THE ART OF MOLECULAR DYNAMICS SIMULATION

The extremely powerful technique of molecular dynamics simulation involves solving the classical many-body problem in contexts relevant to the study of matter at the atomistic level. Since there is no alternative approach capable of handling this broad range of problems at the required level of detail, molecular dynamics methods have proved themselves indispensable in both pure and applied research. This book is a blend of tutorial and recipe collection, providing both an introduction to the subject for beginners and a reference manual for more experienced practitioners. It is organized as a series of case studies that take the reader through each of the steps from formulating the problem, developing the necessary software, and then using the programs to make actual measurements.

This second edition has been extensively revised and enlarged. It contains a substantial amount of new material and the software used in the case studies has been completely rewritten.

Dennis Rapaport received his B.Sc. and M.Sc. degrees in physics from the University of Melbourne, and his Ph.D. in theoretical physics from King's College, University of London. He is a Professor of Physics at Bar-Ilan University and is currently departmental chairman. He has held visiting appointments at Cornell University and IBM in New York, is an Adjunct Professor at the University of Georgia and a Fellow of the American Physical Society. His interest in computer modeling emerged during his undergraduate years and his present research interests include both the methodology of molecular dynamics simulation and its application to a variety of fields.

THE ART OF MOLECULAR DYNAMICS SIMULATION

Second Edition

D. C. RAPAPORT

CAMBRIDGE
UNIVERSITY PRESS

PUBLISHED BY THE PRESS SYNDICATE OF THE UNIVERSITY OF CAMBRIDGE
The Pitt Building, Trumpington Street, Cambridge, United Kingdom

CAMBRIDGE UNIVERSITY PRESS
The Edinburgh Building, Cambridge CB2 2RU, UK
40 West 20th Street, New York, NY 10011-4211, USA
477 Williamstown Road, Port Melbourne, VIC 3207, Australia
Ruiz de Alarcón 13, 28014 Madrid, Spain
Dock House, The Waterfront, Cape Town 8001, South Africa

http://www.cambridge.org

First published 1995
Reprinted 1998
First paperback edition 1997
Reprinted 1998, 2001, 2002
Second edition 2004

Printed in the United Kingdom at the University Press, Cambridge

Typeface TimesNewRoman 9.5/13pt. *System* Advent 3B2 [UPH]

A catalog record for this book is available from the British Library

ISBN 0 521 82568 7 hardback

Contents

Preface to the first edition

Molecular dynamics simulation provides the methodology for detailed microscopic modeling on the molecular scale. After all, the nature of matter is to be found in the structure and motion of its constituent building blocks, and the dynamics is contained in the solution to the N-body problem. Given that the classical N-body problem lacks a general analytical solution, the only path open is the numerical one. Scientists engaged in studying matter at this level require computational tools to allow them to follow the movement of individual molecules and it is this need that the molecular dynamics approach aims to fulfill.

The all-important question that arises repeatedly in numerous contexts is the relation between the bulk properties of matter – be it in the liquid, solid, or gaseous state – and the underlying interactions among the constituent atoms or molecules. Rather than attempting to deduce microscopic behavior directly from experiment, the molecular dynamics method – MD for short – follows the constructive approach in that it tries to reproduce the behavior using model systems. The continually increasing power of computers makes it possible to pose questions of greater complexity, with a realistic expectation of obtaining meaningful answers; the inescapable conclusion is that MD will – if it hasn't already – become an indispensable part of the theorist's toolbox. Applications of MD are to be found in physics, chemistry, biochemistry, materials science, and in branches of engineering.

This is a recipe book. More precisely, it is a combination of an introduction to MD for the beginner, and a cookbook and reference manual for the more experienced practitioner. The hope is that through the use of a series of case studies, in which real problems are studied, both goals can be achieved. The book can be read from cover to cover to explore the principles and capabilities of MD, or it can be used in cookbook style – with a certain amount of cross-referencing – to obtain the recipe for a particular kind of computation. Some familiarity with classical and statistical mechanics, numerical methods and computer programming is assumed.

The case studies take the reader through all the stages from initial problem statement to the presentation of the results of the calculation. The link between these endpoints is the computer program – the recipe. The results of the simulations are 'experimental' observations, in the sense that the simulation is an experiment conducted on an actual, albeit highly idealized, substance. Some of these observations amount to mere measurement, while others can include the discovery of qualitatively novel effects; the custom of referring to MD simulation as computer experimentation is most certainly justified.

Computer programs are an important part of any MD project and feature prominently among the recipes. The view that programs are best kept out of sight along with the plumbing is seriously outdated, and program listings are integrated into the text, with the same status as mathematical equations. After all, a computer program is merely the statement of an algorithm (supplemented by a myriad details to assist the computer in performing its task), and an algorithm is a mathematical procedure. Without the details of the programs, the recipe oriented goal would not have been met: there are many vital, but often subtle, details that only emerge when the program is actually written, so that the program text is an essential part of any recipe and is meant to be read.

Given the near ubiquity of MD, the choice of material had to be restricted to avoid a volume of encyclopedic size. The focus is on the simplest of models, since these form the basis of almost all later developments. Even what constitutes a simple model is open to debate, and here a modest bias on the part of the (physicist) author may be discerned. The emphasis is on showing that MD can reproduce known physical phenomena at a qualitative and semiquantitative level, but without fine-tuning potential functions, molecular structures, or other parameters, for precise quantitative agreement with experiment. Exercises such as demonstrating the solid–fluid phase transition in a system of soft-disk atoms, observing the local ordering in a simple model for water, and following the gyrations of a highly idealized polymer chain, are all far more rewarding experiences for the beginner than detailed computations of specific heats or viscosities across the entire state space of the system. Quantitative detail is not neglected, however, although here some aspects will obviously appeal to more limited segments of the audience.

The model systems to be introduced in these pages can be readily extended and adapted to problems of current interest; suggestions for further work of this kind accompany the case studies, and can serve as exercises (or even research projects) in courses devoted to simulation. The same holds true for the computational techniques. We cover a variety of methods, but not all combinations of methods and problems. In some cases all that is required is a simple modification or combination of the material covered, but in other cases more extensive efforts are called for – the literature continues to report such methodological developments. While

MD can hardly be regarded as a new technique, neither can it be regarded as a fully matured method, and thus there are often several ways of approaching a particular problem, with little agreement on which is to be preferred. It is not our intent to pass judgment, and examples based on alternative methods are included.

The practical side of MD is no less important than the theoretical. A true appreciation of the capabilities and shortcomings of the various methods, an understanding of the assumptions used in the models, and a feeling for what kinds of problem are realistic candidates for MD treatment can only be obtained from experience. This is something that even users of commercial and other packaged software should be aware of. The bottom line is that the reader should be prepared to use this book like any other recipe book: off to the kitchen and start cooking!

January, 1995   Dennis C. Rapaport

Preface to the second edition

The second edition of *The Art of Molecular Dynamics Simulation* is an enlarged and updated version of the first. The principal differences between the two editions are the inclusion of a substantial amount of new material, both as additional chapters and within existing chapters, and a complete revision of all the software used in the case studies to reflect a more modern programming style. This style change is a consequence of the population shift in the research community. At the time the first edition was written older versions of the Fortran language were still in widespread use; despite this fact, C was chosen as the programming language for the book in preference to Fortran, but in a form that would appear familiar to Fortran programmers of the era. Now that C – and related languages – are in widespread use, and Fortran has even evolved to become more like C, the expressive capabilities of C can be employed to the full, resulting in software that is easier to follow. The power of desktop computers has also increased by a large factor since the case studies of the first edition were developed; in recognition of this fact some of the studies consider larger systems, reflecting a shifting view of what is considered a 'short' computation. Other minor changes and corrections have been incorporated throughout the text. The exhortation to employ this volume as a cookbook remains unchanged.

January, 2003   D.C.R.

About the software

Software availability

Readers interested in downloading the software described in this book in a computer-readable form for personal, noncommercial use should visit the Cambridge University Press web site at *http://uk.cambridge.org*, where the home page for this book and the software can be found; a listing of the programs included in the software package appears in the Appendix. Additional material related to the book, as well as contact information, can be found at the author's website – *http://www.ph.biu.ac.il/~rapaport*.

Legal matters

The programs appearing in this book are provided for educational purposes only. Neither the author nor the publisher warrants that these programs are free from error or suitable for particular applications, and both disclaim all liability from any consequences arising out of their use.

1

Introduction

1.1 Historical background

The origins of molecular dynamics – MD – are rooted in the atomism of antiquity. The ingredients, while of more recent vintage, are not exactly new. The theoretical underpinnings amount to little more than Newton's laws of motion. The significance of the solution to the many-body problem was appreciated by Laplace [del51]: 'Given for one instant an intelligence which could comprehend all the forces by which nature is animated and the respective situation of the beings who compose it – an intelligence sufficiently vast to submit these data to analysis – it would embrace in the same formula the movements of the greatest bodies of the universe and those of the lightest atom; for it, nothing would be uncertain and the future, as the past, would be present to its eyes'. And the concept of the computer, without which there would be no MD, dates back at least as far as Babbage, even though the more spectacular hardware developments continue to this day. Thus MD is a methodology whose appearance was a foregone conclusion, and indeed not many years passed after digital computers first appeared before the first cautious MD steps were taken [ald57, gib60, rah64].

The N-body problem originated in the dynamics of the solar system, and the general problem turns out to be insoluble for three or more bodies. Once the atomic nature of matter became firmly established, quantum mechanics took charge of the microscopic world, and the situation became even more complicated because even the constituent particles seemed endowed with a rather ill-defined existence. But a great deal of the behavior of matter in its various states can still be understood in classical (meaning nonquantum) terms, and so it is that the classical N-body problem is also central to understanding matter at the microscopic level. And it is the task of the numerical solution of this problem that MD addresses.

For systems in thermal equilibrium, theory, in the form of statistical mechanics, has met with a considerable measure of success, particularly from the conceptual

1

point of view. Statistical mechanics provides a formal description – based on the partition function – of a system in equilibrium; however, with a few notable exceptions, there are no quantitative answers unless severe approximations are introduced, and even then it is necessary to assume large (essentially infinite) systems. Once out of equilibrium, theory has very little to say. Simulations of various kinds, including MD, help fill the gaps on the equilibrium side, but in the more general case it is only by means of simulation – principally MD – that progress is possible.

From the outset, the role of computers in scientific research has been a central one, both in experiment and in theory. For the theoretician, the computer has provided a new paradigm of understanding. Rather than attempting to obtain simplified closed-form expressions that describe behavior by resorting to (often uncontrolled) approximation, the computer is now able to examine the original system directly. While there are no analytic formulae to summarize the results neatly, all aspects of the behavior are open for inspection.

1.2 Computer simulation

Science requires both observation and comprehension. Without observation there are no facts to be comprehended; without comprehension science is mere documentation. The basis for comprehension is theory, and the language of theoretical science is mathematics. Theory is constructed on a foundation of hypothesis; the fewer the hypotheses needed to explain existing observations and predict new phenomena, the more 'elegant' the theory – Occam's razor.

The question arises as to how simulation is related to physical theory. University education abounds with elegant theoretical manipulation and is a repository for highly idealized problems that are amenable to closed-form solution. Despite the almost 'unreasonable applicability' of mathematics in science [wig60], the fact is that there is usually a chasm between the statement of a theory and the ability to extract quantitative information useful in interpreting experiment. In the real world, exact solutions are the notable exception. Theory therefore relies heavily on approximation, both analytical and numerical, but this is often uncontrolled and so reliability may be difficult to establish. Thus it might be said that simulation rests on the basic theoretical foundations, but tries to avoid much of the approximation normally associated with theory, replacing it by a more elaborate calculational effort. Where theory and simulation differ is in regard to cost. Theory requires few resources beyond the cerebral and is therefore 'cheap'; simulation needs the hardware and, despite plummeting prices, a computer system for tackling problems at the forefront of any field can still prove costly.

Simulation also draws from experiment. Experimental practice rests on a long (occasionally blemished) tradition; computer simulation, because of its novelty, is still somewhat more haphazard, but methodologies are gradually evolving. The output of any simulation should be treated by the same statistical methods used in the analysis of experiments. In addition to estimating the reliability of the results (on the assumption that the measurements have been made correctly) there is also the issue of adequate sampling. This is particularly important when attempting to observe 'rare' events: quantitative studies of such events require that the entire occurrence be reproduced as many times as necessary to assure adequate sampling – if computer resources cannot accommodate this requirement it is presumptuous to expect reliable results.

What distinguishes computer simulation in general from other forms of computation, if such a distinction can be made, is the manner in which the computer is used: instead of merely performing a calculation, the computer becomes the virtual laboratory in which a system is studied – a numerical experiment. The analogy can be carried even further; the results emerging from a simulation may be entirely unexpected, in that they may not be at all apparent from the original formulation of the model. A wide variety of modeling techniques have been developed over the years, and those relevant for work at the molecular level include, in addition to MD, classical Monte Carlo [all87, lan00], quantum based techniques involving path-integral [ber86c, gil90] and Monte Carlo methods [sch92], and MD combined with electron density-function theory [rem90, tuc94], as well as discrete approaches such as cellular automata and the lattice–Boltzmann method [doo91].

Although the goal of science is understanding, it is not always obvious what constitutes 'understanding'. In the simulational context, understanding is achieved once a plausible model is able to reproduce and predict experimental observation. Subsequent study may lead to improvements in the model, or to its replacement, in order to explain further experiments, but this is no different from the way in which science is practiced in the broader context. Clearly, there is no inherent virtue in an excessively complex model if there is no way of establishing that all its features are essential for the desired results (Occam again). The practical consequence of this policy is that, despite any temptation to do otherwise, features should be added gradually. This helps with quality control in the notoriously treacherous world of computer programming; since the outcome of a simulation often cannot be predicted with enough confidence to allow full validation of the computation, the incremental approach becomes a practical necessity.

Simulation plays an important role in education. It takes little imagination to see how interactive computer demonstrations of natural phenomena can enrich any scientific presentation. Whether as an adjunct to experiment, a means of enhancing theoretical discussion, or a tool for creating hypothetical worlds, simulation is without peer. Especially in a conceptually difficult field such as physics, simulation can be used to help overcome some of the more counterintuitive concepts encountered even at a relatively elementary level. As to the role of MD, it can bring to life the entire invisible universe of the atom, an experience no less rewarding for the experienced scientist than for the utter tyro. But, as with education in general, simulation must be kept honest, because seeing is believing, and animated displays can be very convincing irrespective of their veracity.

1.3 Molecular dynamics

Foundations

The theoretical basis for MD embodies many of the important results produced by the great names of analytical mechanics – Euler, Hamilton, Lagrange, Newton. Their contributions are now to be found in introductory mechanics texts (such as [gol80]). Some of these results contain fundamental observations about the apparent workings of nature; others are elegant reformulations that spawn further theoretical development. The simplest form of MD, that of structureless particles, involves little more than Newton's second law. Rigid molecules require the use of the Euler equations, perhaps expressed in terms of Hamilton's quaternions. Molecules with internal degrees of freedom, but that are also subject to structural constraints, might involve the Lagrange method for incorporating geometric constraints into the dynamical equations. Normal equilibrium MD corresponds to the microcanonical ensemble of statistical mechanics, but in certain cases properties at constant temperature (and sometimes pressure) are required; there are ways of modifying the equations of motion to produce such systems, but of course the individual trajectories no longer represent the solution of Newton's equations.

The equations of motion can only be solved numerically. Because of the nature of the interatomic interaction, exemplified by the Lennard-Jones potential with a strongly repulsive core, atomic trajectories are unstable in the sense that an infinitesimal perturbation will grow at an exponential rate[†], and it is fruitless to seek more than moderate accuracy in the trajectories, even over limited periods of time. Thus a comparatively low-order numerical integration method often suffices; whether or not this is adequate emerges from the results, but the reproducibility of MD measurements speaks for itself. Where softer interactions are involved, such as harmonic springs or torsional interactions, either or both of which are often used for modeling molecules with internal degrees of freedom, a higher-order integrator, as well as a smaller timestep than before, may be more appropriate to accommodate the fast internal motion. The numerical treatment of constraints introduces an additional consideration, namely that the constraints themselves must be preserved to much higher accuracy than is provided by the integration method, and methods

---

† This is discussed in §3.8.

exist that address this problem. All these issues, and more, are covered in later chapters.

While MD is utterly dependent on the now ubiquitous computer, an invention of the twentieth century, it pays little heed to the two greatest developments that occurred in physics in the very same century – relativity and quantum mechanics. Special relativity proscribes information transfer at speeds greater than that of light; MD simulation assumes forces whose nature implies an infinite speed of propagation. Quantum mechanics has at its base the uncertainty principle; MD requires – and provides – complete information about position and momentum at all times. In practice, the phenomena studied by MD simulation are those where relativistic effects are not observed and quantum effects can, if necessary, be incorporated as semiclassical corrections – quantum theory shows how this should be done [mai81]. But, strictly speaking, MD deals with a world that, while intuitively appealing to late nineteenth-century science, not to mention antiquity, has little concern for anything that is 'nonclassical'. This fact has in no way diminished the power and effectiveness of the method.

Relation to statistical mechanics

Statistical mechanics (for example [mcq76]) deals with ensemble averages. For the canonical ensemble, in which the temperature T and number of particles N_m are fixed, the equilibrium average of some quantity G is expressed in terms of phase-space integrals involving the potential energy $U(r_1, \ldots r_{N_m})$,

$$
\langle G \rangle = \frac{\displaystyle\int G(r_1, \ldots r_{N_m}) e^{-\beta U(r_1, \ldots r_{N_m})} \, dr_1 \cdots r_{N_m}}{\displaystyle\int e^{-\beta U(r_1, \ldots r_{N_m})} \, dr_1 \cdots r_{N_m}}
\tag{1.3.1}
$$

where $\{r_i \,|\, i = 1, \ldots N_m\}$ are the coordinates, $\beta = 1/k_B T$, and k_B is the Boltzmann constant. This average corresponds to a series of measurements over an ensemble of independent systems.

The ergodic hypothesis relates the ensemble average to measurements carried out for a single equilibrium system during the course of its natural evolution – both kinds of measurement should produce the same result. Molecular dynamics simulation follows the dynamics of a single system and produces averages of the form

$$
\langle G \rangle = \frac{1}{M} \sum_{\mu=1}^{M} G_\mu(r_1, \ldots r_{N_m})
\tag{1.3.2}
$$

over a series of M measurements made as the system evolves. Assuming that the sampling is sufficiently thorough to capture the typical behavior, the two kinds of averaging will be identical. The observation that MD corresponds to the micro-canonical (constant energy) ensemble, rather than to the canonical (constant temperature) ensemble, will be addressed when it appears likely to cause problems.

Relation to other classical simulation methods

The basic Monte Carlo method [lan00] begins by replacing the phase-space integrals in (1.3.1) by sums over states

$$\langle G \rangle = \frac{\sum_s G(s) e^{-\beta U(s)}}{\sum_s e^{-\beta U(s)}} \tag{1.3.3}$$

Then, by a judicious weighting of the states included in the sum, which for the general case results in

$$\langle G \rangle = \frac{\sum_s W(s)^{-1} G(s) e^{-\beta U(s)}}{\sum_s W(s)^{-1} e^{-\beta U(s)}} \tag{1.3.4}$$

where $W(s)$ is the probability with which states are chosen, (1.3.4) can be reduced to a simple average over the S states examined, namely,

$$\langle G \rangle = \frac{1}{S} \sum_{s=1}^{S} G(s) \tag{1.3.5}$$

Clearly, we require

$$W(s) = e^{-\beta U(s)} \tag{1.3.6}$$

for this to be true, and much of the art of Monte Carlo is to ensure that states are actually produced with this probability; the approach is called importance sampling. The Monte Carlo method considers only configuration space, having eliminated the momentum part of phase space. Since there are no dynamics, it can only be used to study systems in equilibrium, although if dynamical processes are represented in terms of collision cross sections it becomes possible to study the consequences of the process, even if not the detailed dynamics [bir94].

Molecular dynamics operates in the continuum, in contrast to lattice-based methods [doo91], such as cellular automata, which are spatially discrete. While the latter are very effective from a computational point of view, they suffer from certain design problems such as the lack of a range of particle velocities, or unwanted effects due to lattice symmetry, and are also not easily extended. The MD approach is computationally demanding, but since it attempts to mimic nature it has few inherent limitations. One further continuum-dynamical method, known as Brownian dynamics [erm80], is based on the Langevin equation; the forces are no longer computed explicitly but are replaced by stochastic quantities that reflect the fluctuating local environment experienced by the molecules.

Applications and achievements

Given the modeling capability of MD and the variety of techniques that have emerged, what kinds of problem can be studied? Certain applications can be eliminated owing to the classical nature of MD. There are also hardware imposed limitations on the amount of computation that can be performed over a given period of time – be it an hour or a month – thus restricting the number of molecules of a given complexity that can be handled, as well as storage limitations having similar consequences (to some extent, the passage of time helps alleviate hardware restrictions).

The phenomena that can be explored must occur on length and time scales that are encompassed by the computation. Some classes of phenomena may require repeated runs based on different sets of initial conditions to sample adequately the kinds of behavior that can develop, adding to the computational demands. Small system size enhances the fluctuations and sets a limit on the measurement accuracy; finite-size effects – even the shape of the simulation region – can also influence certain results. Rare events present additional problems of observation and measurement.

Liquids represent the state of matter most frequently studied by MD methods. This is due to historical reasons, since both solids and gases have well-developed theoretical foundations, but there is no general theory of liquids. For solids, theory begins by assuming that the atomic constituents undergo small oscillations about fixed lattice positions; for gases, independent atoms are assumed and interactions are introduced as weak perturbations. In the case of liquids, however, the interactions are as important as in the solid state, but there is no underlying ordered structure to begin with.

The following list includes a somewhat random and far from complete assortment of ways in which MD simulation is used:

- Fundamental studies: equilibration, tests of molecular chaos, kinetic theory, diffusion, transport properties, size dependence, tests of models and potential functions.
- Phase transitions: first- and second-order, phase coexistence, order parameters, critical phenomena.
- Collective behavior: decay of space and time correlation functions, coupling of translational and rotational motion, vibration, spectroscopic measurements, orientational order, dielectric properties.

- Complex fluids: structure and dynamics of glasses, molecular liquids, pure water and aqueous solutions, liquid crystals, ionic liquids, fluid interfaces, films and monolayers.
- Polymers: chains, rings and branched molecules, equilibrium conformation, relaxation and transport processes.
- Solids: defect formation and migration, fracture, grain boundaries, structural transformations, radiation damage, elastic and plastic mechanical properties, friction, shock waves, molecular crystals, epitaxial growth.
- Biomolecules: structure and dynamics of proteins, protein folding, micelles, membranes, docking of molecules.
- Fluid dynamics: laminar flow, boundary layers, rheology of non-Newtonian fluids, unstable flow.

And there is much more.

The elements involved in an MD study, the way the problem is formulated, and the relation to the real world can be used to classify MD problems into various categories. Examples of this classification include whether the interactions are short- or long-ranged; whether the system is thermally and mechanically isolated or open to outside influence; whether, if in equilibrium, normal dynamical laws are used or the equations of motion are modified to produce a particular statistical mechanical ensemble; whether the constituent particles are simple structureless atoms or more complex molecules and, if the latter, whether the molecules are rigid or flexible; whether simple interactions are represented by continuous potential functions or by step potentials; whether interactions involve just pairs of particles or multiparticle contributions as well; and so on and so on.

Despite the successes, many challenges remain. Multiple phases introduce the issue of interfaces that often have a thickness comparable to the typical simulated region size. Inhomogeneities such as density or temperature gradients can be difficult to maintain in small systems, given the magnitude of the inherent fluctuations. Slow relaxation processes, such as those typical of the glassy state, diffusion that is hindered by structure as in polymer melts, and the very gradual appearance of spontaneously forming spatial organization, are all examples of problems involving temporal scales many orders of magnitude larger than those associated with the underlying molecular motion.

1.4 Organization

Case studies

Case studies are used throughout. The typical case study begins with a review of the theoretical background used for formulating the computational approach. The computation is then described, either by means of a complete listing of the functions that make up the program, or as a series of additions and modifications to an earlier program. Essential but often neglected details such as the initial conditions, organization of the input and output, accuracy, convergence and efficiency are also addressed.

Results obtained from running each program are shown. These sometimes reproduce published results, although no particular effort is made to achieve a similar level of accuracy since our goal is one of demonstration, not of compiling a collection of definitive measurements. Suggested extensions and assorted other projects are included as exercises for the reader.

We begin with the simplest possible example, to demonstrate that MD actually works. Later chapters extend the basic model in a variety of directions, improve the computational methods, deal with various kinds of measurement and introduce new models for more complex problems. The programs themselves are constructed incrementally, with most case studies building on programs introduced earlier. In order to avoid a combinatorial explosion, the directions explored in each chapter tend to be relatively independent, but in more ambitious MD applications it is quite likely that combinations of the various techniques will be needed. Some care is necessary here, because what appears obvious and trivial for simple atoms may, for example, require particular attention for molecules subject to constraints – each case must be treated individually.

Itinerary

Chapter 2 introduces the MD approach using the simplest possible example, and demonstrates how the system behaves in practice; general issues of programming style and organization that are used throughout the book are also introduced here. In Chapter 3 we discuss the methodology for simulating monatomic systems, the algorithms used, and the considerations involved in efficient and accurate computation. Chapter 4 focuses on measuring the thermodynamic and structural properties of systems in equilibrium; some of these properties correspond to what can be measured in the laboratory, while others provide a microscopic perspective unique to simulation. The dynamical properties of equilibrium systems are the subject of Chapter 5, including transport coefficients and the correlation functions that are associated with space- and time-dependent processes.

More complex systems and environments form the subject of subsequent chapters. Modifications of the dynamics to allow systems to be studied under conditions of constant temperature and pressure, as opposed to the constant energy and volume implicit in the basic MD approach, are covered in Chapter 6. In Chapter 7 we discuss further methods for measuring transport properties, both by modeling the relevant process directly and by using a modified form of the dynamics designed for systems not in thermal equilibrium. The dynamics of rigid molecules forms the subject of Chapter 8; methods for handling the general problem are described and a model for water is treated in some detail. Flexible molecules are discussed in Chapter 9 and a model for surfactants examined. Molecules possessing internal degrees of freedom, but also subject to geometric constraints that provide a certain amount of rigidity, are analyzed in Chapter 10, together with a model used for simulating alkane chains. An alternative route to dealing with molecules having internal degrees of freedom, based on treating the internal coordinates directly, is described in Chapter 11. Approaches used for three-body and many-body interactions are introduced in Chapter 12. Specialized methods for treating long-range forces involving Ewald sums and multipole expansions are discussed in Chapter 13.

Chapter 14 describes an alternative approach to MD based on step potentials, rather than on the continuous potentials of earlier chapters; this calls for entirely different computational techniques. In Chapter 15 we focus on the study of time-dependent behavior and demonstrate the ability of MD to reproduce phenomena normally associated with macroscopic hydrodynamics. The methods developed for MD can also be applied to studying the dynamics of granular materials; a short introduction to this subject appears in Chapter 16. The special considerations that are involved in implementing MD computations on parallel and vector supercomputers form the subject of Chapter 17. Chapter 18 deals with a range of software topics not covered by the case studies. And, finally, some closing thoughts on where MD may be headed appear in Chapter 19. A concise alphabetical summary of the variables used in the software and a list of the programs that are available for use with the book appear in the Appendix.

1.5 Further reading

A great deal of information about MD methodology and applications is scattered throughout the scientific literature, and references to material relevant to the subjects covered here will appear in the appropriate places. Three volumes of conference proceedings include pedagogical expositions of various aspects of MD simulation [cic86a, cat90, all93b] and a monograph on liquid simulation covers both MD and Monte Carlo techniques [all87]. Another book devoted in part to MD is [hoo91]. Three evenly spaced reviews of the role of simulation in statistical mechanics are [bee66, woo76, abr86]. Two extensive literature surveys on liquid simulation [lev84, lev92] and a collection of reprints [cic87] are also available.

2

Basic molecular dynamics

2.1 Introduction

This chapter provides the introductory appetizer and aims to leave the reader new to MD with a feeling for what the subject is all about. Later chapters will address the techniques in detail; here the goal is to demonstrate a working example with a minimum of fuss and so convince the beginner that MD is not only straightforward but also that it works successfully. Of course, the technique for evaluating the forces discussed here is not particularly efficient from a computational point of view and the model is about the simplest there is. Such matters will be rectified later. The general program organization and stylistic conventions used in case studies throughout the book are also introduced.

2.2 Soft-disk fluid

Interactions and equations of motion

The most rudimentary microscopic model for a substance capable of existing in any of the three most familiar states of matter – solid, liquid and gas – is based on spherical particles that interact with one another; in the interest of brevity such particles will be referred to as atoms (albeit without hint of their quantum origins). The interactions, again at the simplest level, occur between pairs of atoms and are responsible for providing the two principal features of an interatomic force. The first is a resistance to compression, hence the interaction repels at close range. The second is to bind the atoms together in the solid and liquid states, and for this the atoms must attract each other over a range of separations. Potential functions exhibiting these characteristics can adopt a variety of forms and, when chosen carefully, actually provide useful models for real substances.

The best known of these potentials, originally proposed for liquid argon, is the Lennard-Jones (LJ) potential [mcq76, mai81]; for a pair of atoms i and j located

11

at r_i and r_j the potential energy is

$$u(r_{ij}) = \begin{cases} 4\epsilon \left[\left(\dfrac{\sigma}{r_{ij}} \right)^{12} - \left(\dfrac{\sigma}{r_{ij}} \right)^{6} \right] & r_{ij} < r_c \\ 0 & r_{ij} \geq r_c \end{cases} \qquad (2.2.1)$$

where $r_{ij} = r_i - r_j$ and $r_{ij} \equiv |r_{ij}|$. The parameter ϵ governs the strength of the interaction and σ defines a length scale; the interaction repels at close range, then attracts, and is eventually cut off at some limiting separation r_c. While the strongly repulsive core arising from (in the language of quantum mechanics) the nonbonded overlap between the electron clouds has a rather arbitrary form, and other powers and functional forms are sometimes used, the attractive tail actually represents the van der Waals interaction due to electron correlations. The interactions involve individual pairs of atoms: each pair is treated independently, with other atoms in the neighborhood having no effect on the force between them.

We will simplify the interaction even further by ignoring the attractive tail and changing (2.2.1) to

$$u(r_{ij}) = \begin{cases} 4\epsilon \left[\left(\dfrac{\sigma}{r_{ij}} \right)^{12} - \left(\dfrac{\sigma}{r_{ij}} \right)^{6} \right] + \epsilon & r_{ij} < r_c = 2^{1/6}\sigma \\ 0 & r_{ij} \geq r_c \end{cases} \qquad (2.2.2)$$

with r_c chosen so that $u(r_c) = 0$. A model fluid constructed using this potential is little more than a collection of colliding balls that are both soft (though the softness is limited) and smooth. All that holds the system together is the container within which the atoms (or balls) are confined. While the kinds of system that can be represented quantitatively by this highly simplified model are limited – typically gases at low density – it does nevertheless have much in common with more de-tailed models, and has a clear advantage in terms of computational simplicity. If certain kinds of behavior can be shown to be insensitive to specific features of the model, in this instance the attractive tail of the potential, then it is clearly prefer-able to eliminate them from the computation in order to reduce the amount of work, and for this reason the soft-sphere system will reappear in many of the case studies.

The force corresponding to $u(r)$ is

$$f = -\nabla u(r) \qquad (2.2.3)$$

so the force that atom j exerts on atom i is

$$f_{ij} = \left(\frac{48\epsilon}{\sigma^2} \right) \left[\left(\frac{\sigma}{r_{ij}} \right)^{14} - \frac{1}{2} \left(\frac{\sigma}{r_{ij}} \right)^{8} \right] r_{ij} \qquad (2.2.4)$$

provided $r_{ij} < r_c$, and zero otherwise. As r increases towards r_c the force drops to zero, so that there is no discontinuity at r_c (in both the force and the potential); ∇f and higher derivatives are discontinuous, though this has no real impact on the numerical solution. The equations of motion follow from Newton's second law,

$$m\ddot{\boldsymbol{r}}_i = \boldsymbol{f}_i = \sum_{\substack{j=1 \\ (j \neq i)}}^{N_m} \boldsymbol{f}_{ij} \qquad (2.2.5)$$

where the sum is over all N_m atoms (or molecules in the monatomic case), excluding i itself, and m is the atomic mass. It is these equations which must be numerically integrated. Newton's third law implies that $\boldsymbol{f}_{ji} = -\boldsymbol{f}_{ij}$, so each atom pair need only be examined once. The amount of work[†] is proportional to N_m^2, so that for models in which r_c is small compared with the size of the container it would obviously be a good idea to determine those atom pairs for which $r_{ij} \leq r_c$ and use this information to reduce the computational effort; we will indeed adopt such an approach in Chapter 3. In the present example, which focuses on just the smallest of systems, we continue with this all-pairs approach.

Dimensionless units

At this point we introduce a set of dimensionless, or reduced, MD units in terms of which all physical quantities will be expressed. There are several reasons for doing this, not the least being the ability to work with numerical values that are not too distant from unity, instead of the extremely small values normally associated with the atomic scale. Another benefit of dimensionless units is that the equations of motion are simplified because some, if not all, of the parameters defining the model are absorbed into the units. The most familiar reason for using such units is related to the general notion of scaling, namely, that a single model can describe a whole class of problems, and once the properties have been measured in dimensionless units they can easily be scaled to the appropriate physical units for each problem of interest. From a strictly practical point of view, the switch to such units removes any risk of encountering values lying outside the range that is representable by the computer hardware.

For MD studies using potentials based on the LJ form (2.2.1) the most suitable dimensionless units are defined by choosing σ, m and ϵ to be the units of length,

---

† Note that for the potential function (2.2.2), or the corresponding force (2.2.4), it is never necessary to evaluate $|\boldsymbol{r}_{ij}|$; only its square is needed, so that the (sometimes costly) square root computation is avoided.

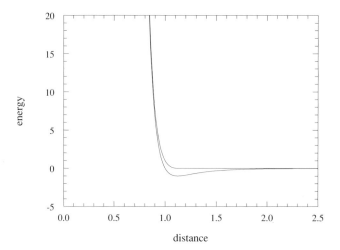

Fig. 2.1. Lennard-Jones and soft-sphere interaction energy (in dimensionless MD units).

mass and energy, respectively, and making the replacements

$$\text{length:} \quad r \rightarrow r\sigma$$
$$\text{energy:} \quad e \rightarrow e\epsilon \qquad\qquad (2.2.6)$$
$$\text{time:} \quad t \rightarrow t\sqrt{m\sigma^2/\epsilon}$$

The resulting form of the equation of motion, now in MD units, is

$$\ddot{\boldsymbol{r}}_i = 48 \sum_{j\,(\neq i)} \left(r_{ij}^{-14} - \tfrac{1}{2} r_{ij}^{-8} \right) \boldsymbol{r}_{ij} \qquad\qquad (2.2.7)$$

The dimensionless kinetic and potential energies, per atom, are

$$E_K = \frac{1}{2N_m} \sum_{i=1}^{N_m} v_i^2 \qquad\qquad (2.2.8)$$

$$E_U = \frac{4}{N_m} \sum_{1 \leq i < j \leq N_m} \left(r_{ij}^{-12} - r_{ij}^{-6} \right) \qquad\qquad (2.2.9)$$

where $\boldsymbol{v}_i$ is the velocity. The functional forms of the LJ and soft-sphere potentials, in MD units, are shown in Figure 2.1.

The unit of temperature is ϵ/k_B, and since each translational degree of freedom contributes $k_B T/2$ to the kinetic energy, the temperature of a d-dimensional ($d = 2$

or 3) system is

$$T = \frac{1}{dN_m} \sum_i v_i^2 \qquad (2.2.10)$$

We have set $k_B = 1$, so that the MD unit of temperature is now also defined. Strictly speaking, of the total dN_m degrees of freedom, d are eliminated because of momentum conservation, but if N_m is not too small this detail can be safely ignored.

If the model is intended to represent liquid argon, the relations between the dimensionless MD units and real physical units are as follows [rah64]:

- Lengths are expressed in terms of $\sigma = 3.4\,\text{Å}$.
- The energy units are specified by $\epsilon/k_B = 120\,\text{K}$, implying that $\epsilon = 120 \times 1.3806 \times 10^{-16}\,\text{erg/atom}^\dagger$.
- Given the mass of an argon atom $m = 39.95 \times 1.6747 \times 10^{-24}$ g, the MD time unit corresponds to 2.161×10^{-12} s; thus a typical timestep size of $\Delta t = 0.005$ used in the numerical integration of the equations of motion corresponds to approximately 10^{-14} s.
- Finally, if N_m atoms occupy a cubic region of edge length L, then a typical liquid density of 0.942 g/cm^3 implies that $L = 4.142 N_m^{1/3}\,\text{Å}$, which in reduced units amounts to $L = 1.218 N_m^{1/3}$.

Suitably chosen dimensionless units will be employed throughout the book. Other quantities, such as the diffusion coefficient and viscosity studied in Chapter 5, will also be expressed using dimensionless units, and these too are readily converted to physical units.

Boundary conditions

Finite and infinite systems are very different, and the question of how large a relatively small system must be to yield results that resemble the behavior of the infinite system faithfully lacks a unique answer. The simulation takes place in a container of some kind, and it is tempting to regard the container walls as rigid boundaries against which atoms collide while trying to escape from the simulation region. In systems of macroscopic size, only a very small fraction of the atoms is close enough to a wall to experience any deviation from the environment prevailing in the interior. Consider, for example, a three-dimensional system with $N_m = 10^{21}$ at liquid density. Since the number of atoms near the walls is of order $N_m^{2/3}$, this amounts to 10^{14} atoms – a mere one in 10^7. But for a more typical MD value of

---

$\dagger$ Several kinds of units are in use for energy; conversion among them is based on standard relations that include 1.3806×10^{-16} erg/atom $= 1.987 \times 10^{-3}$ kcal/mole $= 8.314$ J/mole.

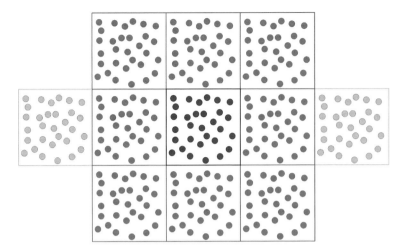

Fig. 2.2. The meaning of periodic boundary conditions (the two-dimensional case is shown).

$N_m = 1000$, roughly 500 atoms are immediately adjacent to the walls, leaving very few interior atoms; if the first two layers are excluded a mere 216 atoms remain. Thus the simulation will fail to capture the typical state of an interior atom and the measurements will reflect this fact. Unless the goal is the study of behavior near real walls, a problem that is actually of considerable importance, walls are best eliminated.

A system that is bounded but free of physical walls can be constructed by resorting to periodic boundary conditions, shown schematically in Figure 2.2. The introduction of periodic boundaries is equivalent to considering an infinite, space-filling array of identical copies of the simulation region. There are two consequences of this periodicity. The first is that an atom that leaves the simulation region through a particular bounding face immediately reenters the region through the opposite face. The second is that atoms lying within a distance r_c of a boundary interact with atoms in an adjacent copy of the system, or, equivalently, with atoms near the opposite boundary – a wraparound effect. Another way of regarding periodic boundaries is to think of mapping the region (topologically, not spatially) onto the equivalent of a torus in four dimensions (a two-dimensional system is mapped onto a torus); then it is obvious that there are no physical boundaries. In this way it is possible to model systems that are effectively bounded but that are nevertheless spatially homogeneous insofar as boundaries are concerned.

The wraparound effect of the periodic boundaries must be taken into account in both the integration of the equations of motion and the interaction computations. After each integration step the coordinates must be examined, and if an atom is found to have moved outside the region its coordinates must be adjusted to bring it back inside. If, for example, the x coordinate is defined to lie between $-L_x/2$ and $L_x/2$, where L_x is the region size in the x direction, the tests (which can be expressed in various equivalent ways) are:

- if $r_{ix} \geq L_x/2$, replace it by $r_{ix} - L_x$;
- otherwise, if $r_{ix} < -L_x/2$, replace it by $r_{ix} + L_x$.

The effect of periodicity on the interaction calculation appears in determining the components of the distance between pairs of atoms; the tests are very similar:

- if $r_{ijx} \geq L_x/2$, replace it by $r_{ijx} - L_x$;
- otherwise, if $r_{ijx} < -L_x/2$, replace it by $r_{ijx} + L_x$.

Periodic wraparound may also have to be considered when analyzing the results of a simulation, as will become apparent later.

Periodic boundaries are most easily handled if the region is rectangular in two dimensions, or a rectangular prism in three. This is not an essential requirement, and any space-filling, convex region can be used, although the boundary computations will not be as simple as those just illustrated. The motivation for choosing alternative region shapes is to enlarge the volume to surface ratio, and thus increase the maximum distance between atoms before periodic ambiguity appears (it is obviously meaningless to speak of interatomic distances that exceed half the region size), the most desirable shape in three dimensions – though not space filling – being the sphere. In two dimensions a hexagon might be used, while in three the truncated octahedron [ada80] is one such candidate. Another reason for choosing more complex region shapes is to allow the modeling of crystalline structures with nonorthogonal axes, for example, a trigonal unit cell; there, too, an alternative region shape, such as a sheared cube, might be worth considering.

Although not an issue in this particular case, the use of periodic boundaries limits the interaction range to no more than half the smallest region dimension – in practice the range is generally much less. Long-range forces require entirely different approaches that will be described in Chapter 13. Problems can also arise if there are strong correlations between atoms separated by distances approaching the region size, because periodic wraparound can then lead to spurious effects. One example is the vibration of an atom producing what are essentially sound waves; the disturbance, if not sufficiently attenuated, can propagate around the system and eventually return to affect the atom itself.

Even with periodic boundaries, finite-size effects are still present, so how big does the system have to be before they can be neglected? The answer depends on the kind of system and the properties of interest. As a minimal requirement, the size should exceed the range of any significant correlations, but there may be more subtle effects even in larger systems. Only detailed numerical study can hope to resolve this question.

Initial state

In order for MD to serve a useful purpose it must be capable of sampling a representative region of the total phase space of the system. An obvious corollary of this requirement is that the results of a simulation of adequate duration are insensitive to the initial state, so that any convenient initial state is allowed. A particularly simple choice is to start with the atoms at the sites of a regular lattice – such as the square or simple cubic lattice – spaced to give the desired density. The initial velocities are assigned random directions and a fixed magnitude based on temperature; they are also adjusted to ensure that the center of mass of the system is at rest, thereby eliminating any overall flow. The speed of equilibration to a state in which there is no memory of this arbitrarily selected initial configuration is normally quite rapid, so that more careful attempts at constructing a 'typical' state are of little benefit.

2.3 Methodology

Integration

Integration of the equations of motion uses the simplest of numerical techniques, the leapfrog method. The origin of the method will be discussed in §3.5; for the present it is sufficient to state that, despite its low order, the method has excellent energy conservation properties and is widely used.

If $h = \Delta t$ denotes the size of the timestep used for the numerical integration, then the integration formulae applied to each component of an atom's coordinates and velocities are

$$v_{i\,x}(t + h/2) = v_{i\,x}(t - h/2) + ha_{i\,x}(t) \tag{2.3.1}$$
$$r_{i\,x}(t + h) = r_{i\,x}(t) + hv_{i\,x}(t + h/2) \tag{2.3.2}$$

The name 'leapfrog' stems from the fact that coordinates and velocities are evaluated at different times; if a velocity estimate is required to correspond to the time at which coordinates are evaluated, then

$$v_{i\,x}(t) = v_{i\,x}(t - h/2) + (h/2)a_{i\,x}(t) \tag{2.3.3}$$

can be used. The local errors introduced at each timestep due to the truncation of what should really be infinite series in h are of order $O(h^4)$ for the coordinates and $O(h^2)$ for velocities.

The leapfrog method can be reformulated in an alternative, algebraically equivalent manner that enables the coordinates and velocities to be evaluated at the same instant in time, avoiding the need for the velocity adjustment in (2.3.3). To do this, the computations are split into two parts: Before computing the acceleration values, update the velocities by a half timestep using the old acceleration values, and then update the coordinates by a full timestep using the intermediate velocity values,

$$v_{i\,x}(t + h/2) = v_{i\,x}(t) + (h/2)a_{i\,x}(t) \qquad (2.3.4)$$
$$r_{i\,x}(t + h) = r_{i\,x}(t) + hv_{i\,x}(t + h/2) \qquad (2.3.5)$$

Now use the new coordinates to compute the latest acceleration values and update the velocities over the second half timestep,

$$v_{i\,x}(t + h) = v_{i\,x}(t + h/2) + (h/2)a_{i\,x}(t + h) \qquad (2.3.6)$$

This two-step procedure[†] is the version of the leapfrog method that will be used throughout the book.

Measurements

The most accessible properties of systems in equilibrium are those introduced in elementary thermodynamics, namely, energy and pressure, each expressed in terms of the independent temperature and density variables T and ρ. Measuring such quantities during an MD simulation is relatively simple, and provides the link between the world of thermodynamics – which predates the recognition of the atomic nature of matter – and the detailed behavior at the microscopic level. However, it is energy rather than temperature that is constant in our MD simulation, so the thermodynamic results are expressed in terms of the average $\langle T \rangle$, rather than T.

In this case study, energy and pressure are the only properties measured. Pressure is defined in terms of the virial expression [han86b] (with $k_B = 1$)

$$PV = N_m T + \frac{1}{d}\left\langle \sum_{i=1}^{N_m} r_i \cdot f_i \right\rangle \qquad (2.3.7)$$

In two dimensions, the region volume V is replaced by the area. For pair potentials, (2.3.7) can be written as a sum over interacting atom pairs, namely,

$$PV = N_m T + \frac{1}{d}\left\langle \sum_{i<j} r_{ij} \cdot f_{ij} \right\rangle \qquad (2.3.8)$$

and for the force (2.2.4) this becomes (in MD units)

$$PV = \frac{1}{d}\left\langle \sum_i v_i^2 + 48 \sum_{i<j}(r_{ij}^{-12} - \tfrac{1}{2}r_{ij}^{-6}) \right\rangle \qquad (2.3.9)$$

While the total energy per atom $E = E_K + E_U$ is conserved, apart from any numerical integration error, quantities such P and T ($= 2E_K/d$) fluctuate, and averages

---

† The first edition used the one-step method of (2.3.1)–(2.3.2).

must be computed over a series of timesteps; such averaging will be included in the program and used for estimating the mean values as well as the statistical measurement errors.

2.4 Programming

Style and conventions

In this section we will be presenting the full listing of the program used in the case study. Not only is the program the tool for getting the job done, it also incorporates a definitive statement of all the computational details. But before addressing these details a few general remarks on matters of organization and programming style are in order. Style, to a considerable degree, is a matter of personal taste; the widely used C language chosen for this work offers a certain amount of flexibility in this respect[†], a boon for some, but a bane for others.

A similar form of organization is used for most programs in the book. Parts of the program discussed in this chapter may seem to be expressed in a more general form than is absolutely necessary; this is to provide a basis for extending the program to handle later case studies. We assume that the reader has a reasonable (and easily acquired) familiarity with the C language. C requires that all variables be defined prior to use; all the definitions will be included, but because the material is presented in a 'functional' manner, rather than as a serial listing of the program text, variables may first appear in the recipe before they are formally defined (this is of course not the case in the program sources). Local variables used within functions are not preserved between calls.

We adopt the convention that all variable names begin with a lower case letter; names formed by joining multiple words use intermediate capitals to clarify meaning. Function names begin with an upper case letter, as do macro definitions specified using #define statements. Constants specified with #define statements are fully capitalized. The format of a C program is also subject to taste. The physical layout used here is fairly standard, with indentation and the positioning of block-delimiting braces used to emphasize the logical structure. The line numbers are of course not part of the program, and are included merely to aid reference.

---

† In the interest of readability, we have tried to avoid some characteristics of C that allow writing extremely concise code (often bordering on the obfuscated); while the experienced C user may perceive their absence, the efficiency of the compiled program is unlikely to be affected in any serious way. As some readers may notice, the software here differs from the first edition in two key respects: (a) Arrays of C structures are used to represent sets of molecular variables, rather than doubly-indexed arrays that represent individual variables (such as atomic coordinates) in which one of the indices is used to select the component of the vector. (b) The conventional C indexing style is used, in which array indices begin at zero, rather than unity as in the original algebraic formulation of the problem. The programming style of the first edition was aimed at making the software more acceptable to Fortran programmers; with the increasing popularity of C, and other programming languages that borrow much of its syntax, not to mention the changing nature of the Fortran language, this is no longer an issue.

Program organization

The main program[♠] of this elementary MD exercise, which forms the basis of most of the subsequent case studies as well, is as follows.

```
int main (int argc, char **argv)
{
  GetNameList (argc, argv);
  PrintNameList (stdout);
  SetParams ();                                        5
  SetupJob ();
  moreCycles = 1;
  while (moreCycles) {
    SingleStep ();
    if (stepCount >= stepLimit) moreCycles = 0;        10
  }
}
```

After the initialization phase (*GetNameList*, *SetParams*, *SetupJob*), in the course of which parameters and other data are input to the program or initialized, and storage arrays allocated, the program enters a loop. Each loop cycle advances the system by a single timestep (*SingleStep*). The loop terminates when *moreCycles* is set to zero; here this occurs after a preset number of timesteps, but in a more general context *moreCycles* can be zeroed once the total processing time exceeds a preset limit, or even by means of an interrupt generated by the user from outside the program when she feels the run has produced enough results (there are also more drastic ways of terminating a program)[†].

The function that handles the processing for a single timestep, including calls to functions that deal with the force evaluation, integration of the equations of motion, adjustments required by periodic boundaries, and property measurements, is

```
void SingleStep ()
{
  ++ stepCount;
  timeNow = stepCount * deltaT;
  LeapfrogStep (1);                                    5
  ApplyBoundaryCond ();
  ComputeForces ();
  LeapfrogStep (2);
  EvalProps ();
  AccumProps (1);                                      10
  if (stepCount % stepAvg == 0) {
```

♠ *pr_02_1* (This is a reference to one of the programs accompanying the book; the full list appears in the Appendix.)

† As a reminder to lapsed C users, *main* is where the program begins, *argc* is the number of arguments passed to the program from the command line (as in Unix), and the array *argv* provides access to the text of each of these arguments.

```
    AccumProps (2);
    PrintSummary (stdout);
    AccumProps (0);
  }                                                              15
}
```

All the work needed for initializing the computation is concentrated in the following function.

```
void SetupJob ()
{
  AllocArrays ();
  stepCount = 0;
  InitCoords ();                                                 5
  InitVels ();
  InitAccels ();
  AccumProps (0);
}
```

Having dealt with the top level functions of the program it is appropriate to insert a few comments on the program structure adopted in these recipes. The order of presentation of this introductory case study reflects the organization of the program: the organization is modular, with separate functions being responsible for distinct portions of the computation. In this initial case study, given the simplicity of the problem the emphasis on organization may appear overdone, but, as indicated earlier, our aim is to provide a more general framework that will be utilized later[†].

The meaning of most program variables should be apparent from their names, with the same being true for functions. Where the meanings are not obvious, or additional remarks are called for, the text will include further details. An alphabetically ordered summary of the globally declared variables appears in the Appendix. Other questions ought to be resolved by examining functions that appear subsequently.

There are many program elements that are common to MD simulations of various kinds. Some of these already appear in this initial case study, others will be introduced later on. Examples include:

- parameter input with completeness and consistency checks;
- runtime array allocation, with array sizes determined by the actual system size;
- initialization of variables;
- the main loop which cycles through the force computations and trajectory integration, and performs data collection at specified intervals;
- the processing and statistical analysis of various kinds of measurement;

[†] On the other hand, in order to avoid the risk of tedium, we have not carried this functional decomposition to the extremes sometimes practiced in professional software development.

- storage of accumulated results and condensed configurational snapshots for later analysis;
- run termination based on various criteria;
- provision for checkpointing (or saving) the current computational state of a long simulation run, both as a safety measure, and to permit the run to be interrupted and continued at some later time.

Computational functions

The function *ComputeForces* encountered in the listing of *SingleStep* is responsible for the interaction computations. Before considering the general form of this function we start with a version suitable for a two-dimensional system in order to allow the gradual introduction of data structures and other elements that will be used throughout the book.

This listing differs from conventional C in that a new kind of floating-point variable, *real*, is introduced. To allow flexibility, *real* can be set to correspond to either single or double precision, known respectively in C as *float* and *double*. Single precision saves storage, whereas double precision provides additional accuracy; as for relative computation speed, this depends on the particular processor hardware, and either precision may be faster, sometimes significantly. Double precision will be used throughout by including the declaration

```
typedef double real;
```

at the beginning of the program.

Many of the quantities involved in the calculations, such as the atomic coordinates, are in fact vectors; the programming style used here will reflect this observation in order to enhance the readability of the software. With this goal in mind we introduce the following C structure type to represent a two-dimensional vector quantity with floating-point components

```
typedef struct {
  real x, y;
} VecR;
```

Organizing the variables associated with each atom or molecule is simplified by the introduction of another structure

```
typedef struct {
  VecR r, rv, ra;
} Mol;
```

in which r, rv and ra correspond, respectively, to the coordinate, velocity and acceleration vectors of the atom. An array of such structures will be introduced later on to represent the state of the system.

The initial version of the function for computing the forces (which are identical to the accelerations in the MD units defined earlier), as well as the potential energy *uSum*, can be written in terms of these vector quantities as

```
void ComputeForces ()
{
  VecR dr;
  real fcVal, rr, rrCut, rri, rri3;
  int j1, j2, n;                                              5

  rrCut = Sqr (rCut);
  for (n = 0; n < nMol; n ++) {
    mol[n].ra.x = 0.;
    mol[n].ra.y = 0.;                                         10
  }
  uSum = 0.;
  for (j1 = 0; j1 < nMol - 1; j1 ++) {
    for (j2 = j1 + 1; j2 < nMol; j2 ++) {
      dr.x = mol[j1].r.x - mol[j2].r.x;                       15
      dr.y = mol[j1].r.y - mol[j2].r.y;
      if (dr.x >= 0.5 * region.x) dr.x -= region.x;
      else if (dr.x < -0.5 * region.x) dr.x += region.x;
      if (dr.y >= 0.5 * region.y) dr.y -= region.y;
      else if (dr.y < -0.5 * region.y) dr.y += region.y;      20
      rr = dr.x * dr.x + dr.y * dr.y;
      if (rr < rrCut) {
        rri = 1. / rr;
        rri3 = rri * rri * rri;
        fcVal = 48. * rri3 * (rri3 - 0.5) * rri;              25
        mol[j1].ra.x += fcVal * dr.x;
        mol[j1].ra.y += fcVal * dr.y;
        mol[j2].ra.x -= fcVal * dr.x;
        mol[j2].ra.y -= fcVal * dr.y;
        uSum += 4. * rri3 * (rri3 - 1.) + 1.;                 30
      }
    }
  }
}
```

Periodic boundaries are included by testing whether any of the components of the interatomic separation vector *dr* exceed half the system size, and if they do, performing a wraparound operation.

While C does not provide the capability for defining new operations, in particular operations associated with vector algebra, it does support the use of macro definitions that can simplify the code considerably. Definitions of this kind will be introduced as necessary, and a complete listing appears in §18.2.

The following definitions can be used for vector addition and subtraction (in two dimensions),

```
#define VAdd(v1, v2, v3)                                          \
    (v1).x = (v2).x + (v3).x,                                     \
    (v1).y = (v2).y + (v3).y
#define VSub(v1, v2, v3)                                          \
    (v1).x = (v2).x - (v3).x,                                     \
    (v1).y = (v2).y - (v3).y
```

where the extra parentheses are a safety measure to cover the possible ways these definitions might be employed in practice. Other vector operations that will be used here, some of which are specialized instances of preceding definitions, are

```
#define VDot(v1, v2)                                              \
    ((v1).x * (v2).x + (v1).y * (v2).y)
#define VSAdd(v1, v2, s3, v3)                                     \
    (v1).x = (v2).x + (s3) * (v3).x,                              \
    (v1).y = (v2).y + (s3) * (v3).y
#define VSet(v, sx, sy)                                           \
    (v).x = sx,                                                   \
    (v).y = sy
#define VSetAll(v, s)         VSet (v, s, s)
#define VZero(v)              VSetAll (v, 0)
#define VVSAdd(v1, s2, v2)    VSAdd (v1, v1, s2, v2)
#define VLenSq(v)             VDot (v, v)
```

The definitions have been constructed in a manner that will minimize the changes required when switching to three dimensions – such as defining the scalar product of two vectors and then using this in defining the squared length of a vector. Finally, the expressions for handling the periodic wraparound can be defined as

```
#define VWrap(v, t)                                               \
    if (v.t >= 0.5 * region.t) v.t -= region.t;                   \
    else if (v.t < -0.5 * region.t) v.t += region.t
#define VWrapAll(v)                                               \
    {VWrap (v, x);                                                \
     VWrap (v, y);}
```

Note that it is implicitly assumed that atoms will not have moved too far outside the region before the periodic wraparound is applied; the above treatment is clearly inadequate for atoms that have traveled so far that this adjustment does not bring them back inside the region. In practice, it should be impossible for an atom to travel such a distance in just a single timestep; thus the alternative, strictly correct but more costly computation based on evaluating

$$(r_{xi} + L_x/2) \pmod{L_x} - L_x/2 \tag{2.4.1}$$

is not used.

Aided by these definitions, as well as by

```
#define Sqr(x)    ((x) * (x))
#define Cube(x)   ((x) * (x) * (x))
#define DO_MOL    for (n = 0; n < nMol; n ++)
```

we arrive at the following revised version of the interaction function, now also including the contribution of the interactions to the virial.

```
void ComputeForces ()
{
  VecR dr;
  real fcVal, rr, rrCut, rri, rri3;
  int j1, j2, n;                                              5

  rrCut = Sqr (rCut);
  DO_MOL VZero (mol[n].ra);
  uSum = 0.;
  virSum = 0.;                                                10
  for (j1 = 0; j1 < nMol - 1; j1 ++) {
    for (j2 = j1 + 1; j2 < nMol; j2 ++) {
      VSub (dr, mol[j1].r, mol[j2].r);
      VWrapAll (dr);
      rr = VLenSq (dr);                                       15
      if (rr < rrCut) {
        rri = 1. / rr;
        rri3 = Cube (rri);
        fcVal = 48. * rri3 * (rri3 - 0.5) * rri;
        VVSAdd (mol[j1].ra, fcVal, dr);                       20
        VVSAdd (mol[j2].ra, - fcVal, dr);
        uSum += 4. * rri3 * (rri3 - 1.) + 1.;
        virSum += fcVal * rr;
      }
    }
  }                                                           25
}
}
```

The code is more concise and transparent, and the use of the vector definitions reduces the scope for typing errors that might otherwise go unnoticed. It should also be noted that by simply changing the definition of the structure *VecR* to

```
typedef struct {
  real x, y, z;
} VecR;
```

and suitably augmenting the vector operations defined above, as well as *VWrapAll*, to include a *z* component, the code can be used without change for three-dimensional computations. The benefits of this kind of approach will be appreciated more as the problems become increasingly complicated. It is worth reiterating that this approach to the force computations involves all $N_m(N_m - 1)/2$ pairs of atoms, and is not the way to carry out serious simulations of this kind; however, a small performance improvement might be achieved here by testing the magnitudes of the individual *dr* components as they are computed to see if they exceed *rCut*, and bypassing the atom pair as soon as this happens.

The function *LeapfrogStep* handles the task of integrating the coordinates and velocities; it appears twice in the listing of *SingleStep*, with the argument *part* determining which portion of the two-step leapfrog process, (2.3.4)–(2.3.5) or (2.3.6), is to be performed.

```
void LeapfrogStep (int part)
{
    int n;

    if (part == 1) {                                                    5
        DO_MOL {
            VVSAdd (mol[n].rv, 0.5 * deltaT, mol[n].ra);
            VVSAdd (mol[n].r, deltaT, mol[n].rv);
        }
    } else {                                                           10
        DO_MOL VVSAdd (mol[n].rv, 0.5 * deltaT, mol[n].ra);
    }
}
```

The function *ApplyBoundaryCond*, called after the first call to *LeapfrogStep*, is responsible for taking care of any periodic wraparound in the updated coordinates.

```
void ApplyBoundaryCond ()
{
    int n;

    DO_MOL VWrapAll (mol[n].r);                                         5
}
```

The brevity of these functions, and their applicability in both two and three dimensions, are a result of the vector definitions introduced previously.

Initial state

Preparation of the initial state uses the following three functions, one for the atomic coordinates, the others for the velocities and accelerations. The number of atoms in the system is expressed in terms of the size of the array of unit cells in which the

atoms are initially arranged, the relevant values appear in `initUcell`, which is a
vector with integer components defined as

```
typedef struct {
   int x, y;
} VecI;
```

Here a simple square lattice (with the option of unequal edge lengths) is used, so
that each unit cell contains just one atom, and the system is centered about the
origin.

```
void InitCoords ()
{
  VecR c, gap;
  int n, nx, ny;

  VDiv (gap, region, initUcell);
  n = 0;
  for (ny = 0; ny < initUcell.y; ny ++) {
    for (nx = 0; nx < initUcell.x; nx ++) {
      VSet (c, nx + 0.5, ny + 0.5);
      VMul (c, c, gap);
      VVSAdd (c, -0.5, region);
      mol[n].r = c;
      ++ n;
    }
  }
}
```

The new vector operations used here are

```
#define VMul(v1, v2, v3)                              \
   (v1).x = (v2).x * (v3).x,                          \
   (v1).y = (v2).y * (v3).y
```

and the corresponding *VDiv*.

The initial velocities are set to a fixed magnitude *velMag* that depends on the
temperature (see below), and after assigning random velocity directions the velocities are adjusted to ensure that the center of mass is stationary. The function *VRand*
(§18.4) serves as a source of uniformly distributed random unit vectors, here in two
dimensions. The accelerations are simply initialized to zero.

```
void InitVels ()
{
  int n;

  VZero (vSum);
  DO_MOL {
```

```
    VRand (&mol[n].rv);
    VScale (mol[n].rv, velMag);
    VVAdd (vSum, mol[n].rv);
  }                                                                    10
  DO_MOL VVSAdd (mol[n].rv, - 1. / nMol, vSum);
}

void InitAccels ()
{                                                                      15
  int n;

  DO_MOL VZero (mol[n].ra);
}
```

New vector operations used here are

```
#define VScale(v, s)                                            \
    (v).x *= s,                                                  \
    (v).y *= s
#define VVAdd(v1, v2)   VAdd (v1, v1, v2)
```

Variables

It is debatable which should be discussed first, the program, or the variables on which it operates. Here we have picked the former in order to provide some motivation for a discussion of the latter.

The scheme we have chosen is that all variables needed by more than one function are declared globally; this implies that they are accessible to all functions[†]. The alternative is to make extensive use of argument lists, perhaps using structures to organize the data transferred between functions; while offering a means of regulating access to variables, it makes the program longer and more tedious to read, so we forgo the practice.

Having settled this issue, what are the global variables used by the program? The list of declarations – each type ordered alphabetically – follows:

```
Mol *mol;
VecR region, vSum;
VecI initUcell;
Prop kinEnergy, pressure, totEnergy;
real deltaT, density, rCut, temperature, timeNow, uSum, velMag,      5
    virSum, vvSum;
int moreCycles, nMol, stepAvg, stepCount, stepEquil, stepLimit;
```

---

† This is not an approach recommended for large software projects because it is difficult to keep track of (and control) which variables are used where.

A new C structure is introduced for representing property measurements that will undergo additional processing,

```
typedef struct {
  real val, sum, sum2;
} Prop;
```

The three elements here are an actual measured value, a sum accumulated over several such measurements in order to evaluate the average, and a sum of squares used in evaluating the standard deviation (more on this below).

The following definition is also included, both to ensure the correct dimensionality of the vectors, and for use in formulae that depend explicitly on whether the system is two- or three-dimensional,

```
#define NDIM   2
```

Most of the names should be self-explanatory. The variable `mol` is actually a pointer to a one-dimensional array that is allocated dynamically at the start of the run and sized according to the value of `nMol`. From a practical point of view, writing `*mol` in the above list of declarations is equivalent to `mol[...]` with a specific array size, except that in the former case the array size is established when the program is run rather than at compilation time[†]. The vector `region` contains the edge lengths of the simulation region. The other quantities, as well as a list of those variables supplied as input to the program, will be covered by the remaining functions below.

All dynamic array allocations are carried out by the function `AllocArrays`. In this example there is just a single array,

```
void AllocArrays ()
{
  AllocMem (mol, nMol, Mol);
}
```

where `AllocMem` is defined as

```
#define AllocMem(a, n, t)   a = (t *) malloc ((n) * sizeof (t))
```

and provides a convenient means of utilizing the standard C memory allocation function `malloc` while ensuring the appropriate type casting.

[†] The advantage of such dynamic allocation (in addition to bypassing any size limitations that some compilers might impose on arrays whose limits are included in the program source) is that it enhances program flexibility by eliminating any arbitrary built-in size assumptions.

Other variables required for the simulation (expressed in MD units when appropriate), excluding those that form part of the input data, are set by the function *SetParams*,

```
void SetParams ()
{
  rCut = pow (2., 1./6.);
  VSCopy (region, 1. / sqrt (density), initUcell);
  nMol = VProd (initUcell);                                            5
  velMag = sqrt (NDIM * (1. - 1. / nMol) * temperature);
}
```

which uses the additional definitions

```
#define VSCopy(v2, s1, v1)                                            \
   (v2).x = (s1) * (v1).x,                                            \
   (v2).y = (s1) * (v1).y
#define VProd(v)   ((v).x * (v).y)
```

The evaluation of *nMol* and `region` assumes just one atom per unit cell, and allowance is made for momentum conservation (which removes *NDIM* degrees of freedom) when computing *velMag* from the temperature.

Measurements

In this introductory case study the emphasis is on demonstrating a minimal working program. The measurements of the basic thermodynamic properties of the system that are included are covered by the following functions. The quantity *vSum* is used to accumulate the total velocity (or momentum, since all atoms have unit mass) of the system; the fact that this should remain exactly zero serves as a simple – but only partial – check on the correctness of the calculation.

The first of the functions computes the velocity and velocity-squared sums and the instantaneous energy and pressure values.

```
void EvalProps ()
{
  real vv;
  int n;
                                                                      5
  VZero (vSum);
  vvSum = 0.;
  DO_MOL {
    VVAdd (vSum, mol[n].rv);
    vv = VLenSq (mol[n].rv);                                          10
    vvSum += vv;
  }
```

```
    kinEnergy.val = 0.5 * vvSum / nMol;
    totEnergy.val = kinEnergy.val + uSum / nMol;
    pressure.val = density * (vvSum + virSum) / (nMol * NDIM);        15
  }
```

The second function collects the results of the measurements, and evaluates
means and standard deviations upon request.

```
void AccumProps (int icode)
{
  if (icode == 0) {
    PropZero (totEnergy);
    PropZero (kinEnergy);
    PropZero (pressure);                                              5
  } else if (icode == 1) {
    PropAccum (totEnergy);
    PropAccum (kinEnergy);
    PropAccum (pressure);                                            10
  } else if (icode == 2) {
    PropAvg (totEnergy, stepAvg);
    PropAvg (kinEnergy, stepAvg);
    PropAvg (pressure, stepAvg);
  }                                                                   15
}
```

Depending on the value of the argument *icode* (0, 1 or 2), *AccumProps* will
initialize the accumulated sums, accumulate the current values, or produce the
final averaged estimates (which overwrite the accumulated values). The following operations[†] are defined for use with the *Prop* structures (*Max* is defined in
§18.2):

```
#define PropZero(v)                                         \
    v.sum = 0.,                                             \
    v.sum2 = 0.
#define PropAccum(v)                                        \
    v.sum += v.val,                                         \     5
    v.sum2 += Sqr (v.val)
#define PropAvg(v, n)                                       \
    v.sum /= n,                                             \
    v.sum2 = sqrt (Max (v.sum2 / n - Sqr (v.sum), 0.))
#define PropEst(v)                                          \    10
    v.sum, v.sum2
```

[†] While the argument of the square-root function evaluated here should never be negative, the *Max* test is included to guard against computer rounding error in cases where the result is close to zero.

Input and output

The function *GetNameList*, called from *main*, reads all the data required to specify the simulation from an input file. It uses a Fortran-style (almost) 'namelist' to group all the data conveniently and automate the input task. It also checks that all requested data items have been provided. For this case study the list of variables is specified in the following way:

```
NameList nameList[] = {
   NameR (deltaT),
   NameR (density),
   NameI (initUcell),
   NameI (stepAvg),                                        5
   NameI (stepEquil),
   NameI (stepLimit),
   NameR (temperature),
};
```

The C macro definitions *NameR* and *NameI* are used to signify real and integer quantities (either single variables, or entire structures with all members of that type). The name of the data file from which the input values are read is derived from the name of the program (if the program happens to be called *md_prog* then the data file should be named *md_prog.in*). The function *PrintNameList*, also called by *main*, outputs an annotated copy of the input data. Full details of these functions and macros appear in §18.5; *VCSum* simply adds the vector components.

Output from the run is produced by

```
void PrintSummary (FILE *fp)
{
   fprintf (fp,
      "%5d %8.4f %7.4f %7.4f %7.4f %7.4f %7.4f %7.4f\n",
      stepCount, timeNow, VCSum (vSum) / nMol, PropEst (totEnergy),    5
      PropEst (kinEnergy), PropEst (pressure));
}
```

Data are written to a file, which in the present case is just the user's terminal because the call to *PrintSummary* in *SingleStep* used the argument *stdout*. By calling this function twice, with different arguments, output can be sent both to the terminal and to a file that logs all the output[†].

---

† The reader unfamiliar with standard C library functions will find *fprintf* – and numerous other functions used later – described in any text on the C language, or in generally available C documentation.

2.5 Results

In this section we present a few of the results that can be obtained from simulations of the two-dimensional soft-disk fluid. In view of the fact that the MD algorithm described here is far from efficient, the results will mostly be confined to short simulation runs of small systems, just to give a foretaste of what is to come. More detailed results based on more extensive computations will appear later.

The input file used in the first demonstration contains the following entries:

```
deltaT          0.005
density         0.8
initUcell       20 20
stepAvg         100
stepEquil       0
stepLimit       10000
temperature     1.
```

The initial configuration is a 20×20 square lattice so that there are a total of 400 atoms. The timestep value `deltaT` is determined by the requirement that energy be conserved by the leapfrog method (to be discussed in §3.5). The initial temperature is $T = 1$; temperature will fluctuate during the run, and no attempt will be made here to set the mean temperature to any particular value.

Conservation laws

The most obvious test that the computation must pass is that of momentum and energy conservation. While the former is intrinsic to the algorithm and, assuming periodic boundaries, its violation would suggest a software error, the latter is sensitive to the choice of integration method and the size of Δt. One quantity that is not conserved is angular momentum; a conservation law requires the system to be invariant under some change, such as translation, but, because of the periodic boundaries, the rotational invariance needed for angular momentum conservation is not applicable. Programming errors can sometimes (but not always) be detected by the violation of a conservation law; when this occurs the effect can be gradual, intermittent, or catastrophic, depending on the cause of error.

In Table 2.1 we show an edited version of the output[†] of the run specified above; the results listed are the sum of the velocity components, the mean energy and kinetic energy per atom, their standard deviations, and the mean pressure. Clearly, energy and momentum are conserved as expected, kinetic energy fluctuates by a

---

[†] Note that the higher-order digits of some of the values listed here – and elsewhere in the book – may vary, depending on the computer, compiler and level of optimization; this is an expected consequence of the trajectory sensitivity, discussed later in this section.

Table 2.1. Edited output from a short MD run.

timestep	$\sum v$	$\langle E \rangle$	$\sigma(E)$	$\langle E_K \rangle$	$\sigma(E_K)$	$\langle P \rangle$
100	0.0000	0.9952	0.0002	0.6555	0.0910	4.5751
200	0.0000	0.9951	0.0001	0.6493	0.0118	4.5802
300	0.0000	0.9951	0.0001	0.6398	0.0168	4.6445
400	0.0000	0.9951	0.0000	0.6476	0.0155	4.5685
500	0.0000	0.9951	0.0000	0.6599	0.0167	4.4682
1000	0.0000	0.9950	0.0000	0.6481	0.0256	4.5489
2000	0.0000	0.9951	0.0001	0.6500	0.0125	4.5370
3000	0.0000	0.9951	0.0001	0.6301	0.0166	4.6898
5000	0.0000	0.9952	0.0001	0.6410	0.0139	4.6254
10000	0.0000	0.9949	0.0001	0.6535	0.0205	4.4886

limited amount, and it is also apparent that as a result of some of the initial kinetic energy being converted to potential energy the temperature of the system (here $T = E_K$) has dropped considerably below the initial setting.

Equilibration

Characterizing equilibrium is by no means an easy task, especially for small systems whose properties fluctuate considerably. Averaging over a series of timesteps will reduce the fluctuations, but different quantities relax to their equilibrium averages at different rates, and this must also be taken into account when trying to establish when the time is ripe to begin making measurements. Fortunately, relaxation is generally quite rapid, but one must always beware of those situations where this is not true. Equilibration can be accelerated by starting the simulation at a higher temperature and later cooling by rescaling the velocities (this is similar, but not identical, to using a larger timestep initially); too high a temperature will, however, lead to numerical instability.

One simple measure of equilibration is the rate at which the velocity distribution converges to its expected final form. Theory [mcq76] predicts the Maxwell distribution

$$f(\boldsymbol{v}) = \frac{\rho}{(2\pi T)^{d/2}}\, e^{-v^2/2T} \tag{2.5.1}$$

(in MD units) which, after angular integration, becomes

$$f(v) \propto v^{d-1} e^{-v^2/2T} \tag{2.5.2}$$

The distribution can be measured by constructing a histogram of the velocity values $\{h_n \mid n = 1, \ldots N_b\}$, where h_n is the number of atoms with velocity magnitude between $(n - 1)\Delta v$ and $n\Delta v$, $\Delta v = v_m/N_b$, and v_m is a suitable upper limit to v. The normalized histogram represents a discrete approximation to $f(v)$.

The function that carries out this computation[♠] is

```
void EvalVelDist ()
{
  real deltaV, histSum;
  int j, n;
                                                                              5
  if (countVel == 0) {
    for (j = 0; j < sizeHistVel; j ++) histVel[j] = 0.;
  }
  deltaV = rangeVel / sizeHistVel;
  DO_MOL {                                                                   10
    j = VLen (mol[n].rv) / deltaV;
    ++ histVel[Min (j, sizeHistVel - 1)];
  }
  ++ countVel;
  if (countVel == limitVel) {                                               15
    histSum = 0.;
    for (j = 0; j < sizeHistVel; j ++) histSum += histVel[j];
    for (j = 0; j < sizeHistVel; j ++) histVel[j] /= histSum;
    PrintVelDist (stdout);
    countVel = 0;                                                           20
  }
}
```

in which the definitions `Min` (§18.2) and

```
#define VLen(v)  sqrt (VDot (v, v))
```

are used. Depending on the value of `countVel`, the function will, in addition to adding the latest results to the accumulated total, either initialize the histogram counts, or carry out the final normalization. Other kinds of analysis in subsequent case studies will involve functions that operate in a similar manner.

In order to use this function storage for the histogram array must be allocated, and a number of additional variables declared and assigned values. The variables are

```
real *histVel, rangeVel;
int countVel, limitVel, sizeHistVel, stepVel;
```

♠ pr_02_2

and those included in the input data must be added to the array `nameList`,

```
NameI (limitVel),
NameR (rangeVel),
NameI (sizeHistVel),
NameI (stepVel),
```

Allocation of the histogram array is included in `AllocArrays`,

```
AllocMem (histVel, sizeHistVel, real);
```

Initialization, in `SetupJob`, requires the additional statement

```
countVel = 0;
```

and the histogram function is called from `SingleStep` by

```
if (stepCount >= stepEquil &&
    (stepCount - stepEquil) % stepVel == 0) EvalVelDist ();
```

Histogram output is provided by the function

```
void PrintVelDist (FILE *fp)
{
  real vBin;
  int n;

  printf ("vdist (%.3f)\n", timeNow);
  for (n = 0; n < sizeHistVel; n ++) {
    vBin = (n + 0.5) * rangeVel / sizeHistVel;
    fprintf (fp, "%8.3f %8.3f\n", vBin, histVel[n]);
  }
}
```

To demonstrate the way in which the velocity distribution evolves over time during the early portion of the simulation we study a system with $N_m = 2500$. Use of a larger system than before produces smoother results, and these are further improved by averaging over five separate runs with different random initial velocities; to simulate a system of this size efficiently we resorted to methods that will be introduced in §3.4, although this has no effect on the results.

The initial velocities are based on random numbers generated using a default initial seed; to change this value introduce a new integer variable `randSeed` (whose default value is arbitrarily set to 17) and in `SetupJob` use this value to initialize a

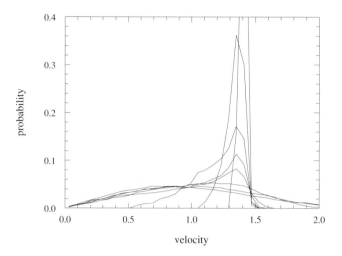

Fig. 2.3. Velocity distribution as a function of time; successively broader graphs are at times 0.02, 0.04, 0.06, 0.08, 0.1, 0.2, 0.4, and 1.0 (the zero-time state – not shown – is a spike at the initial velocity $\sqrt{2}$).

different random sequence by the call

```
InitRand (randSeed);
```

Also add

```
NameI (randSeed),
```

to the array `nameList`. The input data are as above, except for

```
deltaT          0.001
initUcell       50 50
limitVel        4
randSeed        17
rangeVel        3.
sizeHistVel     50
stepVel         5
```

and `randSeed` is different for each run. The results are shown in Figure 2.3; the final distribution develops rapidly and is reached within about 0.4 time units. From results of this kind it is clear that there is no need to assign an initial velocity distribution carefully – the system takes care of this matter on its own (for very small systems there will be deviations from the theoretical distribution [ray91]).

The Boltzmann *H*-function occupies an important position in the development of statistical mechanics [hua63]. It is defined as

$$H(t) = \int f(\boldsymbol{v}, t) \log f(\boldsymbol{v}, t) \, d\boldsymbol{v} \tag{2.5.3}$$

and it can be proved that $\langle dH/dt \rangle \leq 0$, with equality only applying when $f(\boldsymbol{v})$ is the Maxwell distribution. In order to compute $H(t)$ we use the velocity histogram $\{h_n\}$ obtained previously; if we neglect constants, $H(t)$ can be approximated by

$$h(t) = \sum_n h_n \log(h_n / v_n^{d-1}) \tag{2.5.4}$$

An additional variable is required for this computation, namely,

```
real hFunction;
```

and the following code must be added to the summary phase of *EvalVelDist* (for the two-dimensional case),

```
hFunction = 0.;
for (j = 0; j < sizeHistVel; j ++) {
   if (histVel[j] > 0.) hFunction += histVel[j] * log (histVel[j] /
      ((j + 0.5) * deltaV));
}
```
⁵

For output, add the extra line to *PrintVelDist*

```
fprintf (fp, "hfun: %8.3f %8.3f\n", timeNow, hFunction);
```

In Figure 2.4 we show the results of this analysis for several densities, using the above system, but with a quarter the number of atoms to enhance the fluctuations. The long-time limit of the *H*-function depends on T (as well as ρ), and since no attempt is made to force the system to a particular temperature the limiting values will differ. Convergence is fastest at high density, while at lower density $h(t)$ does not begin to change until atoms come within interaction range. Finite systems lack the monotonicity suggested by the theorem, but the overall trend is clear and, strictly speaking, the theorem only addresses average quantities. A computation of this kind was carried out in the early days of MD [ald58]; Boltzmann would presumably have found the results much to his liking.

Thermodynamics

To provide a glimpse of what can be done, we show a few measurements made during some short test runs using as input data,

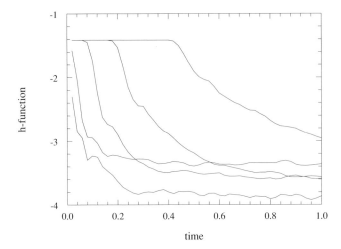

Fig. 2.4. Time dependence of the Boltzmann *H*-function (neglecting constants) starting from an ordered state, at densities 0.2–1.0; convergence ·is faster · at higher
density.

deltaT	0.005
density	0.8
initUcell	20 20
stepAvg	1000
stepEquil	1000
stepLimit	3000
temperature	1.

Various values of `density` are used; any data items not explicitly shown take values specified previously. The output is summarized in Table 2.2.

It is unlikely that the temperature (here just $\langle E_K \rangle$) is the one wanted, and the
value will certainly not be the one used to create the initial state. To obtain a particular $\langle T \rangle$, the velocities must be adjusted over a series of timesteps until the system
settles down at the correct state point. The actual velocity rescaling should be based
on $\langle T \rangle$, and not on the instantaneous T values that may be subject to considerable fluctuation. Though not apparent here, the energy can gradually drift upward
because of the numerical error in the leapfrog method; the drift rate for a given
temperature depends on Δt and is negligible for sufficiently small values. We will
return to these matters in Chapter 3.

Table 2.2. Measurements from soft-disk simulations at different densities: total energy, kinetic energy and pressure are shown.

ρ	$\langle E \rangle$	$\langle E_K \rangle$	$\sigma(E_K)$	$\langle P \rangle$	$\sigma(P)$
0.4	0.9935	0.917	0.014	0.803	0.056
0.6	0.9936	0.823	0.016	1.955	0.099
0.8	0.9952	0.645	0.022	4.578	0.165

Trajectories

The first opportunity for using MD to provide results that are unobtainable by other means is in the study of the trajectories followed by individual atoms. Clearly, a single trajectory conveys very little information, but if the trajectories of groups of nearby atoms are examined a clear picture emerges of the different behavior in the solid, liquid and gaseous states of matter. In the solid phase the atoms are confined to small vibrations around the sites of a lattice, the gas is distinguished by trajectories that are ballistic over relatively long distances, while the liquid is characterized by generally small steps, occasional rearrangement, and no long-range positional order. The differences in the trajectories are reflected at the macroscopic level by the values of the diffusion coefficient. Diffusion is just the mean-square atomic displacement (after allowing for periodic wraparound in the MD case), and is one example of a transport process that MD can examine directly; we will return to this in Chapter 5.

The best way to observe these features is by running an MD simulation interactively and watching the trajectories as they develop for different T and ρ. Trajectories can be shown on a computer display screen by simply drawing a line between the atomic positions every few timesteps; whenever a periodic boundary is crossed simply interrupt the trajectory drawing and restart it from the opposite boundary. Suitable graphics functions are readily added to the program; all that is required, apart from setting up the display functions and arranging for atomic coordinates to be converted to screen coordinates, is the decision as to how frequently the display should be updated. Typical trajectories obtained in the solid and dense fluid phases appear in Figure 2.5.

An example of a simple interactive MD simulation is shown in Figure 2.6. Here the user interface permits realtime control of the computation, including the choice of system size, altering the values of T and ρ, and changing the display update rate. The details involved in writing such programs depend on the computer and software environment; this two-dimensional example is described in [rap97], although

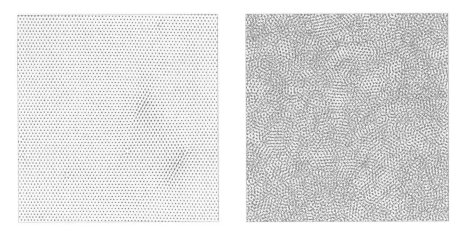

Fig. 2.5. Trajectory plots at densities of 1.05 and 0.85 showing the difference between solid and dense fluid phases, namely, localized and diffusing trajectories.

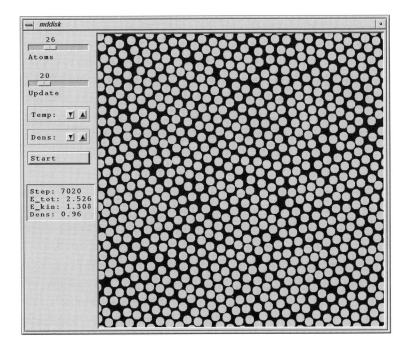

Fig. 2.6. Example of an interactive simulation.

a little more effort would be required for the corresponding three-dimensional case. Visualization plays an essential role in many kinds of problem, and the ability to interact with the simulation while in progress can prove to be of considerable value.

2.6 Further study

2.1 Compare the observed velocity distribution with the theoretical result (2.5.2).

2.2 Check that the correct limiting values of $H(t)$, defined in (2.5.3), are obtained.

2.3 Extend the graphics capability of the interactive MD program so that trajectories can be displayed.

3

Simulating simple systems

3.1 Introduction

In this chapter we focus on a number of techniques used in MD simulation, primarily the methods for computing the interactions and integrating the equations of motion. The goal is to generate the atomic trajectories; subsequent chapters will deal with the all-important question of analyzing this raw 'experimental' data. We continue to work with the simplest atomic systems, in other words, with monatomic fluids based on the LJ potential, not only because we want to introduce the methodology gradually, but also because a lot of the actual qualitative (and even quantitative) behavior of many-body systems is already present in this simplest of models. Models of this kind are widely used in MD studies of basic many-body behavior, examples of which will be encountered in later chapters.

3.2 Equations of motion

While Newton's second law suffices for the dynamics of the simple atomic fluid discussed in this chapter, later chapters will require more complex forms of the equations of motion. The Lagrangian formulation of classical mechanics provides a general basis for dealing with these more advanced problems, and we begin with a brief summary of the relevant results. There are, of course, other ways of approaching the subject, and we will also make passing reference to Hamilton's equations. A full treatment of the subject can be found in textbooks on classical mechanics, for example [gol80].

Lagrange equations of motion

The starting point is Hamilton's variational principle, which concisely summarizes most of classical mechanics into the statement that the phase-space trajectory followed by a mechanical system is the one for which the time integral $\int \mathcal{L}\,dt$ is an extremum, where $\mathcal{L}$ is the Lagrangian. Given a set of N independent generalized coordinates and velocities $\{q_i, \dot{q}_i\}$ that describe the state of a conservative system (one in which all forces derive from some potential energy function U), so that $\mathcal{L} = \mathcal{L}(\{q_i\}, \{\dot{q}_i\}, t)$, then $\mathcal{L}$ can be shown to satisfy the Lagrange equations

$$\frac{d}{dt}\left(\frac{\partial \mathcal{L}}{\partial \dot{q}_i}\right) - \frac{\partial \mathcal{L}}{\partial q_i} = 0, \qquad i = 1, \ldots N \tag{3.2.1}$$

These equations form the starting point for many of the subsequent developments. Newton's second law is a simple consequence of this result, where, if q_i denotes a component of the cartesian coordinates for one of the atoms (and assuming identical masses m),

$$\mathcal{L} = \tfrac{1}{2}m \sum_i \dot{q}_i^2 - U\left(\{q_i\}\right) \tag{3.2.2}$$

so that (3.2.1) becomes

$$m\ddot{q}_i = -\frac{\partial U}{\partial q_i} = f_i \tag{3.2.3}$$

where f_i is the corresponding force component.

Lagrange equations with constraints

There are situations where it is desirable to define the dynamics in ways which cannot be based just on forces obtained from some potential function. For example, in the case of partially rigid molecules the lengths of interatomic bonds should be kept constant. Such restrictions on the dynamics are called constraints and their effect on the equations of motion is the appearance of extra terms that play the role of internal forces, although these terms have an entirely different origin. Here we outline the general framework; the details depend on the problem, and examples will be encountered in Chapters 6 and 10.

Hamilton's principle can be extended to systems with constraints having the general form

$$\sum_k a_{lk}\dot{q}_k + a_l = 0, \qquad l = 1, \ldots M \tag{3.2.4}$$

This includes the special case of holonomic constraints for which there exist relations between the coordinates of form

$$g_l\left(\{q_k\}, t\right) = 0 \tag{3.2.5}$$

in which case

$$a_{lk} = \frac{\partial g_l}{\partial q_k}, \qquad a_l = \frac{\partial g_l}{\partial t} \tag{3.2.6}$$

The resulting Lagrange equations are

$$\frac{d}{dt}\left(\frac{\partial \mathcal{L}}{\partial \dot{q}_i}\right) - \frac{\partial \mathcal{L}}{\partial q_i} = \sum_l \lambda_l a_{li}, \qquad i = 1, \dots N \tag{3.2.7}$$

where the time evolution of the M Lagrange multipliers $\{\lambda_l\}$ is evaluated along with the N coordinates: there is a total of $N + M$ equations with a similar number of unknowns. The sum on the right-hand side of (3.2.7) can be regarded as a generalized force, equivalent in its effect to the imposed constraints.

Hamilton equations of motion

An alternative formulation of the equations of motion sometimes appears in the MD literature. Replace the generalized velocities $\{\dot{q}_i\}$ in the Lagrange formulation by generalized momenta

$$p_i = \partial \mathcal{L}/\partial \dot{q}_i \tag{3.2.8}$$

(if the coordinates are cartesian, then $p_i = m\dot{q}_i$) and consider the Hamiltonian $\mathcal{H} = \mathcal{H}\big(\{q_i\}, \{p_i\}, t\big)$ defined by

$$\mathcal{H} = \sum_i \dot{q}_i p_i - \mathcal{L} \tag{3.2.9}$$

The two first-order equations of motion associated with each coordinate are

$$\dot{q}_i = \frac{\partial \mathcal{H}}{\partial p_i}, \qquad \dot{p}_i = -\frac{\partial \mathcal{H}}{\partial q_i} \tag{3.2.10}$$

If $\mathcal{H}$ has no explicit time dependence, then $\dot{\mathcal{H}} = 0$, and $\mathcal{H}$ – the total energy – is a conserved quantity.

3.3 Potential functions

Origins

Modeling of matter at the microscopic level is based on a comprehensive description of the constituent particles. Although such a description must in principle be based on quantum mechanics, MD generally adopts a classical point of view, typically representing atoms or molecules as point masses interacting through forces that depend on the separation of these objects. More complex applications are likely to require extended molecular structures, in which case the forces will also depend on relative orientation. The quantum picture of interactions arising from overlapping electron clouds has been transformed into a system of masses coupled by exotic 'springs'. The justification for this antithesis of quantum mechanics is that not only does it work, but it appears to work surprisingly well; on the other hand, the rigorous quantum mechanical description is still hard pressed in dealing with even the smallest systems.

Obviously the structural models and potential functions used in classical MD simulation should not be taken too literally, and the potentials are often referred to as effective potentials in order to clarify their status. The classical approximation to the quantum mechanical description of a molecule and its interactions is not derived directly from 'first principles', but, rather, is the result of adapting both structure and potential function to a variety of different kinds of information; these include the results of quantum mechanical energy calculations, experimental data obtained by thermodynamic and various kinds of spectroscopic means, the structure of the crystalline state, measurements of transport properties, collision studies using molecular beams, and so on [hir54, mai81]. These models undergo refinement as new comparisons between simulation and experiment become available, and whenever the evidence against a particular model becomes overwhelming, a revised or even an entirely new model must be developed. (From a strictly theoretical point of view the interactions between molecules can always be written in terms of a multipole expansion [pri84]; if most of the important behavior can be confined to the leading-order terms, then this could be used as the basis for a model potential. While such a systematic approach is appealing, it is not generally used in practice.)

It is always the simplest models that are tried first. Atoms are modeled as point particles interacting through pair potentials. Molecules are represented by atoms with orientation-dependent forces, or as extended structures, each containing several interaction sites; the molecules may be rigid, flexible, or somewhere in between, and if there are internal degrees of freedom there will be internal forces as well. The purpose of this book is not to discuss the design of molecular models; we will make use of existing models and – from a pedagogical viewpoint – the simpler the model the better. Our aim is to demonstrate the general methodology by example, not to review the enormous body of literature devoted to many different kinds of model developed for specific applications. As far as MD is concerned the complexity of the model has little effect on the nature of the computation, merely on the amount of work involved.

Example potentials

The most familiar pair interaction is the LJ potential, introduced in (2.2.1). It has been used quite successfully for liquid argon [rah64, ver67] (although there are better potentials [bar71, mai81]), and is also often used as a generic potential for qualitative explorations not involving specific substances. The LJ interaction is characterized by its strongly repulsive core and weakly attractive tail. To keep computation to a reasonable level the interaction is truncated at a relatively short range; at a typical cutoff distance of $r_c = 2.5$ (in MD units) the interaction energy is just 0.016 of the well depth.

The discontinuity at r_c affects both the apparent energy conservation and the actual atomic motion, with atoms separated by a distance close to r_c sometimes moving repeatedly in and out of interaction range. The discontinuity can be smeared out by changing the form of the potential function slightly, although this must be done carefully since it is the potential that defines the model. For example, a potential function $u(r)$ can be modified to eliminate the discontinuity in both $u(r)$ and the force $-u'(r)$ by replacing it with

$$u_1(r) = u(r) - u(r_c) - \left.\frac{du(r)}{dr}\right|_{r=r_c} (r - r_c) \qquad (3.3.1)$$

This modification applies across the entire interaction range; an example of an alternative that confines the change to the vicinity of r_c involves the use of a cubic spline polynomial (as in §12.3) that interpolates smoothly and differentiably between the value of $u(r)$ at $r = r_c - \delta r$ and zero at r_c.

A slight change to the LJ interaction leads to a potential that is entirely repulsive in nature and very short-ranged (2.2.2). The particles represented by this potential are little more than soft spheres (in three dimensions, or disks in two), although softness is confined to a very narrow range of separations and the spheres rapidly tend to become hard as they are driven together. (Another version of the 'soft-sphere' interaction retains just the r^{-12} term of the LJ potential, again with a cutoff at which $u(r)$ is discontinuous; we will not consider this variant here.) A system subject to the original LJ potential can exist in the solid, liquid, or gaseous states; the attractive part of the potential is used to bind the system when in the solid and liquid states, and the repulsive part prevents collapse. When the attractive interaction is eliminated, the behavior is determined primarily by density; at high density the soft-sphere system is packed into a crystalline state, but once melted, unlike the LJ case where there is also a liquid–gas phase transition, the liquid and gas states are thermodynamically indistinguishable.

Other functional forms can be used for interactions between atoms, and between small molecules in cases where spherical symmetry applies [mai81]. Some prove more suitable than others for particular problems. There is even an alternative to the LJ potential for use in simple cases, namely, a function in which the r^{-12} term is replaced by $Ae^{-\alpha r}$; while such a potential produces a softer central core, the repulsive part contributes over a longer range. But since the subject is MD, not the construction of potential functions, we will not pursue this subject any further.

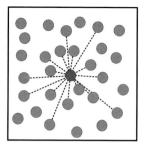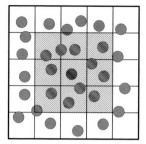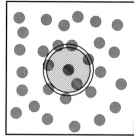

Fig. 3.1. The different approaches to computing interactions: all pairs, cell subdivision (the cell size exceeds the interaction range), and neighbor lists (the concentric circles show the interaction range and the extra area covered by the neighbor list for one of the atoms).

Interactions suitable for describing other kinds of molecule will be introduced in subsequent chapters.

3.4 Interaction computations

All-pairs method

This method was introduced in §2.4. It is the simplest to implement, but extremely inefficient when the interaction range r_c is small compared with the linear size of the simulation region. All pairs of atoms must be examined because, owing to the continual rearrangement that characterizes the fluid state, it is not known in advance which atoms actually interact. Although testing whether atoms are separated by less than r_c is only a part of the overall interaction computation, the fact that the amount of computation needed grows as $O(N_m^2)$ rules out the method for all but the smallest values of N_m. Two techniques for reducing this growth rate to a more acceptable $O(N_m)$ level, often used in tandem, will be discussed here; to within a numerical factor this clearly represents the lower bound for the amount of work required to process all N_m atoms. A schematic summary of the methods appears in Figure 3.1.

Cell subdivision

Cell subdivision [sch73, hoc74] provides a means of organizing the information about atom positions into a form that avoids most of the unnecessary work and reduces the computational effort to the $O(N_m)$ level. Imagine that the simulation region is divided into a lattice of small cells, and that the cell edges all exceed r_c in length. Then if atoms are assigned to cells on the basis of their current positions it is obvious that interactions are only possible between atoms that are either in the same cell or in immediately adjacent cells; if neither of these conditions are met, then the atoms must be at least r_c apart. Because of symmetry only half the neighboring cells need be considered; thus a total of 14 neighboring cells must be examined in three dimensions, and five in two dimensions (these numbers include the cell itself). The wraparound effect due to periodic boundaries is readily incorporated into this scheme. Clearly, the region size must be at least $4r_c$ for the method to be useful, but this requirement is usually met. It is not essential that the cell edges exceed r_c, but if this condition is not satisfied, further cells, not merely nearest neighbors, will have to be included [que73].

The program for the cell-based force calculation involves a form of data organization known as a linked list [knu68]. Rather than accessing data sequentially, the linked list associates a pointer p_n with each data item x_n, the purpose of which is to provide a nonsequential path through the data. Each linked list requires a separate pointer f to access the first data item, and the item terminating the list must have a special pointer value, such as -1, that cannot be mistaken for anything else. Thus $f = a$ points to x_a as the first item in the list, $p_a = b$ points to x_b as the second item, and so on, until a pointer value $p_z = -1$ is encountered, terminating the list. (This kind of data organization will reappear in other contexts in subsequent chapters.)

In the cell algorithm, linked lists are used to associate atoms with the cells in which they reside at any given instant; a separate list is required for each cell. The reason for using linked lists is to economize on storage. It is not known in advance how many atoms occupy each cell, since the number can be anywhere between zero and a value determined by the highest possible packing density; the use of sequential tables that list the atoms in each cell, while guaranteeing sufficient storage so that any cell can be maximally occupied, is extremely wasteful. The linked list approach does not have this problem because of the way the cell occupancy data are organized; the total storage required for all the linked lists is fixed and known in advance.

The additional variables required to support the cell method are

```
VecI cells;
int *cellList;
```

and memory allocation for the array *cellList* that will hold all the information associated with the linked lists is added to the function *AllocArrays*,

```
AllocMem (cellList, VProd (cells) + nMol, int);
```

where, in three dimensions,

```
#define VProd(v)   ((v).x * (v).y * (v).z)
```

and, of course,

```
#define NDIM  3

typedef struct {
  real x, y, z;
} VecR;                                                    5

typedef struct {
  int x, y, z;
} VecI;
```

Rather than use separate arrays for the two kinds of pointer, namely, those between atoms in the same cell (p) and those to the initial atom in the list belonging to each cell (f), the first $nMol$ elements in `cellList` are used for the former and the remainder for the latter. How the list elements are accessed will be clarified by the program listing. The size of the cell array is determined in *SetParams* (using the three-dimensional version of *VSCopy*),

```
VSCopy (cells, 1. / rCut, region);
```

We have tacitly assumed that it is most efficient to use the smallest cells (exceeding r_c in size) possible; only when the density is sufficiently low that the mean cell occupancy drops substantially below unity is it worth considering using larger (and hence fewer) cells.

The force computation function, including cell assignment, allowance for periodic boundaries, energy and virial calculation, is as follows♠ (this is the three-dimensional version).

```
void ComputeForces ()
{
  VecR dr, invWid, rs, shift;
  VecI cc, m1v, m2v, vOff[] = OFFSET_VALS;
  real fcVal, rr, rrCut, rri, rri3, uVal;               5
  int c, j1, j2, m1, m1x, m1y, m1z, m2, n, offset;

  rrCut = Sqr (rCut);
  VDiv (invWid, cells, region);
  for (n = nMol; n < nMol + VProd (cells); n ++) cellList[n] = -1;   10
  DO_MOL {
```

♠ pr_03_1

```
        VSAdd (rs, mol[n].r, 0.5, region);
        VMul (cc, rs, invWid);
        c = VLinear (cc, cells) + nMol;
        cellList[n] = cellList[c];                                      15
        cellList[c] = n;
    }
    DO_MOL VZero (mol[n].ra);
    uSum = 0.;
    virSum = 0.;                                                        20
    for (m1z = 0; m1z < cells.z; m1z ++) {
      for (m1y = 0; m1y < cells.y; m1y ++) {
        for (m1x = 0; m1x < cells.x; m1x ++) {
          VSet (m1v, m1x, m1y, m1z);
          m1 = VLinear (m1v, cells) + nMol;                             25
          for (offset = 0; offset < N_OFFSET; offset ++) {
            VAdd (m2v, m1v, vOff[offset]);
            VZero (shift);
            VCellWrapAll ();
            m2 = VLinear (m2v, cells) + nMol;                           30
            DO_CELL (j1, m1) {
              DO_CELL (j2, m2) {
                if (m1 != m2 || j2 < j1) {
                  VSub (dr, mol[j1].r, mol[j2].r);
                  VVSub (dr, shift);                                    35
                  rr = VLenSq (dr);
                  if (rr < rrCut) {
                    rri = 1. / rr;
                    rri3 = Cube (rri);
                    fcVal = 48. * rri3 * (rri3 - 0.5) * rri;            40
                    uVal = 4. * rri3 * (rri3 - 1.) + 1.;
                    VVSAdd (mol[j1].ra, fcVal, dr);
                    VVSAdd (mol[j2].ra, - fcVal, dr);
                    uSum += uVal;
                    virSum += fcVal * rr;                               45
                  }
                }
              }
            }
          }
        }                                                              50
      }
    }
  }
}
```

The above function is longer and more intricate than the all-pairs version in §2.4, but, as we have already indicated, it is incomparably faster for systems of beyond minimal size; within the limits set by numerical rounding it will of course produce the same answers. The basic organization involves scanning cell pairs, namely each cell with itself and with half its neighbors (not only those sharing a common face, but also those with a shared edge or corner); for each pair of cells the atoms contained in each are also paired to determine which of them lie within interaction range. Part of the code is devoted to the special handling of cells adjacent to one or more of the periodic boundaries, and there is an implicit assumption that there are at least three cells in each direction (otherwise the same neighbor will be accessed on both sides).

As indicated above, the array `cellList` plays a dual role: the first part of the array consists of pointers linking different atoms belonging to the same cell, while the remaining elements, one per cell, point to the first atom in each cell; the value −1 indicates the final atom in the list belonging to a cell and an empty cell respectively. If there are roughly as many cells as there are atoms this array requires close to two elements per atom.

It is important to note that there is no check made to ensure that the cell assignments are 'legal'. If there is any risk that a component of `cc` might lie outside the cell array, an indication that something is seriously wrong with the computation since it implies an atom has escaped from the system (this is more likely to happen when using hard walls rather than periodic boundaries), a check of this kind is easily inserted.

Several new constants and vector operations appear in this listing. The operation

```
#define VLinear(p, s)                                    \
   (((p).z * (s).y + (p).y) * (s).x + (p).x)
```

combines the components of a `VecI` quantity into an offset into a singly-indexed array (the array `cells` is inherently three dimensional, but for computational efficiency it is represented as a one-dimensional array that is accessed by just a single index). Another vector operation is

```
#define VVSub(v1, v2)   VSub (v1, v1, v2)
```

There is also an array `vOff` that specifies the offsets of each of the 14 neighbor cells. The array covers half the neighboring cells, together with the cell itself; its size and contents are specified as

```
#define N_OFFSET  14
#define OFFSET_VALS                                      \
   {{0,0,0}, {1,0,0}, {1,1,0}, {0,1,0}, {-1,1,0}, {0,0,1}, \
    {1,0,1}, {1,1,1}, {0,1,1}, {-1,1,1}, {-1,0,1},        \
    {-1,-1,1}, {0,-1,1}, {1,-1,1}}
```

Dealing with periodic boundaries is handled by the definitions

```
#define VCellWrap(t)                                              \
    if (m2v.t >= cells.t) {                                      \
      m2v.t = 0;                                                 \
      shift.t = region.t;                                        \
    } else if (m2v.t < 0) {                                      \      5
      m2v.t = cells.t - 1;                                       \
      shift.t = - region.t;                                      \
    }
#define VCellWrapAll()                                            \
    {VCellWrap (x);                                              \      10
    VCellWrap (y);                                               \
    VCellWrap (z);}
```

Finally, a loop over all atoms belonging to a particular cell is expressed concisely as

```
#define DO_CELL(j, m)                                            \
    for (j = cellList[m]; j >= 0; j = cellList[j])
```

The corresponding two-dimensional version of the function is easily derived by removing all references to the z components and reducing the number of cell offsets to just the first five.

In view of the fact that the majority of the work in this function is carried out inside a highly nested set of loops, it hardly comes as a surprise to learn that there are different ways of organizing the computation. The method used here is to scan over cells, then over offsets, and only then over cell contents; alternatives include scanning over relative cell offsets and then over cells, or scanning the atoms in the outermost loop, with inner loops that scan the neighboring cells of the cell containing the atom together with their contents. Some computer architectures may be sensitive to the method chosen (§17.6), otherwise it is a matter of convenience. Since the cells are often used as part of the neighbor-list method, this issue is usually not critical.

Neighbor-list method

Only a small fraction of the atoms examined by the cell method – an average of $4\pi/81 \approx 0.16$ in three dimensions, $\pi/9 \approx 0.35$ in two – lie within interaction range. If we construct a list of such pairs from those found by the cell method, but in order to allow this list to be useful over several successive timesteps we replace r_c in the test of interatomic separation by

$$r_n = r_c + \Delta r \tag{3.4.1}$$

then it should be possible to benefit from this reduced neighborhood size [ver67]. The success of the approach relies on the slowly changing microscopic environment, which implies that the list of neighbors remains valid over a number of timesteps, typically 10–20, even for relatively small Δr. The fact that the list contains atom pairs that lie outside the interaction range ensures that over this sequence of timesteps no new interacting pairs can appear that are not already listed. The only disadvantage is the additional storage needed for the list of pairs; once, this might have proved an obstacle, but modern computers usually have sufficient memory for all except (possibly) the very largest of systems.

The value of Δr is inversely related to the rate at which the list must be rebuilt, and it also determines the number of extra noninteracting pairs that are included in the list; it therefore has a certain influence on both processing time and storage requirements. The decision to refresh the neighbor list is based on monitoring the maximum velocity at each timestep and waiting until

$$\sum_{steps}\left(\max_i |v_i|\right) > \frac{\Delta r}{2\Delta t} \tag{3.4.2}$$

before doing the refreshing. This criterion, which is equivalent to examining atomic displacements, errs slightly on the conservative side, since it combines contributions from different atoms, but it guarantees that no interacting pairs are ever missed because atoms cannot approach from r_n to r_c during the elapsed time interval; a more precise test could be based on the accumulated motions of individual atoms, but, because the refreshing is already infrequent, the saving will be minimal. Typically, for the fastest computation at liquid densities, $\Delta r \approx 0.3-0.4$.

The neighbor list can be represented in various ways, one of which is a simple table of atom pairs – the method used here♠. An alternative method, used in §12.2, employs a separate list of neighbors for each atom; all lists are stored in a single array with a separate set of indices specifying the range of list entries for each atom. In either instance, the cell method is used to build the neighbor list, with the cell size now being determined by the distance r_n rather than r_c (if the system is too small – relative to r_n – for the cell method to work, then the more costly all-pairs approach must be used to build the list).

The new variables required by the neighbor-list method are

```
real dispHi, rNebrShell;
int *nebrTab, nebrNow, nebrTabFac, nebrTabLen, nebrTabMax;
```

♠ pr_03_2

The quantities that are set in *SetParams* are

```
VSCopy (cells, 1. / (rCut + rNebrShell), region);
nebrTabMax = nebrTabFac * nMol;
```

the initialization in *SetupJob* is

```
nebrNow = 1;
```

and additional input data items

```
NameI (nebrTabFac),
NameR (rNebrShell),
```

are required. The variable *nebrTabFac* determines how much storage should be provided for the neighbor list (per atom), and *rNebrShell* is the variable corresponding to Δr. The memory allocation in *AllocArrays* for the neighbor list is

```
AllocMem (nebrTab, 2 * nebrTabMax, int);
```

The decision as to when to refresh the neighbor list is based on information about the maximum possible movement of the atoms; this is monitored in *EvalProps* by adding

```
real vvMax;
...
vvMax = 0.;
DO_MOL {
  ...                                                      5
  vvMax = Max (vvMax, vv);
}
dispHi += sqrt (vvMax) * deltaT;
if (dispHi > 0.5 * rNebrShell) nebrNow = 1;
```

If a refresh is due, it is carried out during the next timestep in *SingleStep*,

```
...
LeapfrogStep (1);
ApplyBoundaryCond ();
if (nebrNow) {
  nebrNow = 0;                                             5
  dispHi = 0.;
  BuildNebrList ();
}
ComputeForces ();
LeapfrogStep (2);                                          10
...
```

Refreshing the neighbor list implies complete reconstruction. The construction function is very similar to the cell version of *ComputeForces*. The difference is that, instead of computing the interactions, potentially interacting pairs are merely recorded in the neighbor list for subsequent processing (each pair is recorded as two consecutive values). Note the safety check to ensure that the neighbor list does not grow beyond the storage available; the constant *ERR_TOO_MANY_NEBRS* denotes a predefined error code (§18.4).

```
void BuildNebrList ()
{
  VecR dr, invWid, rs, shift;
  VecI cc, m1v, m2v, vOff[] = OFFSET_VALS;
  real rrNebr;                                                           5
  int c, j1, j2, m1, m1x, m1y, m1z, m2, n, offset;

  rrNebr = Sqr (rCut + rNebrShell);
  VDiv (invWid, cells, region);
  for (n = nMol; n < nMol + VProd (cells); n ++) cellList[n] = -1;      10
  DO_MOL {
    VSAdd (rs, mol[n].r, 0.5, region);
    VMul (cc, rs, invWid);
    c = VLinear (cc, cells) + nMol;
    cellList[n] = cellList[c];                                          15
    cellList[c] = n;
  }
  nebrTabLen = 0;
  for (m1z = 0; m1z < cells.z; m1z ++) {
    for (m1y = 0; m1y < cells.y; m1y ++) {                              20
      for (m1x = 0; m1x < cells.x; m1x ++) {
        VSet (m1v, m1x, m1y, m1z);
        m1 = VLinear (m1v, cells) + nMol;
        for (offset = 0; offset < N_OFFSET; offset ++) {
          VAdd (m2v, m1v, vOff[offset]);                                25
          VZero (shift);
          VCellWrapAll ();
          m2 = VLinear (m2v, cells) + nMol;
          DO_CELL (j1, m1) {
            DO_CELL (j2, m2) {                                          30
              if (m1 != m2 || j2 < j1) {
                VSub (dr, mol[j1].r, mol[j2].r);
                VVSub (dr, shift);
                if (VLenSq (dr) < rrNebr) {
                  if (nebrTabLen >= nebrTabMax)                         35
                    ErrExit (ERR_TOO_MANY_NEBRS);
                  nebrTab[2 * nebrTabLen] = j1;
                  nebrTab[2 * nebrTabLen + 1] = j2;
                  ++ nebrTabLen;
                }                                                       40
              }
            }
          }
```

```
          }
        }
      }                                                                45
     }
    }
  }
```

The neighbor list can now be used to compute the interactions; the following function is also partly derived from the cell version of *ComputeForces*.

```
void ComputeForces ()
{
  VecR dr;
  real fcVal, rr, rrCut, rri, rri3, uVal;
  int j1, j2, n;                                                        5

  rrCut = Sqr (rCut);
  DO_MOL VZero (mol[n].ra);
  uSum = 0.;
  virSum = 0.;                                                          10
  for (n = 0; n < nebrTabLen; n ++) {
    j1 = nebrTab[2 * n];
    j2 = nebrTab[2 * n + 1];
    VSub (dr, mol[j1].r, mol[j2].r);
    VWrapAll (dr);                                                      15
    rr = VLenSq (dr);
    if (rr < rrCut) {
      rri = 1. / rr;
      rri3 = Cube (rri);
      fcVal = 48. * rri3 * (rri3 - 0.5) * rri;                          20
      uVal = 4. * rri3 * (rri3 - 1.) + 1.;
      VVSAdd (mol[j1].ra, fcVal, dr);
      VVSAdd (mol[j2].ra, - fcVal, dr);
      uSum += uVal;
      virSum += fcVal * rr;                                             25
    }
  }
}
```

The check for coordinate wraparound associated with periodic boundaries, using the function *ApplyBoundaryCond* (§2.4), is really only necessary when the neighbor list is about to be refreshed, or when properties that depend on the atomic coordinates are to be evaluated[†].

---

† While only a minor detail here, since the extra work is minimal, it becomes a more significant issue when distributed processing is involved (§17.4).

Further methods

For completeness, we make brief mention of two additional techniques that can prove useful, although they are not in widespread use; both aim at reducing the amount of work required for the interaction calculations.

The replication method simplifies the calculation of interactions across periodic boundaries by introducing copies of all atoms that are within a distance r_n of any region boundary, and placing them just outside the simulation region adjacent to the opposite boundary. If these replica atoms are included in the force computation the wraparound checks are no longer required, but the cell array will have to be enlarged to include the region that the replica atoms can occupy. The set of replica atoms need only be rebuilt (taking care to avoid storage overflow) when the neighbor list is refreshed, but the coordinates of these atoms are updated at each timestep. This technique proves particularly useful for distributed and vector processing (Chapter 17), but when the computations are carried out on a single processor the gain is small and usually barely justifies the effort.

The multiple-timestep method is available for medium-range forces that extend beyond several mean atomic spacings (but excluding long-range forces of the Coulomb type which require special treatment – Chapter 13) [str78]. Pairs of interacting neighbors are divided into groups on the basis of their separation, and the contributions of more distant groups are evaluated at less frequent intervals. While the method has proved useful, it is essential to verify that this approximation does not adversely affect the behavior being studied.

Force tabulation

In most MD simulations the bulk of the computation time is spent computing interactions, and every effort is made to ensure that this is done as efficiently as possible. As an alternative to direct evaluation, interactions can be computed using a simple table lookup, possibly accompanied by interpolation for additional accuracy (there are also situations where the potential only exists in tabular form). Which method is faster depends on the complexity of the potential function. For the LJ case direct evaluation is likely to be more efficient, but for a potential involving, for example, exponential functions, tabulating the entire function, or at least certain parts of it, could improve performance.

The value of tabulation can depend on the computer hardware in ways that are not obvious. Just to give one example, several floating-point computations can often be carried out in the time required merely to retrieve one item at random from a large table. So, for extensive simulations, some empirical investigation of this subject should prove worthwhile. If the potential function also depends on molecular orientation the lookup table becomes multidimensional, and storage limitations may prevent construction of a table with adequate resolution.

3.5 Integration methods

Selection criteria

A variety of different numerical methods is available, at least in principle, for integrating the equations of motion [pre92]. Most can be quickly dismissed for the simple reason that the heaviest component of the computation is the force evaluation, and any integration method requiring more than one such calculation per timestep is wasteful, unless it can deliver a proportionate increase in the size of the timestep Δt while maintaining the same accuracy. However, because of the strongly repulsive force at short distances in the typical LJ-based potential, there is, in effect, an upper bound to Δt, so that the well-known Runge–Kutta methods are unable to enlarge the timestep beyond this limit. The same holds true for adaptive methods that change Δt dynamically to maintain a specified level of accuracy; the fact that each atom experiences a rapidly changing environment due to the local rearrangement of its neighborhood will defeat such an approach. Only two classes of method have achieved widespread use, one a low-order leapfrog technique, the other involving a predictor–corrector approach; both appear in various different but equivalent forms.

Obtaining a high degree of accuracy in the trajectories is neither a realistic nor a practical goal. As we will see below, the sharply repulsive potentials result in trajectories for which even the most minute numerical errors grow exponentially with time, rapidly overwhelming the power-law type of local error introduced by any of the numerical integrators. This is not merely a mathematical curiosity, it also corresponds to what happens in nature, and the issue of trajectory accuracy beyond several average 'collision times' is not a meaningful one. So the criteria for choosing a numerical method focus on energy conservation and on the ability to reproduce certain time- and space-dependent correlations to a sufficient degree of accuracy.

Leapfrog-type methods

Two very simple numerical schemes that are widely used in MD are known as the leapfrog and Verlet methods [bee76, ber86b]; they are completely equivalent algebraically. In their simplest form the methods yield coordinates that are accurate to third order in Δt, and, from the point of view of energy conservation when LJ-type potentials are involved, tend to be considerably better than the higher-order methods discussed subsequently. Their storage requirements are also minimal.

The derivation of the Verlet formula (described much earlier by Delambre [lev93]) follows immediately from the Taylor expansion of the coordinate variable – typically $x(t)$,

$$x(t + h) = x(t) + h\dot{x}(t) + (h^2/2)\ddot{x}(t) + O(h^3) \qquad (3.5.1)$$

where t is the current time, and $h \equiv \Delta t$. Here, $\dot{x}(t)$ is the velocity component and $\ddot{x}(t)$ the acceleration. Note that although $\ddot{x}(t)$ has been expressed as a function of t, it is actually a known function – via the force law – of the coordinates at time t. If we add the corresponding expansion for $x(t - h)$ to (3.5.1) and rearrange, we obtain

$$x(t + h) = 2x(t) - x(t - h) + h^2\ddot{x}(t) + O(h^4) \qquad (3.5.2)$$

The truncation error is of order $O(h^4)$ because the h^3 terms cancel[†]. The velocity is not directly involved in the solution, but if required it can be obtained from

$$\dot{x}(t) = [x(t + h) - x(t - h)]/2h + O(h^2) \qquad (3.5.3)$$

with higher-order expressions based on values from earlier timesteps available if needed, though rarely used.

The (highly intuitive [fey63]) leapfrog method is equally simple to derive. Rewrite the Taylor expansion as

$$x(t + h) = x(t) + h[\dot{x}(t) + (h/2)\ddot{x}(t)] + O(h^3) \qquad (3.5.4)$$

The term multiplying h is just $\dot{x}(t + h/2)$, so (3.5.4) becomes (3.5.6) below. The result (3.5.5) is obtained by subtracting from $\dot{x}(t + h/2)$ the corresponding expression for $\dot{x}(t - h/2)$. The leapfrog integration formulae are then

$$\dot{x}(t + h/2) = \dot{x}(t - h/2) + h\ddot{x}(t) \qquad (3.5.5)$$
$$x(t + h) = x(t) + h\dot{x}(t + h/2) \qquad (3.5.6)$$

The fact that coordinates and velocities are evaluated at different times does not present a problem; if an estimate for $\dot{x}(t)$ is required there is a simple connection that can be expressed in either of two ways,

$$\dot{x}(t) = \dot{x}(t \mp h/2) \pm (h/2)\ddot{x}(t) \qquad (3.5.7)$$

The initial conditions can be handled in a similar manner, although a minor inaccuracy in describing the starting state, namely, the distinction between $\dot{x}(0)$ and $\dot{x}(h/2)$, is often ignored. The implementation of this method in a slightly more convenient two-step form – that avoids having coordinates and velocities at different

---

† A possible disadvantage of (3.5.2) is that at low machine precision the h^2 term multiplying the acceleration may prove a source of inaccuracy.

times – appeared in §2.3, and corresponds to

$$\dot{x}(t + h/2) = \dot{x}(t) + (h/2)\ddot{x}(t) \qquad\qquad (3.5.8)$$
$$x(t + h) = x(t) + h\dot{x}(t + h/2) \qquad\qquad (3.5.9)$$

followed by

$$\dot{x}(t + h) = \dot{x}(t + h/2) + (h/2)\ddot{x}(t + h) \qquad\qquad (3.5.10)$$

Predictor–corrector methods

Predictor–corrector (PC) methods [gea71, bee76, ber86b] are multiple-value methods, in the sense that they make use of several items of information computed at
one or more earlier timesteps. In the two most familiar forms of the method there
is a choice between using the acceleration values at a series of previous timesteps –
the multistep (Adams) approach – or using the higher derivatives of the acceleration at the current timestep (the Nordsieck method). For methods accurate to a
given power of h the two forms can be shown to be algebraically equivalent. The
methods are of higher order than leapfrog, but entail a certain amount of extra computation and require storage for the additional variables associated with each atom.
We will focus just on multistep methods, because derivatives of the acceleration –
quantities that are not natural participants in Newtonian dynamics – are absent. The
advantage of using higher derivatives is that h can easily be changed in the course
of the calculation; this is never done in MD.

Since the origin of the numerical coefficients appearing in the PC formulae may
seem a little mysterious we include a brief summary of the derivation. The goal is
to solve the second-order differential equation

$$\ddot{x} = f(x, \dot{x}, t) \qquad\qquad (3.5.11)$$

with $P()$ and $C()$ denoting the formulae used in the predictor and corrector steps
of the calculation. The predictor step for time $t + h$ is simply an extrapolation of
values computed at earlier times $t, t - h, \ldots$, namely,

$$P(x): \quad x(t + h) = x(t) + h\dot{x}(t) + h^2 \sum_{i=1}^{k-1} \alpha_i f(t + [1 - i]h) \qquad (3.5.12)$$

and, for a given value of k, this Adams–Bashforth formula (which contains the
same information as a Taylor expansion) provides exact results for $x(t) = t^q$ provided $q \le k$; in the general case the local error is $O(h^{k+1})$. In order for this to be

true the coefficients $\{\alpha_i\}$ must satisfy the set of $k - 1$ equations

$$\sum_{i=1}^{k-1}(1 - i)^q \alpha_i = \frac{1}{(q + 1)(q + 2)}, \qquad q = 0, \ldots k - 2 \qquad (3.5.13)$$

These and the subsequent sets of linear equations are readily solved; the coefficients are all rational fractions. A similar result holds for $\dot{x}$,

$$P(\dot{x}): \quad h\dot{x}(t + h) = x(t + h) - x(t) + h^2 \sum_{i=1}^{k-1} \alpha_i' f(t + [1 - i]h) \qquad (3.5.14)$$

with coefficients that satisfy equations

$$\sum_{i=1}^{k-1}(1 - i)^q \alpha_i' = \frac{1}{q + 2} \qquad (3.5.15)$$

After computing the value of $f(t + h)$ using the predicted values of x and $\dot{x}$, the corrections are made with the aid of the Adams–Moulton formula (which was originally formulated as a separate implicit method, but subsequently adopted for use as a means of refining the predicted estimate),

$$C(x): \quad x(t + h) = x(t) + h\dot{x}(t) + h^2 \sum_{i=1}^{k-1} \beta_i f(t + [2 - i]h) \qquad (3.5.16)$$

$$C(\dot{x}): \quad h\dot{x}(t + h) = x(t + h) - x(t) + h^2 \sum_{i=1}^{k-1} \beta_i' f(t + [2 - i]h) \qquad (3.5.17)$$

with coefficients obtained from

$$\sum_{i=1}^{k-1}(2 - i)^q \beta_i = \frac{1}{(q + 1)(q + 2)}, \qquad \sum_{i=1}^{k-1}(2 - i)^q \beta_i' = \frac{1}{q + 2} \qquad (3.5.18)$$

Note that the predicted values do not appear in the corrector formulae, except for their involvement in evaluating f. The coefficients $(\alpha_i, \ldots)$ obtained by solving these equations for $k = 4$ and 5 appear in Table 3.1, and those for $k = 4$ are embedded in the integration functions described below. The results are readily adapted to the multivariable MD situation: the first part of the processing involves applying the predictor step to all the variables (atomic coordinates and velocities), followed by the force computation based on the predicted values, and finally the corrector step.

While most of the dynamical problems studied here can be expressed as second-order differential equations, there are cases where first-order equations are required. Analogous PC formulae are available for the equation

$$\dot{x} = f(x, t) \qquad (3.5.19)$$

Table 3.1. PC coefficients for second-order equations.

$k = 4 \, (\times \, 1/24)$	1	2	3	
$P(x)$:	19	−10	3	
$P(\dot{x})$:	27	−22	7	
$C(x)$:	3	10	−1	
$C(\dot{x})$:	7	6	−1	
$k = 5 \, (\times \, 1/360)$	1	2	3	4
$P(x)$:	323	−264	159	−38
$P(\dot{x})$:	502	−621	396	−97
$C(x)$:	38	171	−36	7
$C(\dot{x})$:	97	114	−39	8

Table 3.2. PC coefficients for first-order equations.

$k = 3 \, (\times \, 1/12)$	1	2	3	
$P(x)$:	23	−16	5	
$C(x)$:	5	8	−1	
$k = 4 \, (\times \, 1/24)$	1	2	3	4
$P(x)$:	55	−59	37	−9
$C(x)$:	9	19	−5	1

The predictor and corrector are

$$P(x): \quad x(t + h) = x(t) + h \sum_{i=1}^{k} \alpha_i f(t + [1 - i]h) \tag{3.5.20}$$

$$C(x): \quad x(t + h) = x(t) + h \sum_{i=1}^{k} \beta_i f(t + [2 - i]h) \tag{3.5.21}$$

with coefficients that satisfy

$$\sum_{i=1}^{k} (1 - i)^q \alpha_i = \frac{1}{q + 1}, \qquad \sum_{i=1}^{k} (2 - i)^q \beta_i = \frac{1}{q + 1} \tag{3.5.22}$$

The resulting coefficients are listed in Table 3.2 and incorporated into programs used in later case studies.

The functions that use the $k = 4$ PC method for integrating the MD equations of motion follow; the predicted velocities are not always required but are included

here for use in those cases where they are. Several definitions will first be introduced to simplify the code. The basic operations involved in the PC method are
contained in the definitions

```
#define PCR4(r, ro, v, a, a1, a2, t)                      \
    r.t = ro.t + deltaT * v.t + wr * (cr[0] * a.t +       \
        cr[1] * a1.t + cr[2] * a2.t)
#define PCV4(r, ro, v, a, a1, a2, t)                      \
    v.t = (r.t - ro.t) / deltaT + wv * (cv[0] * a.t +     \          5
        cv[1] * a1.t + cv[2] * a2.t)
```

and, for the particular case considered here,

```
#define PR(t)                                             \
    PCR4 (mol[n].r, mol[n].r, mol[n].rv, mol[n].ra,       \
        mol[n].ra1, mol[n].ra2, t)
#define PRV(t)                                            \
    PCV4 (mol[n].r, mol[n].ro, mol[n].rv, mol[n].ra,      \          5
        mol[n].ra1, mol[n].ra2, t)
#define CR(t)                                             \
    PCR4 (mol[n].r, mol[n].ro, mol[n].rvo, mol[n].ra,     \
        mol[n].ra1, mol[n].ra2, t)
#define CRV(t)                                            \          10
    PCV4 (mol[n].r, mol[n].ro, mol[n].rv, mol[n].ra,      \
        mol[n].ra1, mol[n].ra2, t)
```

where additional quantities $ra1$, $ra2$, ro and rvo, all of type $VecR$, have been
added to the definition of the Mol structure; these quantities are used to hold the
acceleration values from two earlier timesteps (times $t - h$ and $t - 2h$), and to
provide temporary storage for the old coordinates and velocities from time t so
that they can be overwritten by the predicted values (if the predicted velocity is not
required all reference to it can be dropped).

The predictor and corrector functions[♠] can then be written in compact form as

```
void PredictorStep ()
{
    real cr[] = {19., -10., 3.}, cv[] = {27., -22., 7.}, div = 24.,
        wr, wv;
    int n;                                                           5

    wr = Sqr (deltaT) / div;
    wv = deltaT / div;
    DO_MOL {
        mol[n].ro = mol[n].r;                                        10
        mol[n].rvo = mol[n].rv;
        PR (x);
        PRV (x);
```

♠ *pr_03_3, pr_03_4*

```
      PR (y);
      PRV (y);                                                              15
      PR (z);
      PRV (z);
      mol[n].ra2 = mol[n].ra1;
      mol[n].ra1 = mol[n].ra;
    }                                                                       20
  }

  void CorrectorStep ()
  {
    real cr[] = {3., 10., -1.}, cv[] = {7., 6., -1.}, div = 24.,            25
      wr, wv;
    int n;

    wr = Sqr (deltaT) / div;
    wv = deltaT / div;                                                      30
    DO_MOL {
      CR (x);
      CRV (x);
      CR (y);
      CRV (y);                                                             35
      CR (z);
      CRV (z);
    }
  }
```

In *SingleStep*, the changes necessary in order to use this method (including removal of the calls to *LeapfrogStep*) are

```
    PredictorStep ();
    ApplyBoundaryCond ();
    ComputeForces ();
    CorrectorStep ();
    ApplyBoundaryCond ();                                                   5
```

The interactions are evaluated using the results of the predictor step, but are not reevaluated following the corrector; as a consequence, those properties of the system that depend on the interactions themselves, such as the pressure, are based on the predicted rather than the corrected values – the mean error should be insignificant. Variations of this method tried in the past include actually doing this second evaluation – at considerable computational cost – and applying the corrector more than once; neither were found to provide noticeable improvement in accuracy and they are not used. Two calls to the periodic boundary function *ApplyBoundaryCond* have been included here: if the neighbor-list method is used the first call serves no useful purpose and can be omitted; the second call is really only necessary if the neighbor list is due for reconstruction, or if the corrected coordinates are needed for evaluating properties of the system, but because only a small amount of computation is involved it is perhaps safer to leave it in place.

Comparison

Because of the greater flexibility and potentially higher local accuracy, PC methods tend to be suited to more complex problems such as rigid bodies or constrained dynamics, where greater accuracy at each timestep is desirable. The leapfrog approach needs less work and reduced storage, but has the disadvantage that it must sometimes be specially adapted for certain kinds of problem; because of its essentially time-reversible nature, the leapfrog method provides better energy conservation with strongly divergent LJ-type potentials at larger Δt, and because of its minimal storage needs it is suitable for extremely large-scale studies where storage can become an important issue. Tests of comparative accuracy will be given in §3.7.

3.6 Initial state

Initial coordinates

If we assume that the purpose of the simulation is to study the equilibrium fluid state, then the nature of the initial configuration should have no influence whatsoever on the outcome of the simulation. In choosing the initial coordinates, the usual method is to position the atoms at the sites of a lattice whose unit cell size is chosen to ensure uniform coverage of the simulation region. Typical lattices used in three dimensions are the face-centered cubic (FCC) and simple cubic, whereas in two dimensions the square and triangular lattices are used; if the goal is the study of the solid state, then this will dictate the lattice selection. There is little point in laboriously constructing a random arrangement of atoms, typically using a Monte Carlo procedure to avoid overlap, since the dynamics will produce the necessary randomization very quickly[†].

The function that generates an FCC arrangement (with the option of unequal edges) follows; there are four atoms per unit cell, and the system is centered at the origin. Examples of other lattices are shown subsequently.

```
void InitCoords ()
{
  VecR c, gap;
  int j, n, nx, ny, nz;

  VDiv (gap, region, initUcell);
```

5

---

† An obvious way of reducing equilibration time is to base the initial state on the final state of a previous run.

```
  n = 0;
  for (nz = 0; nz < initUcell.z; nz ++) {
    for (ny = 0; ny < initUcell.y; ny ++) {
      for (nx = 0; nx < initUcell.x; nx ++) {                        10
        VSet (c, nx + 0.25, ny + 0.25, nz + 0.25);
        VMul (c, c, gap);
        VVSAdd (c, -0.5, region);
        for (j = 0; j < 4; j ++) {
          mol[n].r = c;                                              15
          if (j != 3) {
            if (j != 0) mol[n].r.x += 0.5 * gap.x;
            if (j != 1) mol[n].r.y += 0.5 * gap.y;
            if (j != 2) mol[n].r.z += 0.5 * gap.z;
          }                                                          20
          ++ n;
        }
      }
    }
  }                                                                  25
}
```

For the FCC lattice, evaluation of the region size in *SetParams* (§2.4) uses the expression

```
  VSCopy (region, 1. / pow (density / 4., 1./3.), initUcell);
```

and the total number of atoms must be changed to

```
  nMol = 4 * VProd (initUcell);
```

Examples of alternative versions of *InitCoords* when other lattice arrangements are used as the initial state, together with other necessary changes, will now be demonstrated. For the simple cubic lattice, where there is only a single atom in each unit cell,

```
void InitCoords ()
{
  VecR c, gap;
  int n, nx, ny, nz;
                                                                     5
  VDiv (gap, region, initUcell);
  n = 0;
  for (nz = 0; nz < initUcell.z; nz ++) {
    for (ny = 0; ny < initUcell.y; ny ++) {
      for (nx = 0; nx < initUcell.x; nx ++) {                        10
        VSet (c, nx + 0.5, ny + 0.5, nz + 0.5);
        VMul (c, c, gap);
        VVSAdd (c, -0.5, region);
```

```
      mol[n].r = c;
      ++ n;
    }
   }
  }
}
```

Minor changes are also required in *SetParams*,

```
  VSCopy (region, 1. / pow (density, 1./3.), initUcell);
  nMol = VProd (initUcell);
```

The body-centered cubic (BCC) lattice has two atoms per unit cell,

```
void InitCoords ()
{
  VecR c, gap;
  int j, n, nx, ny, nz;

  VDiv (gap, region, initUcell);
  n = 0;
  for (nz = 0; nz < initUcell.z; nz ++) {
    for (ny = 0; ny < initUcell.y; ny ++) {
      for (nx = 0; nx < initUcell.x; nx ++) {
        VSet (c, nx + 0.25, ny + 0.25, nz + 0.25);
        VMul (c, c, gap);
        VVSAdd (c, -0.5, region);
        for (j = 0; j < 2; j ++) {
          mol[n].r = c;
          if (j == 1) VVSAdd (mol[n].r, 0.5, gap);
          ++ n;
        }
      }
    }
  }
}
```

and in *SetParams*,

```
  VSCopy (region, 1. / pow (density / 2., 1./3.), initUcell);
  nMol = 2 * VProd (initUcell);
```

The diamond lattice calls for a slightly more complicated version of the FCC code since the lattice is most readily defined as two staggered FCC lattices, one of which is offset along the diagonal by a quarter unit cell.

```
void InitCoords ()
{
  VecR c, gap;
```

```
  real subShift;
  int j, m, n, nx, ny, nz;                                                 5

  VDiv (gap, region, initUcell);
  n = 0;
  for (nz = 0; nz < initUcell.z; nz ++) {
    for (ny = 0; ny < initUcell.y; ny ++) {                                10
      for (nx = 0; nx < initUcell.x; nx ++) {
        VSet (c, nx + 0.125, ny + 0.125, nz + 0.125);
        VMul (c, c, gap);
        VVSAdd (c, -0.5, region);
        for (m = 0; m < 2; m ++) {                                        15
          subShift = (m == 1) ? 0.25 : 0.;
          for (j = 0; j < 4; j ++) {
            VSAdd (mol[n].r, c, subShift, gap);
            if (j != 3) {
              if (j != 0) mol[n].r.x += 0.5 * gap.x;                      20
              if (j != 1) mol[n].r.y += 0.5 * gap.y;
              if (j != 2) mol[n].r.z += 0.5 * gap.z;
            }
            ++ n;
          }                                                               25
        }
      }
    }
  }
}                                                                         30
```

The changes to *SetParams* are as for the BCC, but with the value 2 replaced by 8.

    Returning to two-dimensional systems, the triangular lattice with two atoms per unit cell requires

```
void InitCoords ()
{
  VecR c, gap;
  int j, n, nx, ny;
                                                                          5
  VDiv (gap, region, initUcell);
  n = 0;
  for (ny = 0; ny < initUcell.y; ny ++) {
    for (nx = 0; nx < initUcell.x; nx ++) {
      VSet (c, nx + 0.5, ny + 0.5);                                       10
      VMul (c, c, gap);
      VVSAdd (c, -0.5, region);
      for (j = 0; j < 2; j ++) {
        mol[n].r = c;
        if (j == 1) VVSAdd (mol[n].r, 0.5, gap);                          15
        ++ n;
      }
    }
}
```

```
    }
  }
```

Because the unit cell shape is not square, the region size must be specified differently in *SetParams*,

```
VSet (region, initUcell.x / sqrt (density * sqrt (3.) / 2.),
    initUcell.y / sqrt (density / (2. * sqrt (3.)))));
nMol = 2 * VProd (initUcell);
```

One further initial arrangement is worth including here, namely, a totally random set of initial coordinates. Though not used in the MD programs, it is useful
during analysis of spatial organization, in order to contrast MD results with those
of random point arrays.

```
void InitCoords ()
{
  real randTab[100];
  int i, n, k;

  for (i = 0; i < 100; i ++) randTab[i] = RandR ();
  DO_MOL {
    for (k = 0; k < NDIM; k ++) {
      i = (int) (100. * RandR ());
      VComp (mol[n].r, k) = (randTab[i] - 0.5) * VComp (region, k);
      randTab[i] = RandR ();
    }
  }
}
```

The function *RandR* (§18.4) serves as a source of uniformly distributed random
values in the range (0, 1). To reduce any possible unwanted correlations in the
random numbers a shuffling scheme is employed; the random values are used as
indices for accessing a table of random numbers, the entries of which are replaced
each time they are used. Referencing a particular component of a vector is done by

```
#define VComp(v, k)                                          \
    *((k == 0) ? &(v).x : ((k == 1) ? &(v).y : &(v).z))
```

Initial velocities

Similar considerations apply to the initial velocities, namely, that rapid equilibration renders the careful fabrication of a Maxwell distribution unnecessary. The simple function *InitVels* of §2.4 can be used, with *VRand* (§18.4) now producing a
random unit vector in three dimensions.

Initialization of integration variables

In addition to setting the initial values for the more obvious physical quantities, the numerical integrator requires its own initializing. For the leapfrog method,

```
void InitAccels ()
{
  int n;

  DO_MOL {                                                    s
    VZero (mol[n].ra);
  }
}
```

and for the PC method add

```
        VZero (mol[n].ra1);
        VZero (mol[n].ra2);
```

While these are of course not the correct values, there is little benefit in doing a more careful job, such as using a self-starting Runge–Kutta method for the first few timesteps. One reason for this is that the trajectories are highly sensitive to computational details such as rounding error (see below), and this has a much stronger influence than the precise details of the initial state; the other is that additional velocity adjustments are usually made early in the run to drive the system to the desired temperature.

Temperature adjustment

Bringing the system to the required average temperature calls for velocity rescaling. If there is a gradual energy drift due to numerical integration error, further velocity adjustments will be required over the course of the run. The drift rate depends on a number of factors – the integration method, potential function, the value of Δt and the ambient temperature.

Since T fluctuates there is no point in making adjustments based on instantaneous estimates. Instead, we can make use of the average $\langle T \rangle$ values that are already available. The temperature adjustment (or velocity rescaling) function below would therefore be called from *SingleStep* immediately following the call *AccumProps(2)* used for summarizing the results.

```
void AdjustTemp ()
{
  real vFac;
  int n;
```

```
                                                                    5
    vvSum = 0.;
    DO_MOL vvSum += VLenSq (mol[n].rv);
    vFac = velMag / sqrt (vvSum / nMol);
    DO_MOL VScale (mol[n].rv, vFac);
  }                                                                 10
```

How frequently this adjustment is required, if at all, must be determined empirically; initially it should be omitted since it may interfere with energy conservation.
If needed, the interval between adjustments would be specified by the variable

```
int stepAdjustTemp;
```

the value included in the input data

```
NameI (stepAdjustTemp),
```

and the adjustment made by a call from *SingleStep*

```
if (stepCount % stepAdjustTemp == 0) AdjustTemp ();
```

A possible alternative would be to automate the scheme, applying the adjustment
whenever the drift exceeds a given threshold.

Forcing the system to have the correct $\langle T \rangle$ during the equilibration phase of the
simulation uses separate estimates of $\langle E_K \rangle$. In *SingleStep* add (after the call to
EvalProps)

```
if (stepCount < stepEquil) AdjustInitTemp ();
```

and introduce the new function

```
void AdjustInitTemp ()
{
  real vFac;
  int n;

                                                                   5
  kinEnInitSum += kinEnergy.val;
  if (stepCount % stepInitlzTemp == 0) {
    kinEnInitSum /= stepInitlzTemp;
    vFac = velMag / sqrt (2. * kinEnInitSum);
    DO_MOL VScale (mol[n].rv, vFac);                               10
    kinEnInitSum = 0.;
  }
}
```

with extra variables

```
real kinEnInitSum;
int stepInitlzTemp;
```

an additional input data item

```
NameI (stepInitlzTemp),
```

and initialization

```
kinEnInitSum = 0.;
```

3.7 Performance measurements

Accuracy

In order to demonstrate the accuracy of the integration methods, and the way in which accuracy depends on Δt, we will carry out a series of measurements of the energy as a function of time for the leapfrog and $k = 4$ PC integrators with several timestep values. We use the soft-sphere interaction rather than LJ in order to avoid any additional fluctuations due to the discontinuity at r_c.

Input data to the calculation are as follows:

```
deltaT            0.00125
density           0.8
initUcell         5 5 5
nebrTabFac        8
rNebrShell        0.4
stepAvg           8000
stepEquil         0
stepInitlzTemp    999999
stepLimit         160000
temperature       1.
```

In the series of runs `deltaT` varies over a 16 : 1 range between 0.001 25 and 0.02, the value of `stepLimit` is chosen to give a total run length of 200 time units (the extreme values being 160 000 and 10 000), and `stepAvg` is set so that a result is output every ten time units. The initial state is a simple cubic lattice, so that $N_m = 125$, and computations are carried out in 64-bit (double) precision. This particular value of `nebrTabFac` is more than adequate for the soft-sphere fluid at moderate density; for an LJ fluid the value depends on r_c, with 50 or larger sometimes being required. A few brief test runs with the actual potential function

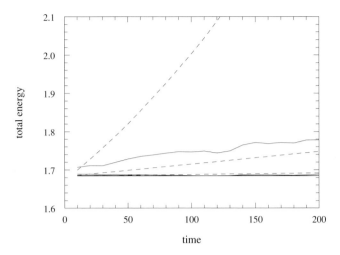

Fig. 3.2. Energy drift for different values of Δt; leapfrog results (solid curves – the first four of which are indistinguishable) are for $\Delta t = 0.001\,25$, 0.0025, 0.005, 0.01 and 0.02; predictor–corrector results (dashed curves) are for $\Delta t = 0.001\,25$, 0.0025 and 0.005.

Table 3.3. Energy conservation for leapfrog (LF) and predictor–corrector (PC) methods.

| Δt: | 0.001 25 | 0.001 25 | 0.0025 | 0.0025 | 0.005 | 0.01 |
t	LF	PC	LF	PC	LF	LF
10	1.6846	1.6848	1.6848	1.6864	1.6858	1.6890
50	1.6846	1.6864	1.6846	1.6991	1.6857	1.6876
100	1.6846	1.6883	1.6848	1.7153	1.6843	1.6853
150	1.6846	1.6902	1.6847	1.7316	1.6853	1.6867
200	1.6847	1.6921	1.6848	1.7482	1.6852	1.6886

in the density range of interest should be sufficient to determine an appropriate value, including a safety margin.

The results are shown in Figure 3.2, where it is clear that the leapfrog method allows a much larger Δt for a given degree of energy conservation [ber86b]. To emphasize the accuracy of the method – from the energy point of view – most of these results are repeated in Table 3.3. To a limited extent, accuracy can be sacrificed in the cause of speed, for example when the goal is a realtime demonstration, but there are limits to the size of Δt if numerical instability is to be avoided. All computations in the case studies will use 64-bit precision (although, in many cases, 32-bit arithmetic could be used without significantly affecting the results).

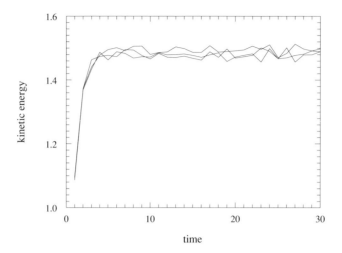

Fig. 3.3. Convergence of mean kinetic energy from different initial states.

Reproducibility

The issue of reproducibility is tied to the rate of approach to equilibrium (or to a stationary final state). In most cases, once the system has equilibrated there will be no memory of the details of the initial state, but problems can arise in cases of very slow convergence, or where there are different metastable states in which the system can become trapped. If we exclude such special circumstances, the averaged results from separate runs should agree to within the limits set by the fluctuations. As a brief demonstration we show how the kinetic energy varies with time for simulations that differ only in the choice of the initial random velocities.

The input data for this test are:

`deltaT`	`0.005`
`density`	`0.8`
`initUcell`	`6 6 6`
`nebrTabFac`	`8`
`randSeed`	`17`
`rNebrShell`	`0.4`
`stepAvg`	`200`
`stepEquil`	`1000`
`stepInitlzTemp`	`200`
`stepLimit`	`6000`
`temperature`	`1.`

The runs use different values of `randSeed` (such as 17, 18 and 19), and the results are shown in Figure 3.3. Convergence of $\langle E_K \rangle$ to its final value is dominated by

Table 3.4. Timing measurements (μs/atom–step) for the different methods.

N_m	pairs	cells	nebrs
64	2.2	2.0	0.7
216	7.0	2.3	0.7
512	16.0	2.2	0.7
1000	–	2.2	0.7

the temperature adjustments that are made while $t < 5$, but the differences between the runs lie within the range of the fluctuations.

Efficiency

Two methods of improving the efficiency of the force computation were described here, cells and neighbor lists. To show that these methods really do provide at least some of the promised benefits we provide a few timing comparisons for the three methods. It is the relative timings and not the the absolute values that are of most interest here[†]. The major cause of any timing anomalies is the size of the cells used in both the cell and neighbor-list methods; the fact that an integral number of cells must fit along each region edge can lead to variations in the mean cell occupancy that will affect performance, especially for smaller systems.

Assorted timing results for soft-sphere systems at $\rho = 0.8$ are shown in Table 3.4. The runs extend over 4000 timesteps each, with $T = 1$ and $\Delta t = 0.005$; leapfrog integration is used (the PC method takes slightly longer). When rNebrShell has the value 0.4 the neighbor list is typically refreshed every 12–15 timesteps. If the theoretical performance expectations are not met in these relatively small systems, allowance should be made for contributions from other parts of the computation as well as the features of the processor architecture[‡].

3.8 Trajectory sensitivity

One particular consequence of the numerical approach deserves special consideration. We have seen how measurements of bulk properties, such as kinetic energy,

---

[†] These particular measurements were made on a 2 GHz Intel Pentium 4 Xeon processor (compiled with GNU C using the optimization option -O3); all case studies were run on this system.

[‡] The way data are organized can sometimes affect performance. Patterns of memory access can have a significant impact on computation speed, especially in modern computers with complex hardware architectures that include mapped and interleaved memory and multilevel caches. Awareness of the issues involved, a subject that can demand some familiarity with the specific processor design, may suggest how to arrange data and organize the loop structure of a program; this is a specialized subject – ignored by most – that will not be covered here.

are reproducible, subject only to well-understood statistical fluctuations. Other equilibrium and steady state properties are similarly well-behaved. When it comes to the trajectories themselves it is an entirely different story: trajectories display an exponential sensitivity to even the most minute perturbation. This implies that trajectories are sensitive to the precision and rounding method used for floating-point arithmetic, and even to the exact sequence of machine instructions in the program. In the absence of infinite precision, there is no way in which two different MD programs, or even the same MD program run on computers of different design, will yield the same trajectories[†]. This is hardly surprising, but it is also irrelevant, since there is no meaningful physical quantity that depends on just a single trajectory realization; all meaningful measurements involve averages that conceal this sensitivity, including the transport properties based on integration along the actual trajectories (§5.2). This extreme sensitivity is the microscopic basis for molecular chaos that plays such an important role in statistical mechanics; though the equations of motion are time reversible, this fact turns out to be unobservable in most practical situations [orb67, lev93].

To actually measure this behavior we consider a system of $2N_m$ atoms in which odd- and even-numbered atoms form independent but identical subsystems that are assigned the same initial coordinates and velocities. One subsystem is slightly perturbed by multiplying its velocities by $1 + \epsilon s$, where ϵ is a small number and s a random value in the range $(-1, 1)$, and we then examine how the root-mean-square coordinate difference

$$\Delta r = \sqrt{\frac{1}{N_m} \sum_{i=1}^{N_m} (r_{2i} - r_{2i-1})^2} \tag{3.8.1}$$

varies with time. Although the study uses soft atoms and a leapfrog method subject to numerical integration error, this error is not the dominant factor, because similar results can also be obtained in hard-sphere studies that are free from integration error.

Only a few simple modifications[♠] to the MD program are required. The atoms are divided into two entirely separate subsystems, and `nMol` is doubled. The

---

† And so the Laplacian vision is laid to rest.
♠ `pr_03_5`

criterion for selecting pairs in *BuildNebrList* is replaced by

```
if ((j1 - j2) % 2 == 0 && (m1 != m2 || j2 < j1))
```

Properties such as the energy can be computed for either or both subsystems by selecting which atoms contribute to the various sums. A simple addition to the innermost loop of *InitCoords* produces consecutive pairs of atoms (n and $n+1$) with the same initial coordinates,

```
mol[n + 1].r = mol[n].r;
n += 2;
```

and *InitVels* is replaced by

```
void InitVels ()
{
  int n;

  VZero (vSum);                                               5
  for (n = 0; n < nMol; n += 2) {
    VRand (&mol[n].rv);
    VScale (mol[n].rv, velMag);
    mol[n + 1].rv = mol[n].rv;
    VVSAdd (vSum, 2., mol[n].rv);                             10
  }
  DO_MOL VVSAdd (mol[n].rv, - 1. / nMol, vSum);
}
```

The trajectory perturbation function is

```
void PerturbTrajDev ()
{
  VecR w;
  int n;
                                                              5
  for (n = 0; n < nMol; n += 2) {
    mol[n + 1].r = mol[n].r;
    VRand (&w);
    VMul (w, w, mol[n].rv);
    VSAdd (mol[n + 1].rv, mol[n].rv, pertTrajDev, w);         10
  }
  countTrajDev = 0;
}
```

and trajectory analysis, allowing for periodic boundaries, is carried out by

```
void MeasureTrajDev ()
{
  VecR dr;
  real dSum;
  int n;                                                               5

  dSum = 0.;
  for (n = 0; n < nMol; n += 2) {
    VSub (dr, mol[n + 1].r, mol[n].r);
    VWrapAll (dr);                                                     10
    dSum += VLenSq (dr);
  }
  valTrajDev[countTrajDev] = sqrt (dSum / (0.5 * nMol));
  ++ countTrajDev;
}                                                                      15
```

The additional code needed in *SingleStep* consists of

```
if (stepCount == stepEquil) PerturbTrajDev ();
if (stepCount > stepEquil &&
    (stepCount - stepEquil) % stepTrajDev == 0) {
  MeasureTrajDev ();
  if (countTrajDev == limitTrajDev) {                                 5
    PrintTrajDev (stdout);
    PerturbTrajDev ();
    BuildNebrList ();
  }
}                                                                     10
```

and output is produced by

```
void PrintTrajDev (FILE *fp)
{
  real tVal;
  int n;
                                                                      5
  for (n = 0; n < limitTrajDev; n ++) {
    tVal = (n + 1) * stepTrajDev * deltaT;
    fprintf (fp, "%.4e %.4e\n", tVal, valTrajDev[n]);
  }
}                                                                     10
```

New variables needed for these measurements are

```
real *valTrajDev, pertTrajDev;
int countTrajDev, limitTrajDev, stepTrajDev;
```

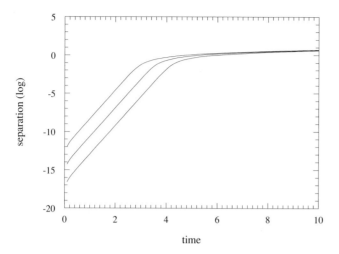

Fig. 3.4. Trajectory divergence for different initial velocity perturbations (the vertical scale is logarithmic).

with additional input data items in `nameList`

```
NameI (limitTrajDev),
NameR (pertTrajDev),
NameI (stepTrajDev),
```

and an array allocated in `AllocArrays`

```
AllocMem (valTrajDev, limitTrajDev, real);
```

The measurements shown in Figure 3.4 are based on a soft-sphere system with an FCC initial state, $N_m = 2048$, $T = 1$, $\rho = 0.8$, and $\Delta t = 0.005$. Other input data items include

```
limitTrajDev        100
pertTrajDev         1.0e-6
stepEquil           3000
stepTrajDev         20
```

Three values of the velocity perturbation `pertTrajDev` are used, namely, 10^{-6}, 10^{-5} and 10^{-4}; just one set of measurements averaged over all atoms is made after allowing sufficient time for equilibration. The linear growth in $\log(\Delta r)$, measured from time zero when the perturbation is applied and, depending on the perturbation, extending to times between 2.5 and 4, corresponds to trajectories that diverge at an exponential rate. Once the size of the deviation reaches the atomic diameter (≈ 1) the more familiar diffusive processes take over.

3.9 Further study

3.1   Implement the cell and neighbor-list methods in two dimensions.

3.2   See whether the $k = 5$ PC method is an improvement over the $k = 4$ method used here.

3.3   Explore the use of PC methods involving derivatives of the acceleration [bee76, ber86b].

3.4   Determine the performance benefits of tabulated interactions.

3.5   How is the computation speed affected by the way the data are organized?

3.6   How is energy conservation affected by smoothing (as in §12.3) the LJ interaction near r_c?

3.7   Investigate the use of multiple-timestep methods.

4

Equilibrium properties of simple fluids

4.1 Introduction

In this chapter we examine the behavior of systems in equilibrium; in particular, we focus on measurements of thermodynamic properties and studies of spatial structure and organization. The treatment of properties associated with the motion of atoms – the dynamical behavior – forms the subject of Chapter 5.

While basic MD simulation methods – formulating and solving the equations of motion – fall into a comparatively limited number of categories, a wide range of techniques is used to analyze the results. Rarely is the wealth of detail embodied in the atomic or molecular trajectories of particular interest in itself, and the issue is how to extract meaningful information from this vast body of data; even a small system of 10^3 structureless atoms followed over a mere 10^4 timesteps can produce up to 6×10^7 numbers, corresponding to a full chronological listing of the atomic coordinates and velocities. A great deal of data averaging and filtration of various kinds is required to reduce this to a manageable and meaningful level; how this is achieved depends on the questions that the simulation is supposed to answer. Much of this processing will be carried out while the simulation is in progress, but some kinds of analysis are best done subsequently, using data saved in the course of the simulation run; the choice of approach is determined by the amount of work and data involved, as well as the need for active user participation in the analysis.

Averages corresponding to thermodynamic quantities in homogeneous systems at equilibrium are the easiest measurements to make. Statistical mechanics relates such MD averages to their thermodynamic counterparts, and the ergodic hypothesis can be invoked to justify equating trajectory averages with ensemble-based thermodynamic properties [mcq76]. However, the fact that statistical mechanics has no knowledge of trajectories means that it is incapable of discussing quantities that are defined in terms of atomic motion – diffusion for example. This is the strength of MD; detailed trajectory histories are available, so that not only can quantities meaningful in a statistical mechanical framework be addressed, but so, too, can any other conceivable quantity.

Some aspects of behavior, such as the structural correlations present in the fluid, ranging from the basic pair correlation function to more subtle correlations involving both position and orientation, or the three-body correlation function, require quite heavy calculations, often rivaling the interaction computation in terms of the amount of work required. Fortunately, such calculations are not needed at each timestep since fluid structure changes only gradually; the rate of change is indeed the criterion for choosing the interval between such measurements.

If the system is spatially inhomogeneous, all quantities, from the simplest thermodynamic values onward, must be based on localized measurements. If the system is also nonstationary over time, long term time averaging is ruled out because it would obliterate the very effects being studied. In short, the more complex the phenomenon the more demanding the measurement task. These topics will be encountered in Chapters 7 and 15.

4.2 Thermodynamic measurements

Relation to statistical mechanics

Measurements of equilibrium properties that are thermodynamic in nature can be regarded as exercises in numerical statistical mechanics. In such instances MD provides an alternative to Monte Carlo, and if no further information is required about the system, computational efficiency alone should determine the choice of technique. While Monte Carlo requires less computation per interacting atom pair, because only the potential energy has to be evaluated, the number of Monte Carlo cycles required to obtain uncorrelated samples (more precisely, a series of samples that are only weakly correlated) may exceed the corresponding number of MD timesteps. The reason for this is that the atomic displacements are randomly chosen in Monte Carlo, and this can be a less efficient way for the system to traverse configuration space than via the cooperative dynamics intrinsic to MD.

Because both the number of atoms and the total energy (assuming that numerical drift has been suppressed) are fixed in the MD simulations encountered so far, the relevant statistical mechanical ensemble for discussing equilibrium behavior is the microcanonical (NVE) one. There is just one minor difference, in that each conserved momentum component removes one degree of freedom, but this is a negligible effect for systems beyond a minimal size.

Error analysis

The measurement process in MD is very similar to experiment. But the experimentalist often has the advantage of knowing that each estimate is independent, allowing well-established statistical methods to be used in the data analysis. With MD, where a series of measurements is carried out in the course of a simulation of limited duration, there is no guarantee that successive estimates are sufficiently unrelated to ensure the reliability of these simple statistical methods. Averages of directly measured quantities may not be the main problem, given an adequate run length, but statistical error estimates are particularly sensitive to correlations between samples.

We assume that the problem has been correctly formulated and implemented; errors in the results can then be categorized as follows. There are systematic errors associated with, for example, finite-size effects, interaction cutoff, and the numerical integration itself; these are an intrinsic part of the computer experiment and are reproducible. There are errors due to inadequate sampling of phase space where, especially near a thermodynamic phase boundary, or in the case of infrequently occurring events, enough of the relevant behavior fails to be sampled; this is symptomatic of poor experimental design. And finally there is statistical error due to random fluctuations in the measurements; under normal circumstances this determines the degree of confidence that can be placed in the results. Only for errors of the last kind is the usual statistical analysis applicable.

Consider a series of M measurements of some fluctuating property A in a system at equilibrium. The mean value is

$$\langle A \rangle = \frac{1}{M} \sum_{\mu=1}^{M} A_\mu \tag{4.2.1}$$

and if each measurement A_μ is independent, with variance

$$\sigma^2(A) = \frac{1}{M} \sum_\mu \left(A_\mu - \langle A \rangle \right)^2 = \langle A^2 \rangle - \langle A \rangle^2 \tag{4.2.2}$$

then the variance of the mean $\langle A \rangle$ is

$$\sigma^2(\langle A \rangle) = \frac{1}{M} \sigma^2(A) \tag{4.2.3}$$

But if, as is usually the case in MD (and other) simulations, the assumed independence of the A_μ is unwarranted, $\sigma^2(\langle A \rangle)$ is liable to be underestimated because the effective number of independent measurements is considerably less than M. How the correlation between measurements affects the results can be seen by rewriting the variance correctly as

$$\sigma^2(\langle A \rangle) = \frac{1}{M} \sigma^2(A) \left[1 + 2 \sum_\mu (1 - \mu/M) \phi_\mu \right] \tag{4.2.4}$$

where ϕ_μ is the autocorrelation function

$$\phi_\mu = \frac{\langle A_\mu A_0 \rangle - \langle A \rangle^2}{\langle A^2 \rangle - \langle A \rangle^2} \tag{4.2.5}$$

A detailed error analysis would involve examining ϕ_μ, but there is little need for this in practice because a much simpler method is available based on block averaging [fly89].

Assuming the A_μ to be correlated, if averages are evaluated over blocks of successive values, then as the block size increases the block averages will be decreasingly correlated; eventually, once the block length exceeds the (unknown) longest correlation time present in the data, the block averages will be independent from a statistical point of view. What is needed is a criterion for choosing the minimal necessary block length: too short a block provides little improvement over the original correlated data, too long a block reduces the number of block averages available for reliable estimation of the variance of the final result.

A very straightforward scheme is based on a series of successive block sizes $b = 1, 2, 4, \ldots$, with the upper bound being set by the total size of the data set. For each b the estimator for the variance can be shown to be

$$\sigma^2(\langle A \rangle_b) = \frac{1}{M_b - 1} \sum_{\beta=1}^{M_b} (A_\beta^2 - \langle A \rangle_b^2) \tag{4.2.6}$$

where M_b is the total number of blocks, A_β a typical block average, and $\langle A \rangle_b$ the overall average. Whenever the current M_b is odd, the last value is simply discarded before doubling the block size. What should happen, assuming that the total measurement period far exceeds the longest correlation time, is that the successive $\sigma^2(\langle A \rangle_b)$ increase until a plateau is eventually reached; the plateau value is the result. In less than ideal situations where the measurement period is too short, or barely adequate in length, the plateau will either not appear at all or will be very narrow; in such cases the variance estimate is unreliable. When the method works successfully the block size at the start of the plateau is an indication of the extent to which the samples are correlated.

Energy

Energy measurements[♠] are the simplest, and here we briefly examine both LJ and soft-sphere systems. The initial state is an FCC lattice, so that $N_m = 500$. Leapfrog integration is used; for the LJ system we use a cutoff $r_c = 2.2$. The temperature fluctuates, and in three dimensions we have $\langle E_K \rangle = 3\langle T \rangle / 2$. Figure 4.1

♠ pr_04_1, pr_04_2

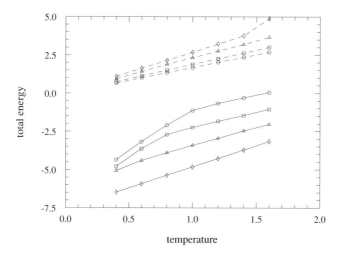

Fig. 4.1. Density and mean-temperature dependence of energy for Lennard-Jones (solid curves) and soft–sphere (dashed) systems, for densities 0.4–1.0.

shows the dependence of E on ρ and $\langle T \rangle$ as measured in a series of runs that include the following input data (the values of `density` and `temperature` are varied):

`deltaT`	`0.005`
`density`	`0.4`
`initUcell`	`5 5 5`
`stepAvg`	`2000`
`stepEquil`	`4000`
`stepInitlzTemp`	`200`
`stepLimit`	`10000`
`temperature`	`0.4`

For careful quantitative studies, the results should be examined closely when deciding on the run length `stepLimit` and the equilibration period `stepEquil`.

In the microcanonical ensemble, thermodynamic quantities based on fluctuations adopt a different form from the canonical ensemble. The most familiar such quantity is the constant-volume specific heat $C_V = (\partial E / \partial T)_V$. It is usually defined in terms of energy fluctuations, namely (with $k_B = 1$),

$$C_V = \frac{N_m}{T^2} \langle \delta E^2 \rangle \tag{4.2.7}$$

where $\langle \delta E^2 \rangle = \langle E^2 \rangle - \langle E \rangle^2$, but while this is appropriate in the canonical ensemble, for MD we have $\langle \delta E^2 \rangle = 0$. Instead, it can be shown [leb67] that the relevant

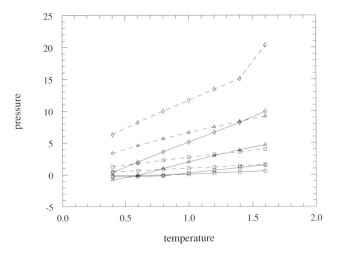

Fig. 4.2. Density and temperature dependence of pressure for Lennard-Jones (solid curves) and soft-sphere (dashed) systems, for densities 0.4–1.0.

fluctuations to consider are those of E_K or E_U individually (they are identical) and that the specific heat is

$$C_V = \frac{3}{2}\left(1 - \frac{2N_m\langle\delta E_K^2\rangle}{3T^2}\right)^{-1} \tag{4.2.8}$$

Either this directly measurable result or numerical differentiation of the $E(T)$ graph – strictly speaking, $E(\langle T\rangle)$ – could be used for estimating C_V.

Equation of state

Pressure is obtained from the virial expression (2.3.8); while it can also be expressed in terms of momentum transferred across an arbitrary plane, there is little reason to resort to such a definition that only uses information from a fraction of the atoms and is therefore subject to larger fluctuations. The virial definition assumes the presence of hard walls responsible for imposing the external pressure, but the result is equally applicable in the case of periodic boundaries [erp77].

Pressure measurements for the runs described above are shown in Figure 4.2. Negative pressure is an indication that the system is being held at too low a density, and in a sufficiently large system separation into distinct liquid (or solid) and vapor phases occurs. A more extensive analysis of this kind would lead to the complete equation of state [nic79]. In the LJ case, when a 'real' substance is being modeled, the values of both E and P can be corrected [vog85] to compensate for the truncation at r_c – as in (4.3.9).

Table 4.1. *Block-averaged estimates of $\sigma(\langle P \rangle)$; b is the block size and M_b the number of blocks.*

b	M_b	$\sigma(\langle P \rangle)$
1	16384	0.0012
2	8192	0.0017
4	4096	0.0023
8	2048	0.0031
16	1024	0.0039
32	512	0.0045
64	256	0.0046
128	128	0.0048
256	64	0.0048
512	32	0.0052
1024	16	0.0060
2048	8	0.0068
4096	4	0.0088

Finite-size effects are already relatively small at $N_m = 500$, at least for positive pressure (and away from the critical point). For example, consider the LJ fluid at $\rho = 0.8$ and $T = 1$. The result obtained in this case is $P = 2.02$, $\sigma(P) = 0.14$, based on a single average over 2000 timesteps. For the case where $N_m = 2048$, an average over 4000 timesteps leads to the same $P = 2.02$, with $\sigma(P) = 0.08$. Thus, even this very rough comparison suggests that size dependence will normally only be an issue if high-quality estimates are required.

The pressure measurements provide an opportunity to demonstrate the block averaging method♠ for estimating the variance of the mean described earlier. Here we consider the soft-sphere fluid with $N_m = 500$, $\rho = 0.8$ and $T = 1$. The pressure measurements are governed by the replacement input data

```
stepAvg          1
stepEquil        1000
stepLimit        17384
```

and the results are sufficient for 12 doublings of the block size starting from $b = 1$. Table 4.1 shows the outcome of this analysis, and reveals that convergence occurs at a block size of 32. The fact that $\sigma(\langle P \rangle) \approx 4\sigma(P)$, the value at $b = 1$, should serve as a reminder that closely spaced measurements are strongly correlated.

♠ `pr_anblockavg` (There are a few supplementary programs in the software package that are not described in the text; this is one of them.)

4.3 Structure

Radial distribution function

The fluid state is characterized by the absence of any permanent structure. There are, nevertheless, well-defined structural correlations that can be measured experimentally to provide important details about the average molecular organization [mcq76, han86b]. The treatment of structural correlation (in the canonical ensemble) begins with the completely general pair distribution function,

$$g(\boldsymbol{r}_1, \boldsymbol{r}_2) = \frac{N_m(N_m - 1) \int e^{-U(\boldsymbol{r}_1, \ldots \boldsymbol{r}_{N_m})/T} \, d\boldsymbol{r}_3 \cdots \boldsymbol{r}_{N_m}}{\rho^2 \int e^{-U(\boldsymbol{r}_1, \ldots \boldsymbol{r}_{N_m})/T} \, d\boldsymbol{r}_1 \cdots \boldsymbol{r}_{N_m}} \tag{4.3.1}$$

where the integral in the denominator is just the partition function (with $k_B = 1$), and the integral in the numerator differs only in that $\boldsymbol{r}_1$ and $\boldsymbol{r}_2$ are excluded from the integration. In the case of spatially homogeneous systems, only relative separation is meaningful, leading to a sum over atom pairs,

$$g(\boldsymbol{r}) = \frac{2V}{N_m^2} \left\langle \sum_{i<j} \delta(\boldsymbol{r} - \boldsymbol{r}_{ij}) \right\rangle \tag{4.3.2}$$

and if the system is also isotropic the function can be averaged over angles without loss of information. The result is the radial distribution function $g(r)$ – RDF for short – a function that describes the spherically averaged local organization around any given atom; $g(r)$ plays a central role in liquid-state physics and all functions that depend on the pair separation, such as potential energy and pressure, can be expressed in terms of integrals involving $g(r)$.

The definition of $g(r)$ implies that $\rho g(r) \, d\boldsymbol{r}$ is proportional to the probability of finding an atom in the volume element $d\boldsymbol{r}$ at a distance r from a given atom, and (in three dimensions) $4\pi\rho g(r)r^2 \Delta r$ is the mean number of atoms in a shell of radius r and thickness Δr surrounding the atom. The RDF is related to the experimentally measurable structure factor $S(\boldsymbol{k})$ by Fourier transformation – $S(\boldsymbol{k})$ is a key quantity in interpreting x-ray scattering measurements. The general result, not assuming isotropy, is

$$S(\boldsymbol{k}) = 1 + \rho \int g(\boldsymbol{r})e^{-i\boldsymbol{k}\cdot\boldsymbol{r}} \, d\boldsymbol{r} \tag{4.3.3}$$

and for isotropic liquids this simplifies to

$$S(\boldsymbol{k}) = 1 + 4\pi\rho \int \frac{\sin kr}{kr} g(r)r^2 \, dr \tag{4.3.4}$$

Equation (4.3.4) supplies an important link between MD simulation and the real world. The MD approach can, of course, provide the answer to any question about structure, such as the nature of spatial correlations between atoms taken three at a time; while this kind of information can prove useful in trying to understand the behavior, comparison is impossible since the corresponding experimental data are unobtainable.

From the definition of $g(r)$ in (4.3.2) it is apparent that the RDF can be measured [rah64, ver68] using a histogram of discretized pair separations. If h_n is the number of atom pairs (i, j) for which

$$(n-1)\Delta r \leq r_{ij} < n\Delta r \tag{4.3.5}$$

then, assuming that Δr is sufficiently small, we have the result[†]

$$g(r_n) = \frac{V h_n}{2\pi N_m^2 r_n^2 \Delta r} \tag{4.3.6}$$

where

$$r_n = \left(n - \tfrac{1}{2}\right)\Delta r \tag{4.3.7}$$

If the RDF measurements extend out to a maximum range r_e the required number of histogram bins is $r_e/\Delta r$. The two-dimensional version is

$$g(r_n) = \frac{A h_n}{\pi N_m^2 r_n \Delta r} \tag{4.3.8}$$

The normalization factors ensure that $g(r \to \infty) = 1$, although periodic boundaries limit the range r_e to no more than half the smallest edge of the simulation region, with wraparound used in evaluating interatomic distances.

The RDF computation has much in common with the interaction calculation, and the cell method can be used if r_e is sufficiently small. Otherwise, all pairs must be considered and this is the version[♠] shown here (where M_PI denotes π); quite accurate results can in fact be obtained from a relatively small number of measurements, so that the overall computational cost is not excessive.

```
void EvalRdf ()
{
  VecR dr;
  real deltaR, normFac, rr;
  int j1, j2, n;                                            5

  if (countRdf == 0) {
    for (n = 0; n < sizeHistRdf; n ++) histRdf[n] = 0.;
  }
  deltaR = rangeRdf / sizeHistRdf;                          10
```

† h_n is $N_m/2$ times the mean number of neighbors in the shell.
♠ pr_04_3

```
    for (j1 = 0; j1 < nMol - 1; j1 ++) {
      for (j2 = j1 + 1; j2 < nMol; j2 ++) {
        VSub (dr, mol[j1].r, mol[j2].r);
        VWrapAll (dr);
        rr = VLenSq (dr);                                           15
        if (rr < Sqr (rangeRdf)) {
          n = sqrt (rr) / deltaR;
          ++ histRdf[n];
        }
      }                                                            20
    }
    ++ countRdf;
    if (countRdf == limitRdf) {
      normFac = VProd (region) / (2. * M_PI * Cube (deltaR) *
        Sqr (nMol) * countRdf);                                    25
      for (n = 0; n < sizeHistRdf; n ++)
        histRdf[n] *= normFac / Sqr (n - 0.5);
      PrintRdf (stdout);
      countRdf = 0;
    }                                                              30
  }
```

In addition to computing the discretized version of the RDF for the current state of the system, the above function also accumulates the average over a series of such 'snapshots', as well as initializing the calculation and producing the final output when sufficient data have been collected; the decision whether to initialize or prepare the final summary is based on the value of *countRdf*. This is the three-dimensional version of the computation; the changes for two dimensions are minor.

New quantities introduced here are

```
real *histRdf, rangeRdf;
int countRdf, limitRdf, sizeHistRdf, stepRdf;
```

and the additional input data items are

```
NameI (limitRdf),
NameR (rangeRdf),
NameI (sizeHistRdf),
NameI (stepRdf),
```

Memory allocation is carried out in *AllocArrays*

```
AllocMem (histRdf, sizeHistRdf, real);
```

and the measurement counter is initialized in *SetupJob*

```
countRdf = 0;
```

The addition to *SingleStep* to request RDF processing is

```
if (stepCount >= stepEquil &&
    (stepCount - stepEquil) %stepRdf == 0) EvalRdf ();
```

while the output function is simply

```
void PrintRdf (FILE *fp)
{
  real rb;
  int n;

  fprintf (fp, "rdf\n");                                    5
  for (n = 0; n < sizeHistRdf; n ++) {
    rb = (n + 0.5) * rangeRdf / sizeHistRdf;
    fprintf (fp, "%8.4f %8.4f\n", rb, histRdf[n]);
  }                                                         10
}
```

The RDF results shown here are obtained from soft-sphere runs that include the following input data (other data items are taken from earlier case studies):

```
initUccll       8 8 8
limitRdf        100
rangeRdf        4.
sizeHistRdf     200
stepEquil       2000
stepInitlzTemp  200
stepLimit       17000
stepRdf         50
temperature     1.
```

An FCC initial state is used, so that $N_m = 2048$. Three density values are used, namely, 0.6, 0.8 and 1.0. The results appearing in Figure 4.3 are those obtained during the last 1000 timesteps of each run; the way in which structure emerges as density increases is clearly visible.

For a simple monatomic fluid $g(r)$ shows how, on average, the neighborhood seen by an atom consists of concentric shells of atoms with well-defined radii. As the density increases, these shells become distorted, an effect reflected in the RDF by additional peaks that appear once the lattice structure of the nascent solid phase begins to make its presence felt. The fact that in the liquid all correlation is lost beyond a few atomic diameters confirms the absence of any long-range positional

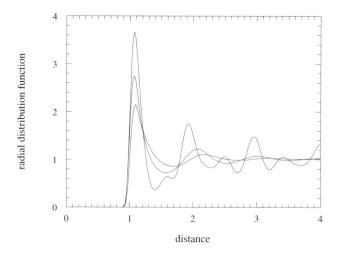

Fig. 4.3. Radial distribution function for soft spheres at densities 0.6–1.0.

order and suggests a picture in which atoms can regard their more distant neighbors as a smeared-out continuum, a useful idealization when trying to construct simple liquid models.

Once the RDF is known, estimates of the errors in the thermodynamic properties due to the interaction cutoff can be determined from the definitions of these quantities in terms of $g(r)$. For example, the error in the potential energy is

$$\Delta E_U = 2\pi\rho \int_{r_c}^{\infty} g(r)u(r)r^2 \, dr \tag{4.3.9}$$

and a related expression exists for the pressure [han86b]. Since $g(r) \approx 1$ at sufficiently large r, the calculation can be simplified; in some cases the error can even be evaluated analytically, such as for the LJ potential, where

$$\Delta E_U = 8\pi\rho \left(\frac{1}{9r_c^9} - \frac{1}{3r_c^3} \right) \tag{4.3.10}$$

Long-range order

The RDF primarily addresses the local structure, but gives little direct information as to whether long-range crystalline order exists. The sharpness of the RDF peaks and the presence of additional peaks at positions indicative of specific lattices provide indirect evidence that is better appreciated once the existence of crystalline order has been established by other means.

Long-range order corresponds to the presence of lattice structure and is the quantity underlying x-ray scattering measurements from crystalline materials. The local density at a point r can be expressed as a sum over atoms,

$$\rho(r) = \sum_{j=1}^{N_m} \delta(r - r_j) \tag{4.3.11}$$

and its Fourier transform is simply

$$\rho(k) = \frac{1}{N_m} \sum_{j=1}^{N_m} e^{-i k \cdot r_j} \tag{4.3.12}$$

In a calculation of $|\rho(k)|$ designed to test for the presence of long-range order, k should be chosen to be a reciprocal lattice vector of the ordered state; this can be any linear combination of the vectors appropriate for the expected FCC lattice, so we choose

$$k = \frac{2\pi}{l}(1, -1, 1) \tag{4.3.13}$$

where l is the unit cell edge. If the system is almost fully ordered $|\rho(k)| \approx 1$, but in the disordered liquid state $|\rho(k)| = O(N_m^{-1/2})$.

The function[*] for evaluating long-range order, assuming (for convenience) all region edges to be the same length, is

```
void EvalLatticeCorr ()
{
  VecR kVec;
  real si, sr, t;
  int n;                                                    5

  kVec.x = 2. * M_PI * initUcell.x / region.x;
  kVec.y = - kVec.x;
  kVec.z = kVec.x;
  sr = 0.;                                                  10
  si = 0.;
  DO_MOL {
    t = VDot (kVec, mol[n].r);
    sr += cos (t);
    si += sin (t);                                          15
  }
  latticeCorr = sqrt (Sqr (sr) + Sqr (si)) / nMol;
}
```

One additional variable is introduced here,

```
real latticeCorr;
```

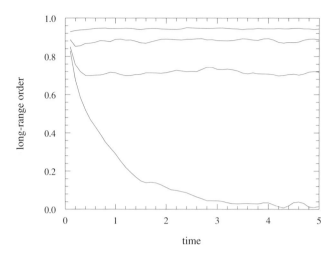

Fig. 4.4.   Time dependence of long-range order in soft-sphere systems that start in an ordered state; the results are for densities 0.8–1.1.

This function is called prior to the call to *PrintSummary* and the output should include the value of *latticeCorr*. No averaging over separate measurements is included, but this could easily be added.

One point must be kept in mind when studying solidification in a finite system, namely, that the best results will be obtained if the region size and shape allow the formation of an integral number of unit cells along each lattice direction. Any mismatch will introduce imperfections of one kind or another into the ordered state, leading to a reduction in the apparent long-range order.

In Figure 4.4 we show how long-range order varies with time during the early stages of runs begun in the ordered state; we use the same system as for the RDF studies but without any initial temperature adjustment. The four density values shown are between 0.8 and 1.1. At the larger densities a moderate to high degree of order persists throughout the observation period (although this is not a guarantee of what might happen over much longer times), whereas at the lowest density the long-range order rapidly vanishes.

4.4 Packing studies

Local structure

There are many reasons for seeking information about local atomic organization that is more detailed than the RDF can provide. In simple fluids the motivation is

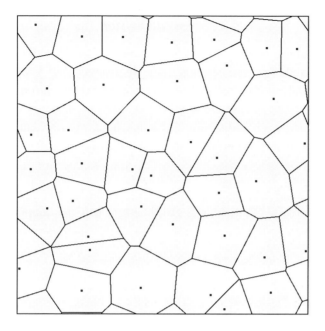

Fig. 4.5. The Voronoi subdivision for a small, random set of points in two dimensions; the region boundaries are periodic.

to understand better how atoms are arranged, and what distinguishes the average packing from the fully ordered crystalline state. In more complex systems the same packing questions can be asked in order to gain more specific information about molecular organization; for example, an estimate of the exposed surface of part of a large molecule can be important for studies of chemical reactivity.

How to describe the spatial organization of what sometimes amounts to little more than a random array of atoms is far from obvious. The most widely used method is based on a Voronoi subdivision [hsu79, cap81, rap83, med90], in which each atom is surrounded by a convex polyhedron constructed using certain pre-scribed rules. The outcome of this construction process is the partitioning of space into a set of polyhedra, with all points that are closer to a particular atom than to any other belonging to its polyhedron. In this way it is possible to define the neighborhood of an atom uniquely, and atoms can then be regarded as adjacent if their polyhedra share a common face. The polyhedra themselves are also of considerable interest since the interactions can influence their geometrical prop-erties. Displaying an image of such a partitioning in three dimensions is not partic-ularly informative, but in Figure 4.5 we show the corresponding two-dimensional result (the name of Dirichlet is associated with this problem) for a random set of points.

Voronoi subdivision

The Voronoi analysis will be carried out separately from the MD run♠ to demonstrate how this kind of postprocessing is done in general; in view of the complexity of the Voronoi analysis it is clearly desirable to keep it distinct from the simulation. Every so often a snapshot containing sufficient information to reproduce the atomic configuration is written to a disk file; this will provide the raw data for subsequent analysis. The following line is added to *main* following the call to *SingleStep*,

```
if (stepCount % stepSnap == 0) PutConfig ();
```

where *stepSnap* specifies the number of timesteps between snapshots and

```
NameI (stepSnap),
```

is added to the input data. The function *PutConfig* is described in §18.6.

Construction of Voronoi polyhedra is an exercise in computational geometry and is by far the longest and most complex of the analysis programs used in these case studies. There are various ways of dealing with this problem [bro78, fin79, tan83]; the version described here computes each polyhedron separately, but does the job with a constant computational effort that is independent of the total number of atoms. Periodic boundaries are assumed.

A concise summary of the method follows. The first step for each atom is to generate a list of its neighbors ordered by distance. A large tetrahedron is then constructed as a generous overestimate of the eventual polyhedron; portions of this polyhedron will be removed in the course of the computation until what remains at the end is the Voronoi polyhedron for that atom. The computation begins with the initial tetrahedron, and carries out the following sequence of operations for each neighbor in the list until none remains that could possibly alter the polyhedron shape:

- compute the bisecting plane between the atom of interest and the neighbor;
- determine which polyhedron vertices lie on the far side of the plane;
- determine which edges and faces are cut by the plane;
- compute the locations of the intercepts of the plane with each cut edge;
- update the description of each cut face and determine which faces are deleted from the polyhedron entirely;
- add the new vertices and edges to the polyhedron, together with the new face;
- remove deleted vertices, edges and faces from the polyhedron description;
- locate the most distant vertices in the new and cut faces.

♠ pr_04_5

When the process terminates, a test is made to ensure that nothing remains of the initial faces; any remnants are symptomatic of a poor choice of initial polyhedron. Measurements made on the resulting polyhedron include vertex, edge and face counts, as well as the volume and surface area.

Assuming that a list of atoms to be tested during the construction of the polyhedron for a particular atom has already been prepared, the following function[♠] shows how the computation is organized.

```
void AnalVorPoly ()
{
  int nf;

  Sort (distSq, siteSeq, nTestSites);                              5
  InitVorPoly ();
  for (curSite = 0; curSite < nTestSites; curSite ++) {
    if (distSq[siteSeq[curSite]] >= 4. * vDistSqMax) break;
    siteB = testSites[siteSeq[curSite]];
    nvDel = 0;                                                     10
    neNew = 0;
    neDel = 0;
    neCut = 0;
    nfDel = 0;
    nfCut = 0;                                                     15
    BisectPlane ();
    if (nvDel > 0) ProcDelVerts ();
    if (neCut > 0) ProcCutEdges ();
    if (nfCut > 0) ProcCutFaces ();
    if (neNew > 0) ProcNewVerts ();                                20
    if (nfCut > 0) ProcNewFace ();
    RemoveOld ();
    if (nfCut > 0) FindDistVerts ();
  }
  for (nf = 0; nf < 4; nf ++)                                      25
      if (face[nf].stat != 0) ErrExit (ERR_SUBDIV_UNFIN);
  PolyGeometry ();
  PolySize ();
}
```

The algorithm

The Voronoi construction task (in common with other exercises in computational geometry) involves a great many details[†]. Because of the rather complex nature of the algorithm these details can be handled in a variety of ways; this is one possible approach. For brevity we omit checks on array overflow and other potential

♠ pr_anvorpol
† The details can be skipped without affecting the continuity of the discussion.

problems, although such safety measures should be included to help detect programming or runtime errors.

The method for determining which atoms can contribute to a particular polyhedron assumes that the region has been subdivided into cells. The atoms required are obtained by first scanning a range of cells around the one containing the atom under examination, then sorting the atoms found into ascending distance order and placing the ordered list of atom indices in the array *siteSeq* (the call to *Sort* in *AnalVorPoly* can use any standard sorting function – see §18.4). Here we only scan neighbor cells, but the range could be extended.

```
void FindTestSites (int na)
{
  VecR dr;
  VecI cn;
  int c, cx, cy, cz, i, ofx, ofy, ofz;                       5

  cx = mol[na].inCell % cells.x;
  cy = (mol[na].inCell / cells.x) % cells.y;
  cz = mol[na].inCell / (cells.x * cells.y);
  nTestSites = 0;                                            10
  for (ofz = -1; ofz <= 1; ofz ++) {
    cn.z = (cz + ofz + cells.z) % cells.z;
    for (ofy = -1; ofy <= 1; ofy ++) {
      cn.y = (cy + ofy + cells.y) % cells.y;
      for (ofx = -1; ofx <= 1; ofx ++) {                     15
        cn.x = (cx + ofx + cells.x) % cells.x;
        c = VLinear (cn, cells) + nMol;
        DO_CELL (i, c) {
          VSub (dr, mol[na].r, mol[i].r);
          VWrapAll (dr);                                     20
          testSites[nTestSites] = i;
          distSq[nTestSites] = VLenSq (dr);
          ++ nTestSites;
        }
      }                                                      25
    }
  }
}
```

Several arrays of structures are used in the program to describe the geometrical details (edges, faces and vertices) of a polyhedron as it is being constructed. The structures themselves are the following:

```
typedef struct {
  int f[2], v[2], stat;
} Edge;

typedef struct {                                             5
  real dist;
```

```
    int fPtr, stat, vFar;
  } Face;

  typedef struct {                                              10
    int e, link, v;
  } Flist;

  typedef struct {
    VecR pos;                                                   15
    real distSq;
    int e[3], stat;
  } Vert;
```

A structure of type *Edge* is associated with each polyhedron edge. In each such structure, the identities of the two vertices joined by the edge are stored in v and the two faces that share the edge appear in f.

With each face is associated a circular linked list (a linked list whose final element points back to the start [knu68]) consisting of elements that are structures of type *Flist* that itemize, in order of appearance, the edges and vertices defining the face boundary; these are stored in e and v, with *link* providing the indices that tie the list together. A structure of type *Face* is also associated with each face. The element *dist* records the shortest distance from the atom to the plane of the face, *fPtr* is the index of the first item in the circular list and *vFar* identifies the furthest vertex in the face from the atom.

In the structure of type *Vert* associated with each vertex, the vertex coordinates are stored in *pos*, the squared distance of each vertex from the atom in *distSq* and the identities of the three edges terminating at the vertex appear in e. To simplify an already complex program, the data representation assumes that there will be exactly three edges attached to each vertex; this excludes certain regular lattice arrangements, as well as the extremely rare case of numerical degeneracy (one candidate for the missing safety checks).

The structures also contain status indicators *stat* which, as the construction progresses, show whether the elements still belong to the polyhedron, have been deleted, or are about to change status. The reader will detect a certain amount of redundancy in the information stored in the structures, but having it all readily accessible simplifies the computation.

The polyhedron used to start the calculation is a tetrahedron. The following function specifies the vertex coordinates and initializes all the data needed to describe the structure of the polyhedron during its subsequent modification.

```
  void InitVorPoly ()
  {
    VecR w, vPosI[] = {{-1., -1., -1.}, {1., -1., -1.}, {0., 2., -1.},
      {0., 0., 3.}};
```

```
    real r2, r6;                                                              5
    int m, n, ne, nf, nv, s,
       vValI[] = {0,2,5,0,1,4,1,2,3,3,4,5},
       eFacesI[] = {0,3,0,1,0,2,1,2,1,3,2,3},
       eVertsI[] = {0,1,1,2,0,2,2,3,1,3,0,3},
       eI[] = {0,1,2,4,3,1,2,3,5,5,4,0},                                     10
       vI[] = {0,1,2,1,3,2,0,2,3,0,3,1};

    r2 = sqrt (2.) * rangeLim;
    r6 = sqrt (6.) * rangeLim;
    siteA = testSites[siteSeq[0]];                                          15
    eLast = 5;
    fLast = 3;
    vLast = 3;
    m = 0;
    for (nv = 0; nv <= vLast; nv ++) {                                      20
      vert[nv].pos = mol[siteA].r;
      VSet (w, r6 / 3., r2 / 3., rangeLim / 3.);
      VMul (w, w, vPosI[nv]);
      VVAdd (vert[nv].pos, w);
      vert[nv].distSq = Sqr (rangeLim);                                     25
      vert[nv].stat = 2;
      for (n = 0; n < 3; n ++) {
        vert[nv].e[n] = vValI[m];
        ++ m;
      }                                                                     30
    }
    vDistSqMax = vert[0].distSq;
    for (ne = 0; ne <= eLast; ne ++) {
      edge[ne].v[0] = eVertsI[2 * ne];
      edge[ne].f[0] = eFacesI[2 * ne];                                      35
      edge[ne].v[1] = eVertsI[2 * ne + 1];
      edge[ne].f[1] = eFacesI[2 * ne + 1];
      edge[ne].stat = 3;
    }
    for (s = 0; s < MAX_FLIST - 1; s ++) flist[s].link = s + 1;             40
    s = 0;
    for (nf = 0; nf <= fLast; nf ++) {
      face[nf].vFar = vI[s];
      face[nf].stat = 3;
      face[nf].fPtr = s;                                                    45
      for (n = 0; n < 3; n ++) {
        flist[s].v = vI[s];
        flist[s].e = eI[s];
        ++ s;
      }                                                                     50
      flist[s - 1].link = face[nf].fPtr;
    }
    fListLast = s - 1;
  }
```

A number of other arrays will be introduced at this point. Elements that have been identified as deleted by the plane currently under consideration are listed in *vDel*, *eDel* and *fDel*, while *eCut* and *fCut* identify edges and faces that are only cut (intersected) by the current plane and *eNew* identifies new edges that are in the process of being added to the polyhedron. Variables such as *fListLast* indicate the last storage element in use.

Bisection of the line between the atom and one of its neighbors to produce a possible new face for the polyhedron is carried out by the following function; the coefficients of the plane equation are placed in the array *fParam*. Allowance is made for periodic boundaries. The vector operations appearing in the function implement the following results:

- the equation of the plane bisecting r_{ij} is

$$(r_i - r_j) \cdot p = (r_i^2 - r_j^2)/2 \qquad (4.4.1)$$

  or, more concisely, $a \cdot p = b$;
- the equation of the edge joining vertices v_1 and v_2 is

$$p = p_{v_1} + \alpha(p_{v_2} - p_{v_1}) \qquad (4.4.2)$$

  where $0 \le \alpha \le 1$;
- the intercept between the plane and the edge, if there is one, occurs when

$$\alpha = \frac{(b - a \cdot p_{v_1})}{a \cdot (p_{v_2} - p_{v_1})} \qquad (4.4.3)$$

```
void BisectPlane ()
{
  VecR dr, shift;
  real d1, d2, d3;
  int nv;                                                          5

  d1 = 0.;
  fParamS = 0.;
  VSub (fParamV, mol[siteB].r, mol[siteA].r);
  VZero (shift);                                                   10
  VShiftAll (fParamV);
  VVAdd (fParamV, shift);
  d1 = VDot (fParamV, mol[siteA].r);
  VAdd (dr, mol[siteB].r, shift);
  fParamS = 0.5 * (VLenSq (dr) - VLenSq (mol[siteA].r));           15
  for (nv = 0; nv <= vLast; nv ++) {
    if (vert[nv].stat != 0) {
      d2 = VDot (fParamV, vert[nv].pos);
      if (d1 != d2) {
        d3 = (fParamS - d1) / (d2 - d1);                           20
        if (d3 > 0. && d3 < 1.) {
```

```
              vDel[nvDel] = nv;
              ++ nvDel;
              vert[nv].stat = 1;
            }                                                    25
          }
        }
      }
    }
```

To handle periodic boundaries we have introduced

```
#define VShift(v, t)                                           \
    if (v.t >= 0.5 * region.t) shift.t -= region.t;            \
    else if (v.t < -0.5 * region.t) shift.t += region.t
#define VShiftAll(v)                                           \
    {VShift (v, x);                                            \     5
     VShift (v, y);                                            \
     VShift (v, z);}
```

Several functions are called in succession from *AnalVorPoly* to deal with those
vertices, edges and faces of the polyhedron that are added, deleted or modified, as
a result of including this new face. The first of these functions determines the edges
and faces affected by the deleted vertices.

```
void ProcDelVerts ()
{
  int e, m, n, nv;

  for (nv = 0; nv < nvDel; nv ++) {                             5
    for (m = 0; m < 3; m ++) {
      e = vert[vDel[nv]].e[m];
      -- edge[e].stat;
      if (edge[e].stat == 2) {
        eCut[neCut] = e;                                       10
        ++ neCut;
      } else {
        eDel[neDel] = e;
        ++ neDel;
      }                                                        15
      for (n = 0; n < 2; n ++) {
        if (face[edge[e].f[n]].stat == 3) {
          fCut[nfCut] = edge[e].f[n];
          ++ nfCut;
          face[edge[e].f[n]].stat = 2;                         20
        }
      }
    }
  }
}                                                              25
```

The next function deals with the edges that have been cut by the plane; the intersection points will become vertices of the polyhedron.

```
void ProcCutEdges ()
{
  VecR dr;
  real d, dt1, dt2;
  int nd, ne, vt1, vt2;                                              5

  for (ne = 0; ne < neCut; ne ++) {
    if (edge[eCut[ne]].stat == 2) {
      edge[eCut[ne]].stat = 3;
      vt1 = edge[eCut[ne]].v[0];
      vt2 = edge[eCut[ne]].v[1];                                     10
      dt1 = VDot (fParamV, vert[vt1].pos);
      dt2 = VDot (fParamV, vert[vt2].pos);
      if (vert[vt1].stat == 1) nd = 0;
      else if (vert[vt2].stat == 1) nd = 1;                          15
      ++ vLast;
      vert[vLast].stat = 2;
      vert[vLast].distSq = 0.;
      d = (fParamS - dt1) / (dt2 - dt1);
      VInterp (vert[vLast].pos, d, vert[vt2].pos, vert[vt1].pos);    20
      VSub (dr, vert[vLast].pos, mol[siteA].r);
      vert[vLast].distSq = VLenSq (dr);
      edge[eCut[ne]].v[nd] = vLast;
      vert[vLast].e[0] = eCut[ne];
      vert[vLast].e[1] = 0;                                          25
      vert[vLast].e[2] = 0;
    }
  }
}
```

Here,

```
#define VInterp(v1, s2, v2, v3)                           \
    VSSAdd (v1, s2, v2, 1. - (s2), v3)
#define VSSAdd(v1, s2, v2, s3, v3)                        \
    (v1).x = (s2) * (v2).x + (s3) * (v3).x,               \
    ...                                                           5
```

The faces cut by the plane are now examined; if a face is not completely eliminated, its lists of boundary edges and vertices are updated to account for the changes.

```
void ProcCutFaces ()
{
  int faceGone, nf, s, s1, s2, s3, s4, v1, v2, vDelCount;

  eLastP = eLast;                                                   5
  ++ fLast;
  for (nf = 0; nf < nfCut; nf ++) {
```

```
    s = face[fCut[nf]].fPtr;
    faceGone = 0;
    while (vert[flist[s].v].stat != 2 && ! faceGone) {           10
      s = flist[s].link;
      if (s == face[fCut[nf]].fPtr) faceGone = 1;
    }
    if (faceGone) {
      fDel[nfDel] = fCut[nf];                                    15
      face[fCut[nf]].stat = 1;
      ++ nfDel;
    } else {
      face[fCut[nf]].stat = 3;
      face[fCut[nf]].fPtr = s;                                   20
      for (s1 = s, s2 = flist[s1].link; vert[flist[s2].v].stat == 2;
         s2 = flist[s1].link) s1 = s2;
      vDelCount = 1;
      for (s3 = s2, s4 = flist[s3].link; vert[flist[s4].v].stat != 2;
         s4 = flist[s3].link) {                                  25
        ++ vDelCount;
        s3 = s4;
      }
      v1 = edge[flist[s1].e].v[0] + edge[flist[s1].e].v[1] -
         flist[s1].v;                                            30
      v2 = edge[flist[s3].e].v[0] + edge[flist[s3].e].v[1] -
         flist[s4].v;
      ++ eLast;
      flist[s3].v = v2;
      if (vDelCount == 1) {                                      35
        ++ fListLast;
        s = fListLast;
        flist[s1].link = s;
        flist[s].link = s2;
        flist[s].v = v1;                                         40
        flist[s].e = eLast;
      } else {
        flist[s2].v = v1;
        flist[s2].e = eLast;
        if (vDelCount > 2) flist[s2].link = s3;                  45
      }
      edge[eLast].v[0] = v1;
      edge[eLast].v[1] = v2;
      edge[eLast].f[0] = fCut[nf];
      edge[eLast].f[1] = fLast;                                  50
      edge[eLast].stat = 2;
      eNew[neNew] = eLast;
      ++ neNew;
    }
  }                                                              55
}
```

A little extra bookkeeping is required for the newly added vertices,

```
void ProcNewVerts ()
{
  int ne, v;

  for (ne = 0; ne < neNew; ne ++) {                              5
    if (eNew[ne] > eLastP) {
      v = edge[eNew[ne]].v[0];
      if (vert[v].e[1] == 0) vert[v].e[1] = eNew[ne];
      else vert[v].e[2] = eNew[ne];
      v = edge[eNew[ne]].v[1];                                   10
      if (vert[v].e[1] == 0) vert[v].e[1] = eNew[ne];
      else vert[v].e[2] = eNew[ne];
    }
  }
}                                                                15
```

and likewise for new faces,

```
void ProcNewFace ()
{
  int e, n, ne, v;

  for (n = 0; n < neNew; n ++) {                                 5
    ++ fListLast;
    if (n == 0) {
      e = eNew[0];
      face[fLast].fPtr = fListLast;
      v = edge[e].v[0];                                          10
    } else {
      ne = 1;
      for (e = eNew[ne]; edge[e].v[0] != v && edge[e].v[1] != v ||
        edge[e].stat == 3; e = eNew[ne]) ++ ne;
    }                                                            15
    flist[fListLast].v = v;
    v = edge[e].v[0] + edge[e].v[1] - v;
    flist[fListLast].e = e;
    edge[e].stat = 3;
  }                                                              20
  face[fLast].stat = 3;
  flist[fListLast].link = face[fLast].fPtr;
  face[fLast].dist = 0.5 * sqrt (distSq[siteSeq[curSite]]);
}
```

Deleted vertices, edges and faces are then flagged appropriately.

```
void RemoveOld ()
{
  int n;
```

```
    for (n = 0; n < nvDel; n ++) vert[vDel[n]].stat = 0;              5
    for (n = 0; n < neDel; n ++) {
      if (edge[eDel[n]].stat == 1) edge[eDel[n]].stat = 0;
    }
    for (n = 0; n < nfDel; n ++) face[fDel[n]].stat = 0;
  }                                                                   10
```

Keeping track of the most distant vertex in each face, as well as the furthest vertex of all, simplifies the task of determining whether a given plane could become a face of the polyhedron. This is done as follows.

```
void FindDistVerts ()
{
  real dd;
  int nf, s;
                                                                     5
  fCut[nfCut] = fLast;
  for (nf = 0; nf < nfCut + 1; nf ++) {
    if (face[fCut[nf]].stat != 0) {
      s = face[fCut[nf]].fPtr;
      dd = vert[flist[s].v].distSq;                                  10
      face[fCut[nf]].vFar = flist[s].v;
      for (s = flist[s].link; s != face[fCut[nf]].fPtr;
          s = flist[s].link) {
        if (vert[flist[s].v].distSq > dd) {
          dd = vert[flist[s].v].distSq;                              15
          face[fCut[nf]].vFar = flist[s].v;
        }
      }
    }
  }                                                                  20
  vDistSqMax = 0.;
  for (nf = 0; nf <= fLast; nf ++) {
    if (face[nf].stat != 0 && vDistSqMax < vert[face[nf].vFar].distSq)
      vDistSqMax = vert[face[nf].vFar].distSq;
  }                                                                  25
}
```

Evaluation of the geometrical properties of the current polyhedron is as follows. Here, the four quantities computed and stored in the array `polyGeom` are the numbers of vertices, edges, faces, and the average number of edges per face. These results are subsequently combined (in `main`) with those from other polyhedra to produce averages for the entire system.

```
void PolyGeometry ()
{
  int n, ne, nf, nv, s;

  for (n = 0; n < 4; n ++) polyGeom[n].val = 0.;                     5
```

```
for (nv = 0; nv <= vLast; nv ++) {
   if (vert[nv].stat != 0) ++ polyGeom[0].val;
}
for (ne = 0; ne <= eLast; ne ++) {
   if (edge[ne].stat != 0) ++ polyGeom[1].val;                    10
}
for (nf = 0; nf <= fLast; nf ++) {
   if (face[nf].stat != 0) {
      ++ polyGeom[2].val;
      ++ polyGeom[3].val;                                         15
      for (s = flist[face[nf].fPtr].link; s != face[nf].fPtr;
         s = flist[s].link) ++ polyGeom[3].val;
   }
}
polyGeom[3].val /= polyGeom[2].val;                               20
}
```

The surface area and volume of the polyhedron are computed by the function below; these results will also be used in producing averages. The area of a single (convex) face f of the polyhedron is just the sum of the areas of the triangles into which it can be decomposed,

$$A_f = \tfrac{1}{2} \sum_i |(\boldsymbol{r}_{i+1} - \boldsymbol{r}_1) \times (\boldsymbol{r}_i - \boldsymbol{r}_1)| \qquad (4.4.4)$$

where the $\boldsymbol{r}_i$ are the vertices of f, and the volume is given by the sum over faces

$$V = \tfrac{1}{3} \sum_f d_f A_f \qquad (4.4.5)$$

where d_f is the distance of the face from the atom position.

```
void PolySize ()
{
   VecR ca, d1, d2, d3;
   real a;
   int nf, s, v1, v2;                                            5

   polyArea.val = 0.;
   polyVol.val = 0.;
   for (nf = 0; nf <= fLast; nf ++) {
      if (face[nf].stat != 0) {                                  10
         s = face[nf].fPtr;
         v1 = flist[s].v;
         s = flist[s].link;
         v2 = flist[s].v;
         VSub (d1, vert[v2].pos, vert[v1].pos);                  15
         VZero (ca);
         for (s = flist[s].link; s != face[nf].fPtr; s = flist[s].link) {
```

```
      v2 = flist[s].v;
      VSub (d2, vert[v2].pos, vert[v1].pos);
      VCross (d3, d1, d2);                                            20
      VVAdd (ca, d3);
      d1 = d2;
    }
    a = VLen (ca);
    polyArea.val += a / 2.;                                          25
    polyVol.val += face[nf].dist * a / 6.;
  }
 }
}
```

Here, *VCross* evaluates the vector cross product,

```
#define VCross(v1, v2, v3)                                           \
    (v1).x = (v2).y * (v3).z - (v2).z * (v3).y,                      \
    (v1).y = (v2).z * (v3).x - (v2).x * (v3).z,                      \
    (v1).z = (v2).x * (v3).y - (v2).y * (v3).x
```

A list of the (global) variables used in the program follows.

```
typedef struct {
  VecR r;
  int inCell;
} Mol;
                                                                     5
Mol *mol;
Edge *edge;
Face *face;
Flist *flist;
Vert *vert;                                                          10
VecR *r, fParamV, region;
VecI cells;
real *distSq, cellRatio, eulerSum, fParamS, fracPolyVol, rangeLim,
    regionVol, timeNow, vDistSqMax;
Prop polyGeom[4], polyArea, polyVol;                                 15
int *cellList, *eCut, *eDel, *eNew, *fCut, *fDel, *siteSeq,
    *testSites, *vDel, blockNum, blockSize, curSite, eLast, eLastP,
    fLast, fListLast, nCell, neCut, neDel, neNew, nfCut, nfDel, nMol,
    nTestSites, nvDel, runId, siteA, siteB, stepCount, vLast;
FILE *fp;                                                            20
```

The variables *blockNum*, *blockSize*, *runId* and *fp* are needed for dealing with the snapshot file. Several parameters are used to set the sizes of the arrays, namely,

```
#define MAX_EDGE    200
#define MAX_FACE    50
#define MAX_FLIST   500
```

```
#define MAX_ITEM    50
#define MAX_VERT    200                                              5
```

The values are larger than necessary (for simplicity, the storage used by deleted items is not reused), but if safety checks are added to the program all risk of array overflow can be avoided.

Array allocation is carried out by a new version of *AllocArrays*.

```
void AllocArrays ()
{
  AllocMem (mol, nMol, Mol);
  AllocMem (distSq, nMol, real);
  AllocMem (siteSeq, nMol, int);                                    5
  AllocMem (testSites, nMol, int);
  AllocMem (cellList, VProd (cells) + nMol, int);
  AllocMem (edge, MAX_EDGE, Edge);
  AllocMem (face, MAX_FACE, Face);
  AllocMem (flist, MAX_FLIST, Flist);                               10
  AllocMem (vert, MAX_VERT, Vert);
  AllocMem (eCut, MAX_ITEM, int);
  ... (similarly eDel, eNew, fCut, fDel, vDel) ...
}
```

The main program for the Voronoi calculation follows. Configurations are read by *GetConfig* (described in §18.6) while the input snapshot file is selected by calling *SetupFiles* (§18.7); the value of *runId* must be supplied when the program is run. The function *SubdivCells* (not shown) contains code borrowed from the cell version of *ComputeForces* (§3.4) that assigns atoms to cells and also saves the cell numbers in *mol[].inCell*.

```
int main (int argc, char **argv)
{
  int n, na;

  runId = atoi (argv[1]);                                           5
  cellRatio = 0.5;
  SetupFiles ();
  blockNum = -1;
  while (GetConfig ()) {
    regionVol = VProd (region);                                     10
    SubdivCells ();
    rangeLim = region.x;
    PropZero (polyArea);
    PropZero (polyVol);
    for (n = 0; n < 4; n ++) PropZero (polyGeom[n]);                15
    for (na = 0; na < nMol; na ++) {
      FindTestSites (na);
      AnalVorPoly ();
      PropAccum (polyArea);
```

```
      PropAccum (polyVol);                                                       20
        for (n = 0; n < 4; n ++) PropAccum (polyGeom[n]);
      }
      fracPolyVol = polyVol.sum / regionVol;
      PropAvg (polyArea, nMol);
      PropAvg (polyVol, nMol);                                                   25
      for (n = 0; n < 4; n ++) PropAvg (polyGeom[n], nMol);
      polyArea.sum /= pow (regionVol, 2./3.);
      polyArea.sum2 /= pow (regionVol, 2./3.);
      polyVol.sum /= regionVol;
      polyVol.sum2 /= regionVol;                                                 30
      eulerSum = polyGeom[0].sum + polyGeom[2].sum - polyGeom[1].sum;
      ... (print the results) ...
    }
  }
```

The first call to *GetConfig* calls *AllocArrays*, and also the following function
to set the cell size; the prescription is somewhat arbitrary, with the parameter
cellRatio used to adjust the number of cells per edge to ensure that the full
complement of neighbors is found.

```
void SetCellSize ()
{
  VSCopy (cells, cellRatio, region);
  nCell = VProd (cells);
}                                                                               5
```

Two quantities evaluated in *main* serve as checks on the computation: the sum
of the volumes of the polyhedra must of course be identical to the region volume,
otherwise it is likely that insufficient neighbors are being examined, and the value
of *eulerSum* should be exactly 2, a familiar result from graph theory.

Results

The sample results shown Table 4.2 are obtained using soft-sphere systems with
864 atoms, started at $T = 1$ but without any temperature adjustment, and at
three different densities. Three sets of configurations are recorded at intervals of
1000 timesteps for use in the analysis (starting after 2000 timesteps). To allow
comparison with the behavior of random systems, results from arrays of 4000 totally random points (generated by a special version of *InitCoords* shown in §3.6)
are included. The trends in the results are clearly visible.

4.5 Cluster analysis

Cluster algorithm

Cluster formation in fluids is a subject of frequent interest, both because clustering
is a real physical process (see also §9.6) and because some models attribute special

Table 4.2. Properties of Voronoi polyhedra for soft-sphere systems and for random sets of points; results shown are mean numbers of vertices n_v, edges n_e and faces n_f per polyhedron, and mean number of edges per face n_{ef}; three sets of results are shown for each case.

density	n_v	n_e	n_f	n_{ef}
1.0	24.199	36.299	14.100	5.145
	24.194	36.292	14.097	5.146
	24.259	36.389	14.130	5.147
0.8	25.310	37.965	14.655	5.173
	25.477	38.215	14.738	5.177
	25.347	38.021	14.674	5.174
0.6	25.819	38.729	14.910	5.183
	25.843	38.764	14.921	5.182
	25.750	38.625	14.875	5.180
random	27.005	40.508	15.502	5.187
	27.021	40.532	15.511	5.189
	27.067	40.600	15.534	5.189

properties to clusters. In either case it is important to be able to identify atoms belonging to common clusters and to measure various cluster properties. Here we focus on clusters appearing in instantaneous snapshots of a soft-sphere system, but a deeper analysis might also need to consider time-dependent behavior, such as cluster growth rates, or cluster lifetimes in systems where attractive interactions actually bind atoms together.

Different criteria are available for determining whether an atom belongs to a cluster. One option is to consider the energy that binds (assuming interactions with an attractive component) an atom to other atoms already in the cluster. An alternative method requiring less computation (which, for attractive pair potentials, is essentially the same) is to base the criterion on the interatomic distance, so that if atom i is already in the cluster, atom j will also be included if $r_{ij} < r_d$, where r_d is the chosen threshold separation; we will adopt this definition here. The value of r_d would typically be based on some energy condition, but this does not affect the technique. If there is no physical reason for preferring a particular value of r_d, the sensitivity of the results to a change in r_d should of course be examined.

This analysis♠ will also be carried out separately from the MD run; the user will want to try different r_d values, so it is more sensible to have the MD configurations available for immediate analysis. The configuration data are input to the analysis

---

♠ `pr_anclust`

program in the same way as in the earlier Voronoi study. Cluster construction begins by determining those atom pairs that are separated by less than r_d; the function
used for this is derived from *BuildNebrList* (§3.4), with a cell size (use of cells
is optional) based on the variable *rClust* that corresponds to r_d.

```
void BuildClusters ()
{
  real rrClust;

  rrClust = Sqr (rClust);                                          5
  ...
  if (VLenSq (dr) < rrClust) AddBondedPair (j1, j2);
  ...
}
```

The tasks of adding atoms to clusters and merging existing clusters that are found
to share a common member are carried out by *AddBondedPair* below. Two structures are defined here, one a modified version of *Mol*, the other to help organize the
cluster work,

```
typedef struct {
  VecR r;
  int inClust;
} Mol;
                                                                   5
typedef struct {
  int head, next, size;
} Clust;
```

In *Mol*, *inClust* records the cluster to which an atom belongs. In *Clust*, *head*
points to the first atom of the cluster, *next* is a pointer from one atom in the cluster
to the next (atoms belonging to a cluster are associated using a linked list) and
size eventually contains the number of atoms in the cluster. Tracing the detailed
logic of this function is left as an exercise for the reader.

```
void AddBondedPair (int j1, int j2)
{
  int cBig, cSmall, m, mp, nc1, nc2;

  nc1 = mol[j1].inClust;                                           5
  nc2 = mol[j2].inClust;
  if (nc1 < 0 && nc2 < 0) {
    mol[j1].inClust = nClust;
    mol[j2].inClust = nClust;
    clust[nClust].size = 2;                                        10
    clust[nClust].head = j1;
    clust[j1].next = j2;
```

```
      clust[j2].next = -1;
      ++ nClust;
   } else if (mol[j1].inClust < 0) {
      mol[j1].inClust = nc2;
      clust[j1].next = clust[nc2].head;
      clust[nc2].head = j1;
      ++ clust[nc2].size;
   } else if (mol[j2].inClust < 0) {
      mol[j2].inClust = nc1;
      clust[j2].next = clust[nc1].head;
      clust[nc1].head = j2;
      ++ clust[nc1].size;
   } else {
      if (nc1 != nc2) {
         cBig = (clust[nc1].size > clust[nc2].size) ? nc1 : nc2;
         cSmall = nc1 + nc2 - cBig;
         for (m = clust[cSmall].head; m >= 0; m = clust[m].next) {
            mol[m].inClust = cBig;
            mp = m;
         }
         clust[mp].next = clust[cBig].head;
         clust[cBig].head = clust[cSmall].head;
         clust[cBig].size += clust[cSmall].size;
         clust[cSmall].size = 0;
      }
   }
}
```

Prior to starting cluster construction a little preparation is required.

```
void InitClusters ()
{
   int n;

   DO_MOL mol[n].inClust = -1;
   nClust = 0;
}
```

After the work is complete the clusters can be reindexed to remove any reference to those clusters that were absorbed by others during construction.

```
void CompressClusters ()
{
   int j, m, nc;

   nc = 0;
   for (j = 0; j < nClust; j ++) {
      if (clust[j].size > 0) {
         clust[nc].head = clust[j].head;
         clust[nc].size = clust[j].size;
```

```
    for (m = clust[nc].head; m >= 0; m = clust[m].next)              10
        mol[m].inClust = nc;
    ++ nc;
  }
}
nClust = nc;                                                          15
}
```

The arrays and variables used here include

```
Mol *mol;
Clust *clust;
VecR region;
VecI cells;
real rClust, timeNow;                                                5
Prop cSize;
int *cellList, bigSize, blockNum, blockSize, nCellEdge, nClust,
    nMol, nSingle, runId, stepCount;
FILE *fp;
```

All the arrays used to hold cluster data are of size $nMol$, to allow for the extreme situation where all atoms form their own clusters. The size of the cell array is based on the number of cells per edge,

```
nCellEdge = region.x / rClust;
```

The array allocation function is

```
void AllocArrays ()
{
  AllocMem (mol, nMol, Mol);
  AllocMem (clust, nMol, Clust);
  AllocMem (cellList, Cube (nCellEdge) + nMol, int);             5
}
```

Once generation is complete the analysis of both geometric and spatial properties of the clusters can be carried out. Spatial properties of the clusters include the radius of gyration and moments of the mass distribution; such studies involve calculations similar to those used for polymer chains in §9.4 and will not be considered here. Other measurements are of a more geometrical flavor; as an example the following function counts the number of isolated atoms, finds the cluster with the most atoms and evaluates the mean and standard deviation of the cluster size distribution.

```
void AnalClusterSize ()
{
  int cBig, nc, ncUse;

  PropZero (cSize);
  ncUse = 0;
  cBig = 0;
  for (nc = 0; nc < nClust; nc ++) {
    cSize.val = clust[nc].size;
    if (cSize.val > clust[cBig].size) cBig = nc;
    if (cSize.val > 1) {
      ++ ncUse;
      PropAccum (cSize);
    }
  }
  bigSize = clust[cBig].size;
  nSingle = nMol - cSize.sum;
  if (ncUse > 0) PropAvg (cSize, ncUse);
}
```

More complex aspects, such as the number of ways in which atoms are linked into a cluster, or the topology of the link network, can also be explored using the data available.

The main program used in the cluster analysis is the following; the values of *runId* and *rClust* must be supplied when the program is run.

```
int main (int argc, char **argv)
{
  runId = atoi (argv[1]);
  rClust = atof (argv[2]);
  SetupFiles ();
  blockNum = -1;
  while (GetConfig ()) {
    InitClusters ();
    BuildClusters ();
    CompressClusters ();
    AnalClusterSize ();
    printf ("%d %d %d %.1f %.1f\n", nSingle, nClust, bigSize,
        PropEst (cSize));
  }
}
```

Measurements

Examples of cluster properties are shown in Table 4.3. The configuration data produced by the $\rho = 0.8$ Voronoi run are used here as well. The results of analyzing three different realizations are shown for various values of the cluster threshold

Table 4.3. Cluster properties for a soft-sphere fluid; results shown are numbers of isolated atoms n_i and multisite clusters n_c, size of the largest cluster s_m and mean cluster size $\langle s \rangle$; three sets of results are shown for each r_d.

r_d	n_i	n_c	s_m	$\langle s \rangle$
1.02	344	148	25	3.5
	313	147	26	3.7
	340	152	23	3.4
1.04	175	117	165	5.9
	173	90	280	7.7
	183	121	98	5.6
1.06	80	39	636	20.1
	84	23	712	33.9
	89	48	623	16.1
1.08	38	4	817	206.5
	31	10	808	83.3
	36	13	796	63.7
1.10	16	3	843	282.7
	16	2	846	424.0
	14	4	840	212.5

separation r_d. Percolation theory can be used to explain the changing behavior as r_d is varied and also to inspire other kinds of cluster analysis [sta92].

4.6 Further study

4.1   Compare specific heats obtained from kinetic energy fluctuations and from dE/dT.

4.2   Examine the errors in energy and pressure due to truncating the LJ interaction.

4.3   Study the soft-sphere equation of state near the melting transition; what kind of transition occurs?

4.4   The possible existence of a hexatic phase in two-dimensional liquids – in which there is long-range orientational order although no translational order – has been explored using MD [abr86]; look into the subject.

4.5   Examine the difference between the LJ and soft-sphere RDFs.

4.6   Extend the structural analysis to consider correlations involving the coordinates of three atoms at a time [vog84, bar88]; for example, study the distribution of angles subtended by pairs of neighbors (suitably defined) of each atom.

4.7   The Voronoi analysis is greatly simplified when applied to systems in two dimensions (see Figure 4.5); generate and analyze some typical soft-disk configurations.

4.8   Examine the cluster distributions for the two-dimensional case from the point of view of percolation theory [hey89].

4.9   Apply cluster analysis to the LJ fluid; here, unlike soft spheres, the binding energy can be computed for each cluster, so that the study of cluster formation takes on physical meaning.

5

Dynamical properties of simple fluids

5.1 Introduction

In this chapter we encounter measurements of a type demonstrating some of the unique capabilities of MD. Because the complete trajectories are available, it is no more difficult to measure time-dependent properties, both in and out of equilibrium, than it is to measure thermodynamic and structural properties at equilibrium. Here we concentrate on properties defined in terms of time-dependent correlation functions at the atomic level – the dynamic structure factor and transport coefficients such as the shear viscosity are examples. Most of the analysis is incorporated into the simulation program, but it would of course be possible (though extremely storage intensive) to store the required trajectory data for subsequent processing.

5.2 Transport coefficients

Background

Transport coefficients describe the material properties of a fluid within the framework of continuum fluid dynamics. Discrete atoms play no role whatsoever in the continuum picture, but this does not seriously limit the enormous range of practical engineering applications of the continuum approach. The most familiar of the transport coefficients are those applicable to simple fluids; these are the diffusion coefficient, the shear and bulk viscosities and the thermal conductivity. Other transport coefficients appear when dealing with more complex fluids, such as those containing more than one species, or those with novel rheological behavior. In many problems the transport coefficients are assumed to be experimentally determined constants, depending only on the temperature and density of the fluid, which themselves are often assumed constant for a given problem, but in more complex situations transport coefficients can depend on local behavior, an example being the dependence of shear viscosity on the velocity gradient.

While statistical mechanics focuses its attention on equilibrium systems, and there is no corresponding general theory for systems away from equilibrium, linear response theory [mcq76, han86b] describes the reaction of an equilibrium system to a small external perturbation and defines generalized 'susceptibilities' that are expressed in terms of various equilibrium correlation functions. The transport coefficients we will be discussing here can be expressed in a similar fashion [hel60, mcq76], despite the fact that there are no obvious mechanical perturbations corresponding to the concentration, velocity and thermal gradients associated with the underlying transport processes; we will return to this subject in §7.4.

Each transport coefficient can be derived directly from one of the continuum equations of fluid dynamics, such as the Navier–Stokes equation, after taking the long wavelength (small k) limit of the Fourier transformed version of the equation. The eventual result of the derivation is a direct relation between a macroscopic transport coefficient and the time integral of a particular microscopic autocorrelation function measured in an equilibrium system; such correlations are not directly accessible to experiment.

The alternative, and from the historical point of view original, definition of a transport coefficient, namely, the constant factor relating the response of a system to an imposed driving force – such as the Newtonian definition of shear viscosity, or Fourier's law of heat transport – implies a nonequilibrium system. Measurements based on these definitions are also feasible within the MD framework; there are, however, certain technical details that must be addressed in order to carry out such simulations, and we will deal with this approach in §7.3.

Diffusion

In a continuous system the diffusion coefficient D is defined by Fick's law relating mass flow to density gradient [mcq76],

$$\rho \boldsymbol{u} = -D\nabla\rho \tag{5.2.1}$$

where $\boldsymbol{u}(\boldsymbol{r}, t)$ is the local velocity and $\rho(\boldsymbol{r}, t)$ the local density or concentration, so that the time evolution of ρ is described by the equation

$$\frac{\partial \rho}{\partial t} = D\nabla^2\rho \tag{5.2.2}$$

This result applies both to the diffusion of one species through another and to self-diffusion within a single species. At the discrete-particle level ρ is just

$$\rho(\boldsymbol{r}, t) = \sum_{j=1}^{N_m} \delta\big(\boldsymbol{r} - \boldsymbol{r}_j(t)\big) \tag{5.2.3}$$

Then, for large t – compared with the 'collision interval', a vague but intuitively obvious period of time where continuous potentials are involved – we have the Einstein expression [mcq76]

$$D = \lim_{t \to \infty} \frac{1}{6N_m t} \left\langle \sum_{j=1}^{N_m} [r_j(t) - r_j(0)]^2 \right\rangle \tag{5.2.4}$$

Note that for a finite system, t cannot become too large because the allowed displacements are bounded; eventually this asymptotic result will break down, so that after reaching a plateau D will begin to drop to zero.

For periodic boundaries we need the 'true' atomic displacements $r'_j(t)$, from which the effects of wraparound have been removed. If we assume that the displacement per timestep is small relative to the system size (as it always is), then the two sets of coordinate components are related by

$$r'_{jx}(t) = r_{jx}(t) + \mathrm{nint}\big([r'_{jx}(t - \Delta t) - r_{jx}(t)]/L_x\big)L_x \tag{5.2.5}$$

where $\mathrm{nint}(x)$ is the nearest integer to x, and $r'_j(0) = r_j(0)$. In (5.2.4), $\langle \ldots \rangle$ denotes the average over a sufficiently large number of (in principle) independent samples.

The alternative Green–Kubo expression for D [mcq76] is based on the integrated velocity autocorrelation function,

$$D = \frac{1}{3N_m} \int_0^\infty \left\langle \sum_{j=1}^{N_m} v_j(t) \cdot v_j(0) \right\rangle dt \tag{5.2.6}$$

The two definitions, (5.2.4) and (5.2.6), can be shown to be completely equivalent.

A reliable estimate of D, as well as the other transport coefficients discussed subsequently, requires that the trajectories be computed relatively accurately for as long as the velocities remain correlated. As pointed out in §3.8, the main source of uncertainty in the trajectories is the strongly repulsive potential and not the truncation error of the numerical method used for solving the differential equations. The former is a real physical effect that influences the velocity correlations in a way that mimics nature, so that the velocities remain correlated until overwhelmed by the noise inherent in the trajectories.

Shear viscosity

The shear viscosity η is defined by the Navier–Stokes equation [mcq76]

$$\rho \left(\frac{\partial}{\partial t} + u \cdot \nabla \right) u = \eta \nabla^2 u + \left(\frac{\eta}{3} + \eta_v \right) \nabla(\nabla \cdot u) - \nabla P \tag{5.2.7}$$

Another transport coefficient also appears in this equation, the bulk viscosity η_v, but it will not be studied here. Theory then leads to an expression analogous to the Einstein diffusion formula (5.2.4) [hel60, mcq76] (however, see [all93a]), namely,

$$\eta = \lim_{t \to \infty} \frac{1}{6TVt} \left\langle \sum_{x<y} \left[\sum_j m_j r_{jx}(t) v_{jy}(t) - \sum_j m_j r_{jx}(0) v_{jy}(0) \right]^2 \right\rangle \quad (5.2.8)$$

where $\sum_{x<y}$ denotes a sum over the three pairs of distinct vector components (xy, yz and zx) used to improve the statistics (and $k_B = 1$). The formula shows how η characterizes the rate at which some component (for example, y) of momentum diffuses in a perpendicular (x) direction. While this result bears a certain formal similarity to the diffusion expression, the conspicuous difference is that here a single sum combines the contributions from all atoms, whereas with diffusion each atom contributes individually – in short, the square of a sum as opposed to a sum of squares. This expression turns out to be unusable with periodic boundaries because they violate the translational invariance assumed in the derivation [all93a].

The alternative Green–Kubo form, based on the integrated autocorrelation function of the pressure tensor, does not experience this problem. The definition is [mcq76]

$$\eta = \frac{V}{3T} \int_0^\infty \left\langle \sum_{x<y} P_{xy}(t) P_{xy}(0) \right\rangle dt \quad (5.2.9)$$

where

$$P_{xy} = \frac{1}{V} \left[\sum_j m_j v_{jx} v_{jy} + \tfrac{1}{2} \sum_{i \neq j} r_{ijx} f_{ijy} \right] \quad (5.2.10)$$

is a component of the pressure tensor (the negative of which is known as the stress tensor). Evaluation of the second term in P_{xy} can be carried out along with the force computation, treating periodic boundaries in the normal way. For pair potentials such as LJ, in which $f_{ij} = f(r_{ij}) \hat{r}_{ij}$, it is clear that $P_{xy} = P_{yx}$. Averaging over vector components is again used to improve the statistics.

Thermal conductivity

The equation for heat transfer derived from Fourier's law [mcq76], assuming that the process involves thermal conduction alone and that there is no convection (implying mass flow), is

$$\rho C_V \frac{\partial E}{\partial t} = \lambda \nabla^2 E \quad (5.2.11)$$

and the resulting diffusion-like formula for the thermal conductivity is

$$\lambda = \lim_{t \to \infty} \frac{1}{6T^2 V t} \left\langle \sum_x \left[\sum_j r_{jx}(t) e_j(t) - \sum_j r_{jx}(0) e_j(0) \right]^2 \right\rangle \tag{5.2.12}$$

where

$$e_j = \tfrac{1}{2} m v_j^2 + \tfrac{1}{2} \sum_{i (\neq j)} u(r_{ij}) - \langle e \rangle \tag{5.2.13}$$

is the instantaneous excess energy of atom j, $\langle e \rangle$ is the mean energy and $\sum_x$ is a sum over vector components. The periodic boundary limitation also applies here, but there is an alternative form based on the integrated heat flux autocorrelation function,

$$\lambda = \frac{V}{3T^2} \int_0^\infty \langle S(t) \cdot S(0) \rangle \, dt \tag{5.2.14}$$

where[†]

$$S = \frac{1}{V} \left[\sum_j e_j v_j + \tfrac{1}{2} \sum_{i \neq j} r_{ij} (f_{ij} \cdot v_j) \right] \tag{5.2.15}$$

5.3 Measuring transport coefficients

Direct evaluation of diffusion

We now turn to the practical side of studying the transport coefficients, beginning with the simplest example, the diffusion coefficient, based on the Einstein definition (5.2.4). The computations will introduce a standardized framework that can be used for all measurements extending over a series of timesteps, with each such calculation including initialization, the actual process of making and accumulating measurements at evenly spaced time intervals, and a final summary. An important feature of these computations is that the samples are overlapped to provide extra results; this calls for additional storage and bookkeeping. While the overlap increases the correlation between successive samples, with similar consequences for error estimates as described in §4.2, it can improve the quality of the results without extending the duration of the run; ideally, overlap should be confined to time intervals over which the correlation between measurements has decayed to a comparatively small value.

The measurements entail following the atomic trajectories over a sufficient number of timesteps to obtain convergence of (5.2.4) to asymptotic behavior. New sets

---

† The expression for S should also include a term proportional to $\langle h \rangle \sum_j v_j$, where $\langle h \rangle$ is the mean enthalpy per atom, but, provided the total momentum is zero, the term can be dropped.

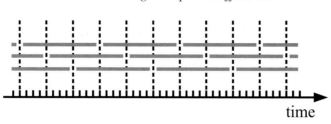

time

Fig. 5.1. Use of overlapped data collection for time-dependent properties; the measurements at any instant contribute to several sets of results (shown as shaded lines), with the dashed lines marking the different time origins.

of diffusion measurements are begun at fixed time intervals, so that several sets of measurements based on different time origins will be in progress simultaneously because of the overlapped measurements – see Figure 5.1. To be more specific, a total of *nValDiffuse* measurements contribute to the set used to produce a single (unaveraged) estimate of *D*, there are *nBuffDiffuse* sets of data being collected at any time (except for very early in the run), each occupying a separate storage buffer, and measurements are made every *stepDiffuse* timesteps. The last of the parameters governing the data collection is *limitDiffuseAv*, which specifies the total number of individual estimates used to produce the averaged value of *D*. Each set of measurements is carried out completely independently and they are combined only when complete, with a variable *count* used to keep track of the number of measurements currently in that set; this will be initialized in a particular way by the function *InitDiffusion* (shown later) to ensure that the measurements are evenly spaced. Given the values of these parameters, the total run length needed for a given number of measurements can easily be determined; only complete sets of results are used and partially filled buffers are discarded at the end of the run.

The functions appearing below measure the squared displacement of each atom from each of several reference points (or origins), one per set, making allowance for periodic boundaries, and produce a series of estimates that for sufficiently large time intervals should converge to *D*. With each set of measurements, or buffer, is associated a structure defined as

```
typedef struct {
  VecR *orgR, *rTrue;
  real *rrDiffuse;
  int count;
} TBuf;
```
5

in which the array *orgR* stores the origins for the set of measurements, *rTrue* accumulates the components of the 'true' displacement of each atom after removing

any wraparound effects, `rrDiffuse` accumulates the mean-square displacements
and `count` is a counter. Similarly named structures, but with different contents,
will be encountered in subsequent time-dependent measurements where multiple
overlapped sets of results must also be collected.

The measurement functions[♦] are as follows, where `Nint` (§18.2) performs the
nint(x) operation.

```
void EvalDiffusion ()
{
  VecR dr;
  int n, nb, ni;
                                                                           5
  for (nb = 0; nb < nBuffDiffuse; nb ++) {
    if (tBuf[nb].count == 0) {
      DO_MOL {
        tBuf[nb].orgR[n] = mol[n].r;
        tBuf[nb].rTrue[n] = mol[n].r;                                      10
      }
    }
    if (tBuf[nb].count >= 0) {
      ni = tBuf[nb].count;
      tBuf[nb].rrDiffuse[ni] = 0.;                                         15
      DO_MOL {
        VSub (dr, tBuf[nb].rTrue[n], mol[n].r);
        VDiv (dr, dr, region);
        dr.x = Nint (dr.x);
        dr.y = Nint (dr.y);                                                20
        dr.z = Nint (dr.z);
        VMul (dr, dr, region);
        VAdd (tBuf[nb].rTrue[n], mol[n].r, dr);
        VSub (dr, tBuf[nb].rTrue[n], tBuf[nb].orgR[n]);
        tBuf[nb].rrDiffuse[ni] += VLenSq (dr);                             25
      }
    }
    ++ tBuf[nb].count;
  }
  AccumDiffusion ();                                                       30
}

void AccumDiffusion ()
{
  real fac;                                                               35
  int j, nb;

  for (nb = 0; nb < nBuffDiffuse; nb ++) {
    if (tBuf[nb].count == nValDiffuse) {
      for (j = 0; j < nValDiffuse; j ++)                                   40
        rrDiffuseAv[j] += tBuf[nb].rrDiffuse[j];
```

```
            tBuf[nb].count = 0;
            ++ countDiffuseAv;
            if (countDiffuseAv == limitDiffuseAv) {
              fac = 1. / (NDIM * 2 * nMol * stepDiffuse *        45
                  deltaT * limitDiffuseAv);
              for (j = 1; j < nValDiffuse; j ++)
                rrDiffuseAv[j] *= fac / j;
              PrintDiffusion (stdout);
              ZeroDiffusion ();                                  50
            }
          }
        }
      }
```

The functions that initialize (*InitDiffusion*) and reset (*ZeroDiffusion*) the
calculation and output the results (*PrintDiffusion*) follow. Note the values initially assigned to *tBuf[].count* that determine the spacing between measurements: using negative initial values delays the start of data collection for each set
of measurements until the appropriate moment.

```
void InitDiffusion ()
{
  int nb;

  for (nb = 0; nb < nBuffDiffuse; nb ++)                        5
      tBuf[nb].count = - nb * nValDiffuse / nBuffDiffuse;
  ZeroDiffusion ();
}

void ZeroDiffusion ()                                          10
{
  int j;

  countDiffuseAv = 0;
  for (j = 0; j < nValDiffuse; j ++) rrDiffuseAv[j] = 0.;       15
}

void PrintDiffusion (FILE *fp)
{
  real tVal;                                                   20
  int j;

  fprintf (fp, "diffusion\n");
  for (j = 0; j < nValDiffuse; j ++) {
    tVal = j * stepDiffuse * deltaT;                           25
    fprintf (fp, "%8.4f %8.4f\n", tVal, rrDiffuseAv[j]);
  }
}
```

Incorporating the above functions into the MD program requires several additions. The measurement function is called from *SingleStep* at regular intervals after equilibration by

```
if (stepCount >= stepEquil &&
    (stepCount - stepEquil) % stepDiffuse == 0) EvalDiffusion ();
```

and program initialization (*SetupJob*) includes

```
InitDiffusion ();
```

The new quantities associated with these calculations are

```
TBuf *tBuf;
real *rrDiffuseAv;
int countDiffuseAv, limitDiffuseAv, nBuffDiffuse, nValDiffuse,
  stepDiffuse;
```

the measurement parameters input to the program are

```
NameI (limitDiffuseAv),
NameI (nBuffDiffuse),
NameI (nValDiffuse),
NameI (stepDiffuse),
```

and the necessary arrays are allocated by *AllocArrays* (note the various array sizes),

```
int nb;
...
AllocMem (rrDiffuseAv, nValDiffuse, real);
AllocMem (tBuf, nBuffDiffuse, TBuf);
for (nb = 0; nb < nBuffDiffuse; nb ++) {
  AllocMem (tBuf[nb].orgR, nMol, VecR);
  AllocMem (tBuf[nb].rTrue, nMol, VecR);
  AllocMem (tBuf[nb].rrDiffuse, nValDiffuse, real);
}
```

Diffusion from the velocity autocorrelation function

The alternative approach to measuring the diffusion coefficient is based on the integrated velocity autocorrelation function (5.2.6). Considerations governing the use of overlapped samples discussed previously also apply; the work itself is organized in a similar way and even the new variables have corresponding names. Each data collection buffer has an associated structure of type *TBuf* whose contents for this particular program are

```
typedef struct {
  VecR *orgVel;
  real *acfVel;
  int count;
} TBuf;                                                              5
```

where *orgVel* stores a copy of the atom velocities at the start of the measurement period and the autocorrelation function is constructed in *acfVel*. The general technique described here forms the basis for studying the remaining transport coefficients later in this section.

The calculation[♦] is carried out by the following functions.

```
void EvalVacf ()
{
  int n, nb, ni;

  for (nb = 0; nb < nBuffAcf; nb ++) {                              5
    if (tBuf[nb].count == 0) {
      DO_MOL tBuf[nb].orgVel[n] = mol[n].rv;
    }
    if (tBuf[nb].count >= 0) {
      ni = tBuf[nb].count;                                          10
      tBuf[nb].acfVel[ni] = 0.;
      DO_MOL tBuf[nb].acfVel[ni] +=
          VDot (tBuf[nb].orgVel[n], mol[n].rv);
    }
    ++ tBuf[nb].count;                                              15
  }
  AccumVacf ();
}

void AccumVacf ()                                                   20
{
  real fac;
  int j, nb;

  for (nb = 0; nb < nBuffAcf; nb ++) {                              25
    if (tBuf[nb].count == nValAcf) {
      for (j = 0; j < nValAcf; j ++)
        avAcfVel[j] += tBuf[nb].acfVel[j];
      tBuf[nb].count = 0;
      ++ countAcfAv;                                                30
      if (countAcfAv == limitAcfAv) {
        fac = stepAcf * deltaT / (NDIM * nMol * limitAcfAv);
        intAcfVel = fac * Integrate (avAcfVel, nValAcf);
        for (j = 1; j < nValAcf; j ++) avAcfVel[j] /= avAcfVel[0];
        avAcfVel[0] = 1.;                                           35
```

```
        PrintVacf (stdout);
        ZeroVacf ();
      }
    }
  }                                                                      40
}
```

The function *Integrate* (§18.4) computes the integral of its first argument using a simple method such as the trapezoidal rule. Other required functions are

```
void InitVacf ()
{
  int nb;

  for (nb = 0; nb < nBuffAcf; nb ++)                                     5
    tBuf[nb].count = - nb * nValAcf / nBuffAcf;
  ZeroVacf ();
}

void ZeroVacf ()                                                        10
{
  int j;

  countAcfAv = 0;
  for (j = 0; j < nValAcf; j ++) avAcfVel[j] = 0.;                      15
}

void PrintVacf (FILE *fp)
{
  real tVal;                                                           20
  int j;

  fprintf (fp, "acf\n");
  for (j = 0; j < nValAcf; j ++) {
    tVal = j * stepAcf * deltaT;                                       25
    fprintf (fp, "%8.4f %8.4f\n", tVal, avAcfVel[j]);
  }
  fprintf (fp, "vel acf integral: %8.3f\n", intAcfVel);
}
```

To incorporate the measurements into the MD program add the following statement to *SingleStep*,

```
if (stepCount >= stepEquil &&
    (stepCount - stepEquil) % stepAcf == 0) EvalVacf ();
```

and a call to the initialization function from *SetupJob*

```
InitVacf ();
```

The additional variables used are

```
TBuf *tBuf;
real *avAcfVel, intAcfVel;
int countAcfAv, limitAcfAv, nBuffAcf, nValAcf, stepAcf;
```

the measurement parameters input to the program are

```
NameI (limitAcfAv),
NameI (nBuffAcf),
NameI (nValAcf),
NameI (stepAcf),
```

and the array allocations (in `AllocArrays`) are

```
AllocMem (avAcfVel, nValAcf, real);
AllocMem (tBuf, nBuffAcf, TBuf);
for (nb = 0; nb < nBuffAcf; nb ++) {
  AllocMem (tBuf[nb].acfVel, nValAcf, real);
  AllocMem (tBuf[nb].orgVel, nMol, VecR);
}
```

s

Shear viscosity and thermal conductivity

These transport coefficient computations[♠] are also based on the appropriate auto-correlation functions and closely follow the treatment used to compute D from the velocity autocorrelation. Indeed, to simplify matters we will assume that all three transport coefficients are computed together and that identical parameters govern the measurements. In order to compute the quantities involved in the autocorrelation functions, certain additions must be made to the interaction calculations.

In the expressions for the pressure tensor (5.2.10) and heat current (5.2.15) used in the definitions of the transport coefficients, sums over products such as $r_{ij\,x}f_{ij\,y}$, $r_{ij\,x}f_{ij\,y}v_{j\,y}$ and $e_j v_{j\,x}$ appear; these terms should be evaluated at the same time as the forces. Additional arrays are needed to save sums of the form $\sum_i r_{ij\,x}f_{ij\,y}$ and the values of e_j separately for each atom; these two arrays are represented by adding extra quantities[†] to the definition of `Mol`,

```
VecR rf[3];
real en;
```

[♠] pr_05_3
[†] The matrix represented by the array `rf` is symmetric because $r_{ij\,x}f_{ij\,y} \equiv r_{ij\,y}f_{ij\,x}$.

The following additions to *ComputeForces* are required:

```
VecR w[3];
int k;
...
DO_MOL {
  mol[n].en = 0.;                                            5
  for (k = 0; k < 3; k ++) VZero (mol[n].rf[k]);
}
for (n = 0; n < nebrTabLen; n ++) {
  ...
  if (rr < rrCut) {                                          10
    ...
    mol[j1].en += uVal;
    mol[j2].en += uVal;
    for (k = 0; k < 3; k ++) w[k] = dr;
    VScale (w[0], fcVal * dr.x);                             15
    VScale (w[1], fcVal * dr.y);
    VScale (w[2], fcVal * dr.z);
    for (k = 0; k < 3; k ++) {
      VVAdd (mol[j1].rf[k], w[k]);
      VVAdd (mol[j2].rf[k], w[k]);                           20
    }
```

New elements that are added to *TBuf* for the autocorrelation function computations are

```
VecR orgTherm, orgVisc;
real *acfTherm, *acfVisc;
```

additional quantities needed are

```
real *avAcfTherm, *avAcfVisc, intAcfTherm, intAcfVisc;
```

and the arrays are allocated (in *AllocArrays*) by

```
AllocMem (avAcfTherm, nValAcf, real);
AllocMem (avAcfVisc, nValAcf, real);
for (nb = 0; nb < nBuffAcf; nb ++) {
  AllocMem (tBuf[nb].acfTherm, nValAcf, real);
  AllocMem (tBuf[nb].acfVisc, nValAcf, real);                5
}
```

The function *EvalVacf* is modified to incorporate the additional data collection:

```
VecR vecTherm, vecVisc;
...
VZero (vecVisc);
VZero (vecTherm);
```

```
DO_MOL {                                                              5
  vecVisc.x += mol[n].rv.y * mol[n].rv.z + 0.5 * mol[n].rf[1].z;
  vecVisc.y += mol[n].rv.z * mol[n].rv.x + 0.5 * mol[n].rf[2].x;
  vecVisc.z += mol[n].rv.x * mol[n].rv.y + 0.5 * mol[n].rf[0].y;
  mol[n].en += VLenSq (mol[n].rv);
  VVSAdd (vecTherm, 0.5 * mol[n].en, mol[n].rv);                     10
  vecTherm.x += 0.5 * VDot(mol[n].rv, mol[n].rf[0]);
  vecTherm.y += 0.5 * VDot(mol[n].rv, mol[n].rf[1]);
  vecTherm.z += 0.5 * VDot(mol[n].rv, mol[n].rf[2]);
}
for (nb = 0; nb < nBuffAcf; nb ++) {                                 15
  ...
  if (tBuf[nb].count == 0) {
    tBuf[nb].orgVisc = vecVisc;
    tBuf[nb].orgTherm = vecTherm;
  }                                                                  20
  tBuf[nb].acfVisc[ni] = VDot (tBuf[nb].orgVisc, vecVisc);
  tBuf[nb].acfTherm[ni] = VDot (tBuf[nb].orgTherm, vecTherm);
}
```

The following additions are made to *AccumVacf* to evaluate the autocorrelation integrals and transport coefficients:

```
if (tBuf[nb].count == nValAcf) {
  ...
  for (j = 0; j < nValAcf; j ++) {
    avAcfVisc[j] += tBuf[nb].acfVisc[j];
    avAcfTherm[j] += tBuf[nb].acfTherm[j];                           5
  }
...
if (countAcfAv == limitAcfAv) {
  ...
  fac = density * stepAcf * deltaT / (3. * temperature *            10
    nMol * limitAcfAv);
  intAcfVisc = fac * Integrate (avAcfVisc, nValAcf);
  for (j = 1; j < nValAcf; j ++) avAcfVisc[j] /= avAcfVisc[0];
  avAcfVisc[0] = 1.;
  fac = density * stepAcf * deltaT / (3. * Sqr (temperature) *      15
    nMol * limitAcfAv);
  intAcfTherm = fac * Integrate (avAcfTherm, nValAcf);
  for (j = 1; j < nValAcf; j ++) avAcfTherm[j] /= avAcfTherm[0];
  avAcfTherm[0] = 1.;
```

Finally, the function *ZeroVacf* requires

```
for (j = 0; j < nValAcf; j ++) {
  avAcfVisc[j] = 0.;
  avAcfTherm[j] = 0.;
}
```

and, in `PrintVacf`, the two quantities `avAcfVisc[j]` and `avAcfTherm[j]` are added to the output loop and the integrated values `intAcfVisc` and `intAcfTherm` are also output.

5.4 Space–time correlation functions

Definitions

The experimental significance of time-dependent correlation functions is that spectroscopic techniques, of which neutron scattering is one example, actually measure the spectra of microscopic dynamical quantities. The MD approach provides equivalent information directly from the trajectories, so that comparison with experiment can be made by carrying out a Fourier analysis of the simulation results – in a sense this amounts to performing the experiment on the model system. Such correlation functions span the entire range of length and time scales, from slow long-wavelength modes at the hydrodynamic limit, right down to the atomic level [boo91].

To link the discrete atomistic picture with the continuum view of a system described in terms of smoothly varying scalar and vector fields, a function such as the number density (for a single species this is just the mass density in dimensionless MD units) at a point r at time t is expressed as a sum over atoms, as in (5.2.3),

$$\rho(r, t) = \sum_j \delta(r - r_j(t)) \tag{5.4.1}$$

In a practical sense, this defines the local density in terms of the average occupancy of a small volume of space situated at r and measured over a short time interval. There are of course fluctuations as atoms enter and leave the volume, but these can be reduced by using larger volumes and/or longer time intervals. The definition satisfies the obvious requirement that matter is conserved,

$$\int \rho(r, t)\, dr = N_m \tag{5.4.2}$$

Space- and time-dependent density correlations are described by means of the van Hove correlation function [han86b], defined as

$$G(r, t) = \frac{1}{N_m} \left\langle \int \rho(r' + r, t)\rho(r', 0)\, dr' \right\rangle \tag{5.4.3}$$

$$= \frac{1}{N_m} \left\langle \sum_{ij} \delta(r + r_i(0) - r_j(t)) \right\rangle \tag{5.4.4}$$

where homogeneity is assumed in order to carry out the r' integration in (5.4.3).

The double summation can be divided into two parts,

$$G(\mathbf{r}, t) = G_s(\mathbf{r}, t) + G_d(\mathbf{r}, t) \tag{5.4.5}$$

where

$$G_s(\mathbf{r}, t) = \frac{1}{N_m} \left\langle \sum_j \delta\big(\mathbf{r} + \mathbf{r}_j(0) - \mathbf{r}_j(t)\big) \right\rangle \tag{5.4.6}$$

is the probability of an atom being displaced by a distance $\mathbf{r}$ during time t and $G_d(\mathbf{r}, t)$ contains the remaining terms of the double sum. The $t = 0$ limits of G_s and G_d are

$$G_s(\mathbf{r}, 0) = \delta(\mathbf{r}) \tag{5.4.7}$$

$$G_d(\mathbf{r}, 0) = \rho g(\mathbf{r}) \tag{5.4.8}$$

In the limits $r \rightarrow \infty$ or $t \rightarrow \infty$, $G_s \rightarrow 1/V$ and $G_d \rightarrow \rho$. There have been attempts in the past to establish a functional relationship between G_s and G_d, but these have been unsuccessful because of the wealth of dynamical detail that must be discarded.

The Fourier transform of the density is

$$\rho(\mathbf{k}, t) = \int \rho(\mathbf{r}, t) e^{-i\mathbf{k} \cdot \mathbf{r}} \, d\mathbf{r} \tag{5.4.9}$$

$$= \sum_j e^{-i\mathbf{k} \cdot \mathbf{r}_j(t)} \tag{5.4.10}$$

and the intermediate scattering function is defined by

$$F(\mathbf{k}, t) = \int G(\mathbf{r}, t) e^{-i\mathbf{k} \cdot \mathbf{r}} \, d\mathbf{r} \tag{5.4.11}$$

$$= \frac{1}{N_m} \langle \rho(\mathbf{k}, t)\rho(-\mathbf{k}, 0) \rangle \tag{5.4.12}$$

Note the connection to the static structure factor, $F(\mathbf{k}, 0) = S(\mathbf{k})$. The dynamic structure factor is defined as

$$S(\mathbf{k}, \omega) = \frac{1}{2\pi} \int_{-\infty}^{\infty} F(\mathbf{k}, t) e^{i\omega t} \, dt \tag{5.4.13}$$

and satisfies the sum rule

$$\int_{-\infty}^{\infty} S(\mathbf{k}, \omega) \, d\omega = S(\mathbf{k}) \tag{5.4.14}$$

If l and τ are the mean free path and collision time (suitably interpreted when dealing with continuous potentials), then the regime $kl \ll 1$ and $\omega\tau \ll 1$, or, equivalently, wavelength $\gg l$ and timescale $\gg \tau$, is where the behavior can be described by continuum fluid mechanics and the underlying atomic nature of the fluid is totally hidden. The ability to use MD to study the behavior across a range of scales provides the bridge between atomistic and macroscopic worlds.

While the local density conveys information about the distribution of atoms, it is equally possible to examine local variations in the motion of the atoms. The definition of the particle current (or momentum current for a single atomic species using MD units) is [han86b]

$$\boldsymbol{\pi}(\boldsymbol{r}, t) = \sum_j \boldsymbol{v}_j \delta(\boldsymbol{r} - \boldsymbol{r}_j(t)) \tag{5.4.15}$$

with Fourier transform

$$\boldsymbol{\pi}(\boldsymbol{k}, t) = \sum_j \boldsymbol{v}_j e^{-i\boldsymbol{k}\cdot\boldsymbol{r}_j(t)} \tag{5.4.16}$$

The spatial correlation functions of the components of the current vector are defined as

$$C_{\alpha\beta}(\boldsymbol{k}, t) = \frac{k^2}{N_m}\langle \pi_\alpha(\boldsymbol{k}, t)\pi_\beta(-\boldsymbol{k}, 0)\rangle \tag{5.4.17}$$

For isotropic fluids, symmetry considerations lead to an expression in terms of longitudinal and transverse currents (the directions are relative to $\boldsymbol{k}$),

$$C_{\alpha\beta}(\boldsymbol{k}, t) = \frac{k_\alpha k_\beta}{k^2}C_L(\boldsymbol{k}, t) + \left(\delta_{\alpha\beta} - \frac{k_\alpha k_\beta}{k^2}\right)C_T(\boldsymbol{k}, t) \tag{5.4.18}$$

and by setting $\boldsymbol{k} = k\hat{\boldsymbol{z}}$ we obtain

$$C_L(k, t) = \frac{k^2}{N_m}\langle \pi_z(\boldsymbol{k}, t)\pi_z(-\boldsymbol{k}, 0)\rangle \tag{5.4.19}$$

$$C_T(k, t) = \frac{k^2}{2N_m}\langle \pi_x(\boldsymbol{k}, t)\pi_x(-\boldsymbol{k}, 0) + \pi_y(\boldsymbol{k}, t)\pi_y(-\boldsymbol{k}, 0)\rangle \tag{5.4.20}$$

The dynamic structure factor is related to the Fourier transform of the longitudinal current

$$S(\boldsymbol{k}, \omega) = \frac{1}{\omega^2}C_L(\boldsymbol{k}, \omega) \tag{5.4.21}$$

In the small k (continuum) limit, the form of $S(\boldsymbol{k}, \omega)$ is known [han86b]. The function is of course symmetric in ω, there are Lorentzian shaped Brillouin peaks at $\omega = \pm v_s k$, where v_s is the adiabatic speed of sound, and there is a Rayleigh peak at $\omega = 0$. The width of each peak is proportional to k^2; the Rayleigh peak width is also proportional to the thermal diffusivity ($\lambda/\rho C_P$) and the Brillouin peak width is proportional to the sound attenuation coefficient (a quantity expressible in terms of transport coefficients and specific heats). Note that wraparound effects can occur for times $t > L/v_s$, where L is the region size; if significant, this sets an upper bound to the timescales that can be examined, and hence a lower limit to ω. The values of k that can be examined are restricted to vectors with components that are integer multiples of $2\pi/L$, so that the larger the region, the smaller the k values that can be reached.

Computational methods

The MD evaluation of $S(k, \omega)$ is based on the Fourier transform of $F(k, t)$. This in turn can be expressed either as the Fourier transform of a discretized form of the van Hove correlation function (5.4.11) – an extension of the method used in §4.3 for $g(r)$ – or as the time correlation of the Fourier-transformed density (5.4.12). The latter is clearly preferable since it requires a great deal less work. Evaluation of $\rho(k, t)$ is based directly on a sum over atoms, as in the study of long-range order in §4.3 (but there only for a single k value). An alternative for very large systems that is not explored here is to first evaluate a coarse-grained density function $\rho(r, t)$ based on a grid with suitable spatial resolution, and then use a discrete (preferably fast-) Fourier transform to obtain $\rho(k, t)$. Some of the detail is lost when grid averages are used, but this affects results at short distances – typically of the order of the grid spacing – while details of the more interesting long-range behavior are preserved.

Further simplification is possible when studying isotropic systems, since the function of interest is the spherically-averaged quantity $S(k, \omega)$, and it is therefore sufficient to consider k vectors in a very limited number of directions. Averaging over several spatially equivalent directions will improve the statistics, so that if we confine our attention to k vectors along the coordinate axes, the computation requires evaluation of $\rho(k, t)$ for the three vectors $k = (k, 0, 0)$, $(0, k, 0)$ and $(0, 0, k)$. If we assume a cubic region, periodic boundaries restrict the allowed values of k to integer multiples of $2\pi/L$.

Program details

We now turn to the details[♦] of computing the density and current correlation functions. The technique of overlapped measurements introduced in §5.3 is also used here, and the variables involved in the data collection are organized in a similar manner.

Because of the considerable amount of data that must be collected, we begin with some remarks on how the computations and data are organized. The calculation starts by evaluating the Fourier sums for the density and the three current components – one longitudinal and two transverse – along each of the three k directions.

♦ pr_05_4

The real and imaginary parts of these complex valued results are stored in the array `valST`, with the index determined by a combination of the direction of k, the value of k and the kind of quantity; there are four of these – the density and the three current components – so storage must be provided for a total of 24 real numbers for each value of k considered. The real parts of the contributions to the correlation functions computed by a single call to `EvalSpacetimeCorr` are placed in the array `acfST` that is part of the structure `TBuf` tailored for this calculation,

```
typedef struct {
  real **acfST, *orgST;
  int count;
} TBuf;
```

The first index of the two-dimensional array `acfST` specifies the kind of correlation, namely the k value and whether the value corresponds to the real part of $C_L(k, t)$, $C_T(k, t)$ or $F(k, t)$; the second index specifies the time offset. The number of different k values used is denoted by `nFunCorr`.

The contributions to all the (overlapped) correlation function measurements in progress at a given instant are evaluated by the following function. A cubic region shape and leapfrog integrator are assumed. The recursion relations

$$\sin n\theta = 2\cos\theta \sin(n-1)\theta - \sin(n-2)\theta \qquad (5.4.22)$$

$$\cos n\theta = 2\cos\theta \cos(n-1)\theta - \cos(n-2)\theta \qquad (5.4.23)$$

are used for evaluating sines and cosines of multiple angles.

```
void EvalSpacetimeCorr ()
{
  real b, c, c0, c1, c2, kVal, s, s1, s2, w;
  int j, k, m, n, nb, nc, ni, nv;

  for (j = 0; j < 24 * nFunCorr; j ++) valST[j] = 0.;      5
  kVal = 2. * M_PI / region.x;
  DO_MOL {
    j = 0;
    for (k = 0; k < 3; k ++) {                             10
      for (m = 0; m < nFunCorr; m ++) {
        if (m == 0) {
          b = kVal * VComp (mol[n].r, k);
          c = cos (b);
          s = sin (b);                                     15
          c0 = c;
        } else if (m == 1) {
          c1 = c;
          s1 = s;
          c = 2. * c0 * c1 - 1.;                           20
          s = 2. * c0 * s1;
```

```
      } else {
        c2 = c1;
        s2 = s1;
        c1 = c;                                              25
        s1 = s;
        c = 2. * c0 * c1 - c2;
        s = 2. * c0 * s1 - s2;
      }
      valST[j ++] += mol[n].rv.x * c;                        30
      valST[j ++] += mol[n].rv.x * s;
      valST[j ++] += mol[n].rv.y * c;
      valST[j ++] += mol[n].rv.y * s;
      valST[j ++] += mol[n].rv.z * c;
      valST[j ++] += mol[n].rv.z * s;                        35
      valST[j ++] += c;
      valST[j ++] += s;
    }
  }
}                                                            40
for (nb = 0; nb < nBuffCorr; nb ++) {
  if (tBuf[nb].count == 0) {
    for (j = 0; j < 24 * nFunCorr; j ++)
      tBuf[nb].orgST[j] = valST[j];
  }                                                          45
  if (tBuf[nb].count >= 0) {
    for (j = 0; j < 3 * nFunCorr; j ++)
      tBuf[nb].acfST[j][tBuf[nb].count] = 0.;
    j = 0;
    for (k = 0; k < 3; k ++) {                               50
      for (m = 0; m < nFunCorr; m ++) {
        for (nc = 0; nc < 4; nc ++) {
          nv = 3 * m + 2;
          if (nc < 3) {
            w = Sqr (kVal * (m + 1));                        55
            -- nv;
            if (nc == k) -- nv;
            else w *= 0.5;
          } else w = 1.;
          tBuf[nb].acfST[nv][tBuf[nb].count] +=             60
            w * (valST[j] * tBuf[nb].orgST[j] +
            valST[j + 1] * tBuf[nb].orgST[j + 1]);
          j += 2;
        }
      }                                                      65
    }
  }
  ++ tBuf[nb].count;
}
AccumSpacetimeCorr ();                                       70
}
```

Accumulating the averages is the task of the following function, where $nValCorr$ is the number of time offsets.

```
void AccumSpacetimeCorr ()
{
  int j, n, nb;

  for (nb = 0; nb < nBuffCorr; nb ++) {                                    5
    if (tBuf[nb].count == nValCorr) {
      for (j = 0; j < 3 * nFunCorr; j ++) {
        for (n = 0; n < nValCorr; n ++)
          avAcfST[j][n] += tBuf[nb].acfST[j][n];
      }                                                                    10
      tBuf[nb].count = 0;
      ++ countCorrAv;
      if (countCorrAv == limitCorrAv) {
        for (j = 0; j < 3 * nFunCorr; j ++) {
          for (n = 0; n < nValCorr; n ++)                                  15
            avAcfST[j][n] /= 3. * nMol * limitCorrAv;
        }
        PrintSpacetimeCorr (stdout);
        ZeroSpacetimeCorr ();
      }                                                                    20
    }
  }
}
```

The additional variables are

```
TBuf *tBuf;
real **avAcfST, *valST;
int countCorrAv, limitCorrAv, nBuffCorr, nFunCorr, nValCorr,
    stepCorr;
```

new input values,

```
NameI (limitCorrAv),
NameI (nBuffCorr),
NameI (nFunCorr),
NameI (nValCorr),
NameI (stepCorr),                                                         5
```

and required array allocations (in $AllocArrays$),

```
AllocMem (valST, 24 * nFunCorr, real);
AllocMem2 (avAcfST, 3 * nFunCorr, nValCorr, real);
AllocMem (tBuf, nBuffCorr, TBuf);
for (nb = 0; nb < nBuffCorr; nb ++) {
  AllocMem (tBuf[nb].orgST, 24 * nFunCorr, real);                         5
```

```
      AllocMem2 (tBuf[nb].acfST, 3 * nFunCorr, nValCorr, real);
   }
```

where `AllocMem2` dynamically allocates a two-dimensional array `a[n1][n2]` of the specified type[†]

```
#define AllocMem2(a, n1, n2, t)                                    \
   AllocMem (a, n1, t *);                                          \
   AllocMem (a[0], n1 * n2, t);                                    \
   for (k = 1; k < n1; k ++) a[k] = a[k - 1] + n2;
```

The calls to the functions that do the analysis (from `SingleStep`) and the initialization (from `SetupJob`) are

```
   if (stepCount > stepEquil && (stepCount - stepEquil) %
      stepCorr == 0) EvalSpacetimeCorr ();

   InitSpacetimeCorr ();
```

Finally, the initialization and output functions are

```
   void InitSpacetimeCorr ()
   {
      int nb;

      for (nb = 0; nb < nBuffCorr; nb ++)                              5
         tBuf[nb].count = - nb * nValCorr / nBuffCorr;
      ZeroSpacetimeCorr ();
   }

   void ZeroSpacetimeCorr ()                                          10
   {
      int j, n;

      countCorrAv = 0;
      for (j = 0; j < 3 * nFunCorr; j ++) {                           15
         for (n = 0; n < nValCorr; n ++) avAcfST[j][n] = 0.;
      }
   }

   void PrintSpacetimeCorr (FILE *fp)                                 20
   {
      real tVal;
      int j, k, n;
      char *header[] = {"cur-long", "cur-trans", "density"};
                                                                      25
      fprintf (fp, "space-time corr\n");
```

[†] The allocation produces a one-dimensional array of size $n1 \times n2$ together with an array of pointers containing the offsets that allow it to be treated as doubly indexed.

```
for (k = 0; k < 3; k ++) {
  fprintf (fp, "%s\n", header[k]);
  for (n = 0; n < nValCorr; n ++) {
    tVal = n * stepCorr * deltaT;                          30
    fprintf (fp, "%7.3f", tVal);
    for (j = 0; j < nFunCorr; j ++)
        fprintf (fp, " %8.4f", avAcfST[3 * j + k][n]);
    fprintf (fp, "\n");
  }                                                         35
}
}
```

Correlation analysis

In order to evaluate $S(k, \omega)$ in (5.4.13) for a given k at a total of n_ω frequency values (including $\omega = 0$), it is necessary to collect $n_\omega + 1$ equally spaced measurements of $F(k, t)$. Since $F(k, t)$ is an even function of t, these results are reflected about $t = 0$, providing a total of $2n_\omega$ values – the first and last measurements are used only once each. Then the Fourier transform is carried out using all $2n_\omega$ values (this would normally be some power of two, to simplify the use of the FFT method), and the discrete form of the function $S(k, \omega)$ appears as the real part of the first n_ω terms [pre92]. The current correlation functions are treated in the same way. This analysis is carried out separately from the run itself; the program for doing this, as well as for tabulating the normalized functions $F(k, t)/F(k, 0)$, will now be described.

The analysis program[♠] shown below first averages the data produced by the run and then either generates the normalized time-dependent correlations (and error estimates, here unused) or the Fourier transforms. The program also demonstrates a general approach to organizing analysis in which only selected data are extracted from the job output file based on the headings accompanying the data. Whether the real-space or Fourier version of the program is run depends on *doFourier*, and for the latter, a windowing function [pre92] is applied if *doWindow* is set; these options are specified on the command line when the program is run, as is the amount of early data to skip, and the name of the file containing a full copy of the program output that is to be processed. Cutoffs for limiting the output are built into the program (but could also be added to the command line if desired).

A complex data type is introduced here

```
typedef struct {
  real R, I;
} Cmplx;
```

♠ pr_anspcor

```
#define CSet(a, x, y)                                          \           5
    a.R = x,                                                   \
    a.I = y
```

The macro *NameVal* (§18.5) is used in locating a selected data item from near the start of the data file[†], while the function *FftComplex* (§18.4) performs a fast Fourier transform (FFT) and overwrites the original data with the result (the length of the processed array must be a power of 2).

```
#define BUFF_LEN  1024
char *header[] = {"cur-long", "cur-trans", "density"},
    *txtCorr = "space-time corr";

int main (int argc, char **argv)                                          5
{
  Cmplx *work;
  real *corrSum[3], *corrSumSq[3], damp, deltaT, deltaTCorr,
      omegaMax, tMax, w, x;
  int doFourier, doWindow, j, k, n, nData, nFunCorr, nSet, nSetSkip,      10
      nv, nValCorr, stepCorr;
  char *bp, *fName, buff[BUFF_LEN];
  FILE *fp;

  n = 1;                                                                  15
  if (-- argc < 1 || ! strcmp (argv[1], "-h")) PrintHelp (argv[0]);
  doFourier = 1;
  doWindow = 0;
  nSetSkip = 1;
  while (-- argc >= 0) {                                                  20
    if (! strcmp (argv[n], "-t")) doFourier = 0;
    else if (! strcmp (argv[n], "-w")) doWindow = 1;
    else if (! strcmp (argv[n], "-s")) nSetSkip = atoi (argv[n] + 2);
    else {
      fName = argv[n];                                                    25
      break;
    }
    ++ n;
  }
  if (argc > 0) PrintHelp (argv[0]);                                      30
  omegaMax = 10.;
  tMax = 5.;
  if ((fp = fopen (fName, "r")) == 0) {
    printf ("no file\n");
    exit (0);                                                             35
  }
  while (1) {
    bp = fgets (buff, BUFF_LEN, fp);
```

[†] The file being processed begins with a copy of the input data used for the run; this is terminated by a line starting with a '-' character, denoted here by *CHAR_MINUS*. Several standard C file and character-string functions are used in this program.

```
    if (*bp == CHAR_MINUS) break;
    NameVal (deltaT);                                                   40
    NameVal (nFunCorr);
    NameVal (nValCorr);
    NameVal (stepCorr);
}
deltaTCorr = stepCorr * deltaT;                                         45
for (j = 0; j < 3; j ++) {
  AllocMem (corrSum[j], nFunCorr * nValCorr, real);
  AllocMem (corrSumSq[j], nFunCorr * nValCorr, real);
  for (n = 0; n < nFunCorr * nValCorr; n ++) {
    corrSum[j][n] = 0.;                                                 50
    corrSumSq[j][n] = 0.;
  }
}
AllocMem (work, 2 * (nValCorr - 1), Cmplx);
nData = 0;                                                             55
nSet = 0;
while (1) {
  if (! (bp = fgets (buff, BUFF_LEN, fp))) break;
  if (! strncmp (bp, txtCorr, strlen (txtCorr))) {
    ++ nSet;                                                           60
    if (nSet < nSetSkip) continue;
    ++ nData;
    for (j = 0; j < 3; j ++) {
      bp = fgets (buff, BUFF_LEN, fp);
      for (n = 0; n < nValCorr; n ++) {                                65
        bp = fgets (buff, BUFF_LEN, fp);
        w = strtod (bp, &bp);
        for (k = 0; k < nFunCorr; k ++) {
          w = strtod (bp, &bp);
          corrSum[j][k * nValCorr + n] += w;                           70
          corrSumSq[j][k * nValCorr + n] += Sqr (w);
        }
      }
    }
  }                                                                    75
}
fclose (fp);
printf ("%d\n", nData);
for (j = 0; j < 3; j ++) {
  for (n = 0; n < nFunCorr * nValCorr; n ++) {                         80
    corrSum[j][n] /= nData;
    corrSumSq[j][n] = sqrt (corrSumSq[j][n] / nData -
        Sqr (corrSum[j][n]));
  }
}                                                                      85
if (doFourier) {
  for (j = 0; j < 3; j ++) {
    for (k = 0; k < nFunCorr; k ++) {
      for (n = 0; n < nValCorr; n ++) {
```

```
            if (doWindow) damp = (nValCorr - n) / (nValCorr + 0.5);        90
            else damp = 1.;
            CSet (work[n], corrSum[j][k * nValCorr + n] * damp, 0.);
          }
          for (n = nValCorr; n < 2 * (nValCorr - 1); n ++)
            work[n] = work[2 * (nValCorr - 1) - n];                        95
          FftComplex (work, 2 * nValCorr - 2);
          for (n = 0; n < nValCorr; n ++)
            corrSum[j][k * nValCorr + n] = work[n].R;
        }
      }                                                                    100
      omegaMax = Min (omegaMax, M_PI / deltaTCorr);
      nv = nValCorr * omegaMax / (M_PI / deltaTCorr);
    } else {
      for (j = 0; j < 3; j ++) {
        for (k = 0; k < nFunCorr; k ++) {                                 105
          for (n = 1; n < nValCorr; n ++)
            corrSum[j][k * nValCorr + n] /= corrSum[j][k * nValCorr];
          corrSum[j][k * nValCorr] = 1.;
        }
      }                                                                    110
      tMax = Min (tMax, (nValCorr - 1) * deltaTCorr);
      nv = nValCorr * tMax / ((nValCorr - 1) * deltaTCorr);
    }
    for (j = 0; j < 3; j ++) {
      printf ("%s\n", header[j]);                                         115
      for (n = 0; n < nv; n ++) {
        if (doFourier) x = n * omegaMax / nv;
        else x = n * deltaTCorr;
        printf ("%9.4f", x);
        for (k = 0; k < nFunCorr; k ++)                                   120
          printf (" %9.4f", corrSum[j][k * nValCorr + n]);
        printf ("\n");
      }
    }
  }                                                                        125

void PrintHelp (char *pName)
{
  printf ("Usage: %s [-t{time_corr}] [-s{skip}n] [-w{window}]"
    " input-file \n", pName);                                             130
  exit (0);
}
```

5.5 Measurements

Velocity autocorrelation function

The velocity autocorrelation functions shown in Figure 5.2 are computed during soft-sphere runs that use the following data,

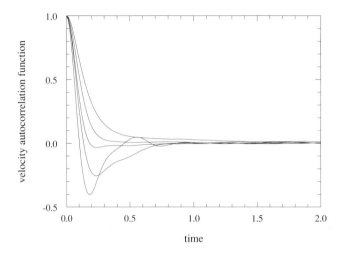

Fig. 5.2. Soft-sphere velocity autocorrelation functions for densities 0.6–1.0.

deltaT	0.005
density	0.6
initUcell	5 5 5
limitAcfAv	200
nBuffAcf	10
nebrTabFac	8
nValAcf	200
rNebrShell	0.4
stepAcf	3
stepAvg	1000
stepEquil	2000
stepInitlzTemp	100
stepLimit	15000
temperature	0.5

with values of `density` between 0.6 and 1.0. Leapfrog integration is used. The initial state is an FCC lattice, so that $N_m = 500$. A single set of results based on 200 sets of partially overlapped measurements is produced during a run of 15 000 timesteps. The negative correlations that are observed at higher densities (both for LJ and hard-sphere systems) were one of the important early revelations of MD [rah64, ald67].

Given the exponential sensitivity of the trajectories to any numerical error, as demonstrated in §3.8, it is important to establish that results such as the velocity autocorrelation function are fully reproducible. Here we show one example to confirm that this is indeed the case. The system used is the same as above, at a density of 0.9, but with different values of Δt; measurement intervals and run lengths are

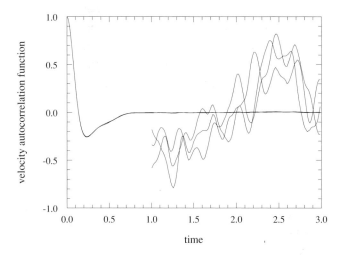

Fig. 5.3. Velocity autocorrelation functions for $\rho = 0.9$ computed using $\Delta t = 0.005$, 0.0025 and 0.00125; the results at later times are also shown with a vertical 100-fold magnification.

adjusted accordingly. Figure 5.3 shows the results; in order to resolve the extremely small differences at later times, the results are replotted with the vertical scale enlarged by a factor of 100.

Transport coefficients

Diffusion coefficient measurements using (5.2.4) for the same system♠ are shown in Figure 5.4. The necessary input data include

limitDiffuseAv	*200*
nBuffDiffuse	*10*
nValDiffuse	*250*
stepDiffuse	*4*
stepLimit	*63000*

For $\rho \geq 0.9$ the values drop to zero and there is no significant diffusion, while at smaller ρ they appear to asymptote to increasingly larger values, although convergence also slows and longer measurements are seen to be necessary at $\rho = 0.6$ and 0.7.

Estimates for D based on (5.2.6) can be obtained by integrating the velocity autocorrelation functions shown in Figure 5.2. The results, without any attempt at

---

♠ *pr_andiffus*

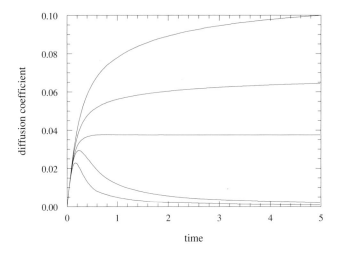

Fig. 5.4.   Diffusion coefficient measurements for densities 0.6–1.0.

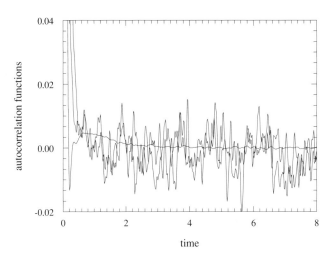

Fig. 5.5.   Velocity, pressure-tensor and heat-current autocorrelation functions; the vertical scale has been expanded to show the noise present in the results.

error estimation, for $\rho = 0.9, 0.8, 0.7$ and 0.6, are $0.001, 0.038, 0.067$ and 0.108 respectively; these can be compared with the D values of Figure 5.4.

In Figure 5.5 we show the three autocorrelation functions whose integrals yield D, η and λ, namely velocity, pressure tensor and heat current. The system used here has $N_m = 864$, $\rho = 0.8$ and $T = 1$. To improve the quality of the results,

Table 5.1. Transport coefficients (c) from the integrated autocorrelation functions.

c	$\langle c \rangle$	$\sigma(c)$	$\sigma(c)/\langle c \rangle$
D	0.0818	0.0015	0.0180
η	1.4468	0.6297	0.4352
λ	5.6299	1.9573	0.3477

the computation is run for almost a half million timesteps, with the following measurement parameters included in the input data:

```
stepEquil      4000
limitAcfAv     500
nBuffAcf       30
nValAcf        600
stepAcf        3
```

This yields 15 sets of autocorrelation results; the leapfrog integrator produces practically no energy drift (a mere one part in 2000) over the entire run. There is a clear difference between the smooth velocity autocorrelation function, which involves separate contributions from each atom, and the other two functions which are comparatively noisy because the entire system must be considered to obtain a single measurement.

Evaluation of the transport coefficients by integrating these autocorrelation functions♠ over the entire range to $t = 9$ leads to the results shown in Table 5.1. The uncertainty in the estimates of η and λ is considerable, but can be reduced once it is realized that the noise in the integrands makes a substantial contribution to the error without improving the estimate of the mean [lev87]. To show the potential for improvement, Table 5.2 lists the results obtained by terminating the integration after the first 70 values, corresponding to $t \approx 1$, as well as after two and three times this number of values. Further information on transport coefficient calculations using these methods can be found in [lev73, sch85, lev87, vog87].

Space–time correlations

Sample results for space–time correlation functions are obtained from a single run of a soft-sphere system with $N_m = 2048$, $T = 0.7$ and $\rho = 0.84$. The state point

---

♠ `pr_antransp`

Table 5.2. Effect of truncating the autocorrelation function integration after n values.

n	c	$\langle c \rangle$	$\sigma(c)$	$\sigma(c)/\langle c \rangle$
70	D	0.0780	0.0006	0.0076
	η	1.6890	0.1877	0.1111
	λ	6.3320	0.6518	0.1029
140	D	0.0807	0.0008	0.0094
	η	1.6904	0.2125	0.1257
	λ	6.3355	1.1238	0.1774
210	D	0.0815	0.0011	0.0129
	η	1.7120	0.3338	0.1950
	λ	5.9660	1.3919	0.2333

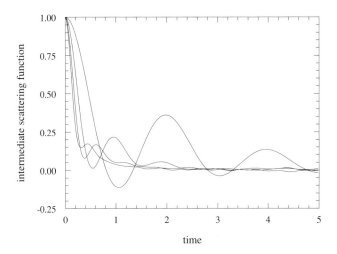

Fig. 5.6. Normalized intermediate scattering function $F(k, t)$ for the four smallest k values; the decay becomes slower as $k \to 0$.

is chosen to be fairly close to published results for the LJ fluid [sch86]. The input data include

```
limitCorrAv    500
nBuffCorr      80
nFunCorr       4
nValCorr       1025
stepCorr       5
```

A run of a little over 4×10^5 timesteps produces 13 sets of results.

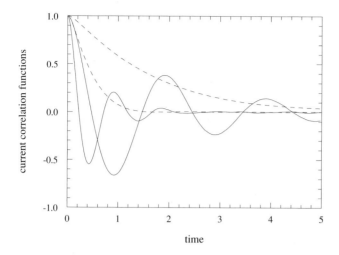

Fig. 5.7. Normalized longitudinal (solid curve) and transverse (dashed) current correlation functions, $C_L(k, t)$ and $C_T(k, t)$, for the two smallest k values.

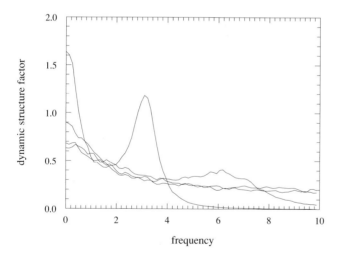

Fig. 5.8. Dynamic structure factor $S(k, \omega)$ for the four smallest k values.

Figure 5.6 shows the normalized intermediate scattering function $F(k, t)$ for the four smallest values of k. The normalized current correlation functions $C_L(k, t)$ and $C_T(k, t)$ are shown in Figure 5.7, with results for the two smallest k values included in both cases. The absence of any structure in C_T is expected for normal liquids that do not support shear waves; the peaks in C_L correspond to sound propagation. Finally, the dynamic structure factor $S(k, \omega)$, with its expected peaks, is shown in Figure 5.8. In general, the soft-sphere results resemble those for the LJ case [sch86] (allowing for the different time units); they can also be compared with hard-sphere results [all83]. Additional correlation functions are treated in [des88].

5.6 Further study

5.1   All the transport coefficient values have been given in reduced MD units; convert them to physical units and compare with the experimental values for argon.

5.2   Extend the diffusion measurements until all converge.

5.3   Compare the transport coefficients for soft-sphere and LJ fluids under similar conditions.

5.4   The large t behavior of the autocorrelation functions is a subject of particular interest [erp85, erp88]; examine the kind of decay that occurs and whether it depends on some power of t (rather than the exponential decay that might have been naïvely expected). What kind of cooperative motion is responsible for this behavior in the case of the velocity autocorrelation function [ald70b]?

5.5   The bulk viscosity is another transport coefficient; measure it using the appropriate autocorrelation function [lev73, lev87].

5.6   New transport coefficients appear when binary fluids are studied; consider the possibilities [vog88].

5.7   Explore alternative, possibly more efficient methods for organizing the space–time correlation computations.

5.8   The space–time correlations show propagating longitudinal modes (these correspond to sound waves); is it possible to observe similar transverse modes (shear waves) at sufficiently high density [lev87]?

6

Alternative ensembles

6.1 Introduction

The equations of motion used in MD are based on Newtonian mechanics; in this way MD mimics nature. If one adopts the purely mechanical point of view there is little more to be said, but if a broader perspective is permitted and MD is regarded as a tool for generating equilibrium states satisfying certain specified requirements, then it is possible to modify the dynamics and address a broader range of problems. But at the outset it must be emphasized that no physical meaning is attributed to the actual dynamics, and the approach is merely one of computational convenience for generating particular equilibrium thermodynamic states, although – and this is not an attempt to extract any such meaning – the deviations of the motion from the truly Newtonian may in fact be extremely small.

Conventional MD differs from most experimental studies in that it is the energy and volume that are fixed, rather than temperature and pressure. In statistical mechanical terms, MD produces microcanonical (NVE) ensemble averages, whereas constant-temperature experiments correspond to the canonical (NVT) ensemble; if constant pressure is imposed as well, as is generally the case in the laboratory, it is the isothermal–isobaric (NPT) ensemble that is the relevant one. While the choice of ensemble is usually one of convenience at the macroscopic level since (away from the critical point) thermal fluctuations are small, for the microscopic systems studied by MD the fluctuations of nonregulated quantities can be sufficiently large to make precise measurement difficult. Modifying the dynamics allows MD to model the equilibrium behavior of such ensembles directly.

We will describe two different approaches to the problem. One employs a feedback mechanism for correcting deviations in the controlled parameter (for example, temperature) from the preset mean value; the value fluctuates, but the size of the fluctuations can be regulated. The other method ensures that the controlled parameter is strictly constant, apart from numerical drift, by augmenting the equations of motion with suitable mechanical constraints; thus temperature can be held constant by introducing a constraint that fixes the kinetic energy.

There are other ways to change the ensemble, such as by coupling the system to a constant-temperature heat bath, or even by simply resetting the kinetic energy at each timestep. The former requires a stochastic mechanism for adjusting velocities in order to reproduce the effect of a heat bath [and80], but this violates the deterministic nature of the dynamics. The latter method is sufficiently crude not to merit consideration, although when it comes to introducing hard walls into the simulation (§7.3) the same idea can be adopted, but the justification is of course entirely different.

6.2 Feedback methods

Controlled temperature

The mechanism for feedback regulation of temperature rests on the idea that because the temperature is proportional to the mean-square velocity it ought to be possible to vary the temperature by adjusting the rate at which time progresses [nos84a]. A new dynamical variable s is introduced into the Lagrangian in a manner that is equivalent to rescaling the unit of time, and extra terms are added in just the way needed to obtain the desired behavior. There are now two distinct time variables: the real, or physical, time t', and a scaled, or virtual, time t; the relation between them is through their differentials,

$$dt = s(t')\, dt' \tag{6.2.1}$$

The Lagrangian for this unusual 'extended' system is written as

$$\mathcal{L} = \tfrac{1}{2} m s^2 \sum_i \dot{r}_i^2 - \sum_{i<j} u(r_{ij}) + \tfrac{1}{2} M_s \dot{s}^2 - n_f T \log s \tag{6.2.2}$$

where T is the required temperature,

$$n_f = 3N_m + 1 \tag{6.2.3}$$

is the number of degrees of freedom (which could be reduced by three to account for momentum conservation), M_s plays the role of a mass that is needed in order to construct an equation of motion for the new 'coordinate' s, and the dot stands for the derivative d/dt – note that (6.2.2) is defined in terms of the virtual time. The Lagrange equations of motion that are obtained by the standard procedure are

$$\ddot{r}_i = \frac{1}{ms^2} f_i - \frac{2\dot{s}}{s} \dot{r}_i \qquad (6.2.4)$$

$$M_s \ddot{s} = ms \sum_i \dot{r}_i^2 - \frac{n_f T}{s} \qquad (6.2.5)$$

Because the relationship between t and t' depends on the entire history of the system, namely,

$$t = \int s(t') \, dt' \qquad (6.2.6)$$

it is more convenient if the equations are transformed to use physical time units [nos84b, hoo85]; henceforth the dot will be used to denote d/dt', and the equations can be rewritten as

$$\ddot{r}_i = \frac{1}{m} f_i - \frac{\dot{s}}{s} \dot{r}_i \qquad (6.2.7)$$

$$\ddot{s} = \frac{\dot{s}^2}{s} + \frac{G_1 s}{M_s} \qquad (6.2.8)$$

where

$$G_1 = m \sum_i \dot{r}_i^2 - n_f T \qquad (6.2.9)$$

The first of these equations of motion resembles the conventional Newtonian equation with an additional frictionlike term, though not true friction because the coefficient can be of either sign; the second equation defines the feedback mechanism by which s is varied to regulate temperature.

The motivation for the $\log s$ term in the Lagrangian can now be appreciated. Assume that it is replaced by a general function $w(s)$; since s is finite, the time average of $\ddot{s}$ must vanish, implying that

$$m \left\langle \frac{1}{s} \sum_i \dot{r}_i^2 \right\rangle = \left\langle \frac{dw}{ds} \right\rangle \qquad (6.2.10)$$

The left-hand side is just $n_f \langle T/s \rangle$, so that if we equate the actual values rather than just the averages we find that $w(s) = n_f T \log s$.

The equilibrium averages of the physical system can be shown to be those of the canonical ensemble at temperature T [nos84a]. In order to establish this result the microcanonical partition function of the extended system is simply integrated over the s variable and what remains is the canonical partition function. The temperature itself is not constant, however, but the negative feedback acting through s ensures that the fluctuations are limited and the mean value is equal to T.

The Hamiltonian of the extended system

$$\mathcal{H} = \tfrac{1}{2}m \sum_i \dot{r}_i^2 + \sum_{i<j} u(r_{ij}) + \tfrac{1}{2}M_s \left(\frac{\dot{s}}{s}\right)^2 + n_f T \log s \qquad (6.2.11)$$

is conserved since there are no time-dependent external forces and this provides a useful check on the accuracy of the numerical solution. The Hamiltonian has no physical significance; its first two terms represent the energy of the physical system, but their sum is free to fluctuate.

The quantity M_s is a parameter whose value must be determined empirically; M_s has no particular physical meaning and is simply a part of the computational technique. In principle, the value of M_s does not affect the final equilibrium results, but it does influence their accuracy and reliability, because if the kinetic energy fluctuations are allowed to become too large it is rather difficult to think of the system existing at a particular temperature. For small variations in s the characteristic period of the fluctuations is [nos84a]

$$\tau_s = 2\pi \sqrt{M_s \langle s \rangle^2 / 2 n_f T} \qquad (6.2.12)$$

and the simulation must extend over many such periods to prevent these fluctuations from influencing the results adversely.

An MD program demonstrating temperature feedback will not be shown separately but will be combined with the version that incorporates pressure feedback as well. This method is described below.

Controlled pressure and temperature

While the connection between time and temperature just introduced is not reminiscent of any physical mechanism, pressure can be adjusted by altering the container volume. In the MD context this is achieved by a uniform isotropic volume change brought about by rescaling the atomic coordinates [and80]. A more useful method emerges if this is combined with the temperature feedback; the appropriate Lagrangian treatment leads to a system whose behavior corresponds to the isothermal–isobaric (NPT) ensemble.

We consider a cubic simulation region with a volume that is allowed to vary. Scaled coordinates r are introduced that span the unit cube and are related to the physical coordinates r' by

$$r = r'/V^{1/3} \qquad (6.2.13)$$

The same scaled time variable introduced previously is also used here, so that now both V and s are treated as supplementary dynamical variables. The Lagrangian for this system is a generalization of (6.2.2), with additional terms designed to ensure the correct pressure feedback mechanism [nos84b] (and once again the dot denotes d/dt),

$$\mathcal{L} = \tfrac{1}{2} m V^{2/3} s^2 \sum_i \dot{r}_i^2 - \sum_{i<j} u(V^{1/3} r_{ij}) + \tfrac{1}{2} M_s \dot{s}^2 + \tfrac{1}{2} M_v \dot{V}^2$$
$$- n_f T \log s - PV$$
(6.2.14)

where P and T are the required (externally imposed) values and M_v is another generalized mass parameter. Roughly speaking, M_v can be regarded as the mass of a piston that could have been used to regulate pressure by altering the volume, but because of the need to avoid explicit walls the effect of a sliding piston is achieved by means of a uniform volume change; a real piston would also introduce undesirable effects such as pressure waves. The first term in the Lagrangian is the kinetic energy, though not that of the physical system obtained by the substitution $\dot{r}' = V^{1/3}\dot{r} + (\dot{V}/3V^{2/3})r$, but a value based on velocities measured relative to the rate at which the region size changes (the $\dot{V}$ term is dropped) [and80]; removal of the flow component of the atomic velocities is essential to ensure the correct definition of temperature.

The Lagrange equations of motion, in scaled variables, are

$$\ddot{r}_i = \frac{1}{mV^{1/3}s^2} f_i - \left(\frac{2\dot{s}}{s} + \frac{2\dot{V}}{3V} \right) \dot{r}_i$$
(6.2.15)

$$M_s \ddot{s} = m V^{2/3} s \sum_i \dot{r}_i^2 - \frac{n_f T}{s}$$
(6.2.16)

$$M_v \ddot{V} = \frac{ms^2}{3V^{1/3}} \sum_i \dot{r}_i^2 + \frac{1}{3V^{2/3}} \sum_{i<j} r_{ij} \cdot f_{ij} - P$$
(6.2.17)

Returning to physical time units, with the dot now denoting d/dt', we obtain

$$\ddot{r}_i = \frac{1}{mV^{1/3}} f_i - \left(\frac{\dot{s}}{s} + \frac{2\dot{V}}{3V} \right) \dot{r}_i$$
(6.2.18)

$$\ddot{s} = \frac{\dot{s}^2}{s} + \frac{G_1 s}{M_s}$$
(6.2.19)

$$\ddot{V} = \frac{\dot{s}\dot{V}}{s} + \frac{G_2 s^2}{3M_v V}$$
(6.2.20)

where

$$G_1 = mV^{2/3} \sum_i \dot{\boldsymbol{r}}_i^2 - n_f T \tag{6.2.21}$$

$$G_2 = mV^{2/3} \sum_i \dot{\boldsymbol{r}}_i^2 + V^{1/3} \sum_{i<j} \boldsymbol{r}_{ij} \cdot \boldsymbol{f}_{ij} - 3PV \tag{6.2.22}$$

Scaled coordinates have been retained since they are more convenient from a computational point of view (see further).

When the dynamics are supplemented by these two extra degrees of freedom, the equilibrium averages of the physical system can be shown to be those of the NPT ensemble [nos84b]. The Hamiltonian is

$$\mathcal{H} = \tfrac{1}{2}mV^{2/3} \sum_i \dot{\boldsymbol{r}}_i^2 + \sum_{i<j} u(V^{1/3} \boldsymbol{r}_{ij}) + \tfrac{1}{2}M_s \left(\frac{\dot{s}}{s}\right)^2$$
$$+ \tfrac{1}{2}M_v \left(\frac{\dot{V}}{s}\right)^2 + n_f T \log s + PV \tag{6.2.23}$$

and, though conserved, it is once again not a physically meaningful quantity. The method of establishing a reasonable value for M_v is again empirical; for small variations in V the characteristic period is [nos83]

$$\tau_v = 2\pi \sqrt{M_v \langle \delta V \rangle^2 / T} \tag{6.2.24}$$

Periodic boundaries are most readily handled when the problem is expressed in terms of scaled coordinates, because the simulation region is then a fixed unit cube; use of physical variables introduces unnecessary complications when handling boundary crossings, because velocities and accelerations must be adjusted as well as coordinates [eva84]. When working with the PC method, the conversion to physical coordinates needed for the interaction calculations can overwrite the scaled coordinates because the predicted values are not needed for the corrector computation. Since the volume varies, provision must be made for an adjustable number of cells for use in the interaction calculations; for simplicity♠ we use the cell method without a neighbor list.

In *SingleStep* the following code is required:

```
PredictorStep ();
PredictorStepPT ();
ApplyBoundaryCond ();
UpdateCellSize ();
UnscaleCoords ();
ComputeForces ();
```

♠ pr_06_1

```
ComputeDerivsPT ();
CorrectorStep ();
CorrectorStepPT ();
ApplyBoundaryCond ();
```
10

The changes to *EvalProps* (assuming for simplicity $n_f = 3N_m$) are

```
totEnergy.val += (0.5 * (massS * Sqr (varSv) +
   massV * Sqr (varVv)) / Sqr (varS) + extPressure * varV) /
   nMol + 3. * temperature * log (varS);
pressure.val = (vvSum + virSum) / (3. * varV);
```

where *totEnergy.val* corresponds to the Hamiltonian (6.2.23). Note the call to *UpdateCellSize* (see below) immediately after the predictor calculation to determine the current cell size.

The new variables needed are

```
real extPressure, g1Sum, g2Sum, massS, massV, varS, varSa,
   varSa1, varSa2, varSo, varSv, varSvo, varV, varVa, varVa1,
   varVa2, varVo, varVv, varVvo;
int maxEdgeCells;
```

where *varS* and *varV* correspond to s and V, and the various suffixes denote derivatives and earlier values (for the PC method) in the same way as for *r*. The extra input data consists of

```
NameR (extPressure),
NameR (massS),
NameR (massV),
```

and additional initialization functions must be called from *SetupJob*,

```
InitFeedbackVars ();
ScaleCoords ();
ScaleVels ();
```

The maximum size of the cell array is set in *SetParams*. The calculation assumes a cubic region, and we allow for a reasonable (but arbitrarily chosen) amount of expansion beyond the initial region size,

```
maxEdgeCells = 1.3 * cells.x;
```

Storage allocation, in *AllocArrays*, to provide a cell array based on this maximum size requires the change

```
AllocMem (cellList, Cube (maxEdgeCells) + nMol, int);
```

Computing the right-hand sides of the feedback equations is as follows.

```
void ComputeDerivsPT ()
{
  real aFac, vFac;
  int n;

  vvSum = 0.;
  DO_MOL vvSum += VLenSq (mol[n].rv);
  vvSum *= pow (varV, 2./3.);
  g1Sum = vvSum - 3. * nMol * temperature;
  g2Sum = vvSum + virSum - 3. * extPressure * varV;
  aFac = pow (varV, -1./3.);
  vFac = - varSv / varS - 2. * varVv / (3. * varV);
  DO_MOL VSSAdd (mol[n].ra, aFac, mol[n].ra, vFac, mol[n].rv);
  varSa = Sqr (varSv) / varS + g1Sum * varS / massS;
  varVa = varSv * varVv / varS + g2Sum * Sqr (varS) /
     (3. * massV * varV);
}
```

The second-order feedback equations for *s* and *V* are solved using the same PC method as the equations of motion.

```
#define PCR4(r, ro, v, a, a1, a2)                                   \
   r = ro + deltaT * v + wr * (cr[0] * a + cr[1] * a1 + cr[2] * a2)
#define PCV4(r, ro, v, a, a1, a2)                                   \
   v = (r - ro) / deltaT + wv * (cv[0] * a + cv[1] * a1 + cv[2] * a2)

void PredictorStepPT ()
{
  real cr[] = {19., -10., 3.}, cv[] = {27., -22., 7.}, div = 24., e,
     wr, wv;

  wr = Sqr (deltaT) / div;
  wv = deltaT / div;
  varSo = varS;
  varSvo = varSv;
  PCR4 (varS, varS, varSv, varSa, varSa1, varSa2);
  PCV4 (varS, varSo, varSv, varSa, varSa1, varSa2);
  varSa2 = varSa1;
  varSa1 = varSa;
  ... (ditto for varV) ...
  e = pow (varV, 1. / NDIM);
  VSetAll (region, e);
}
```

```
void CorrectorStepPT ()
{                                                                      25
  real cr[] = {3., 10., -1.}, cv[] = {7., 6., -1.}, div = 24., e,
    wr, wv;

  wr = Sqr (deltaT) / div;
  wv = deltaT / div;                                                   30
  PCR4 (varS, varSo, varSvo, varSa, varSa1, varSa2);
  PCV4 (varS, varSo, varSvo, varSa, varSa1, varSa2);
  ... (ditto for varV) ...
  e = pow (varV, 1. / NDIM);
  VSetAll (region, e);                                                 35
}
```

Initialization of the extra variables is as follows.

```
void InitFeedbackVars ()
{
  varS = 1.;
  varV = Cube (region.x);
  varSv = 0.;                                                          5
  varSa = 0.;
  varSa1 = 0.;
  varSa2 = 0.;
  ... (ditto for varV...) ...
}                                                                      10
```

Coordinate and velocity rescaling (just one of the functions is shown) and cell size adjustment are handled by

```
void ScaleCoords ()
{
  real fac;
  int n;
                                                                       5
  fac = pow (varV, -1. / 3.);
  DO_MOL VScale (mol[n].r, fac);
}

void UpdateCellSize ()                                                 10
{
  VSCopy (cells, 1. / rCut, region);
  cells.x = Min (cells.x, maxEdgeCells);
  cells.y = Min (cells.y, maxEdgeCells);
  cells.z = Min (cells.z, maxEdgeCells);                               15
}
```

The function *EvalProps* needs a minor addition to allow for the scaled velocities; the form of the kinetic energy term in (6.2.23) determines the way *vvSum* must
be calculated,

```
vvSum *= pow (varV, 2./3.);
```

and the definition of *VWrap* used in *ApplyBoundaryCond* must be altered to use
scaled coordinates (the function itself is unchanged),

```
#define VWrap(v, t)                                              \
    if (v.t >= 0.5) v.t -= 1.;                                   \
    else if (v.t < -0.5) v.t += 1.
```

Output of the instantaneous region edge length *region.x* should be added to
PrintSummary.

The results shown here are based on a soft-sphere system with input data

```
deltaT              0.001
density             0.8
extPressure         6.5
initUcell           5 5 5
massS               0.1
massV               0.01
stepAvg             200
stepEquil           0
stepLimit           20000
temperature         1.
```

The values used for M_v (*massV*) are 0.01, 0.1 and 1.0, with just the single value
used for M_s (*massS*); the value of Δt depends to some extent on both these mass
parameters, but here the drift in the total Hamiltonian (6.2.23) over the entire run
is less than one part in 5000. The initial state is an FCC lattice, so that $N_m = 500$.

The fluctuating kinetic energy for the case $M_v = 1$ is shown in Figure 6.1.
The average value over 18 000 timesteps, ignoring the first 2000 timesteps, is the
expected $\langle E_K \rangle = 1.5000$, with $\sigma(E_K) = 0.0013$.

Pressure could be displayed similarly. A more revealing result, however, is the
way the region edge length varies with time; this is shown for different values of
M_v in Figure 6.2. The corresponding pressure results for all three values of M_v are
listed in Table 6.1. It is clear that in the case of $M_v = 1$ the run is not long enough,
so that even though the apparent $\sigma(P)$ is the smallest, the estimate of $\langle P \rangle$ could
be incorrect due to inadequate sampling. The additional degrees of freedom used
in the feedback methods introduce their own timescales that must be taken into
account in determining the run length.

Table 6.1. Pressure estimates.

M_v	$\langle P \rangle$	$\sigma(P)$
1.00	6.543	0.142
0.10	6.559	0.278
0.01	6.465	0.344

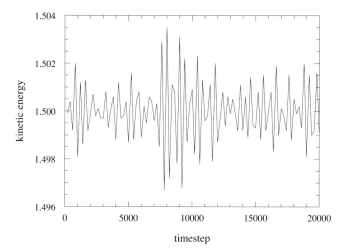

Fig. 6.1. Kinetic energy fluctuations for PT-feedback simulation with $M_v = 1$.

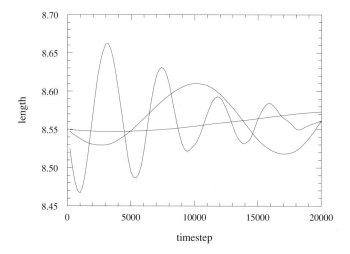

Fig. 6.2. Region edge fluctuations for $M_v = 0.01, 0.1$ and 1.0; the frequency is higher for smaller M_v.

Controlled pressure with variable region shape

In the above treatment of pressure feedback the simulation volume retained its cubic form, so that changes consist of uniform contractions and expansions. The method is readily extended to the case of a simulation region in which the lengths and directions of the edges are allowed to vary independently, subject to uniform external pressure [par80, nos83] (an even more general case where the applied stress components are specified separately can also be handled [par81]). The more flexible approach allows for the size and shape changes needed to accommodate lattice formation on freezing and for the study of structural phase transitions between different crystalline states. We will outline the mathematical formulation of the problem (omitting temperature feedback), but it will not be treated as a case study.

Once again, scaled coordinates are introduced, but they are now defined using a more general linear transformation

$$r' = Hr \tag{6.2.25}$$

where the transformation matrix

$$H = (h_{\mu v}) = (c_1, c_2, c_3) \tag{6.2.26}$$

is defined in terms of the vectors $\{c_\mu\}$ specifying the edges of the MD region, and the volume is

$$V = c_1 \cdot c_2 \times c_3 = \det H \tag{6.2.27}$$

A metric tensor

$$G = H^T H \tag{6.2.28}$$

can be introduced, so that

$$r'^2_{ij} = r^T_{ij} G r_{ij} \tag{6.2.29}$$

The scaled coordinates span the unit cube and periodic images have coordinates $H(r + (n_x, n_y, n_z))$. Note the standard relation between spatial derivatives in the two coordinate systems,

$$\partial/\partial r' = (H^T)^{-1} \partial/\partial r \tag{6.2.30}$$

The distortion of the simulation region is limited by the requirement that the interaction range does not exceed half the smallest region dimension.

The Lagrangian for this system is

$$\mathcal{L} = \tfrac{1}{2}m \sum_i \dot{r}^T_i G \dot{r}_i - \sum_{i<j} u(H r_{ij}) + \tfrac{1}{2}M_v \sum_{\mu v} \dot{h}^2_{\mu v} - PV \tag{6.2.31}$$

so that the Lagrange equations for the coordinates are

$$\ddot{r}_i = H^{-1} f_i / m - G^{-1} \dot{G} \dot{r}_i \tag{6.2.32}$$

In the isotropic case this reduces to the earlier result, because $H^{-1} = V^{1/3} I$ and $G^{-1} \dot{G} = 2\dot{V}/3V$. The Lagrange equations for the components of H are

$$M_v \ddot{h}_{\mu\nu} = m \sum_i [H \dot{r}_i]_\mu \dot{r}_{i\nu} + \sum_{i<j} f_{ij\mu} r_{ij\nu} - P \frac{\partial V}{\partial h_{\mu\nu}} \tag{6.2.33}$$

If we introduce an additional matrix

$$\begin{aligned} U = (u_{\mu\nu}) &= \left(\frac{\partial V}{\partial h_{\mu\nu}} \right) \\ &= V (H^{-1})^T \\ &= (c_2 \times c_3, \ c_3 \times c_1, \ c_1 \times c_2) \end{aligned} \tag{6.2.34}$$

then because

$$r_{i\nu} = V^{-1} \sum_\mu u_{\mu\nu} [H r_i]_\mu \tag{6.2.35}$$

the equation of motion (6.2.33) can be expressed concisely as

$$M_v \ddot{H} = (P - PI)U \tag{6.2.36}$$

where P is the pressure tensor.

6.3 Constraint methods

Constant temperature

The alternative to feedback control is the use of mechanical constraints. Enforcing constant temperature amounts to introducing a nonholonomic constraint into the equations of motion in order to fix the kinetic energy; in effect, this serves as a mathematical thermostat [hoo82, eva83a]. The justification for this arises not from Hamilton's variational principle, but from another formulation of mechanics known as Gauss's principle of least constraint [eva83b], which states that the quantity

$$\sum_i m_i (\ddot{r}_i - f_i / m_i)^2 \tag{6.3.1}$$

is minimized by the constrained motion. If the nonholonomic constraints are nonlinear but homogeneous functions of velocity (as is the case here), the results are formally the same as those the variational principle would have produced [ray72]. The equilibrium properties of this isothermal system can be shown to be those of the canonical ensemble [eva84], but the dynamics must once again be interpreted with care since the motion is no longer Newtonian.

Since there are $3N_m$ degrees of freedom (we ignore the three lost to momentum conservation), the constraint equation designed to ensure constant temperature is (assuming that all m_i are equal)

$$N_m E_K = \tfrac{1}{2} m \sum_{i=1}^{N_m} \dot{r}_i^2 \tag{6.3.2}$$

The constrained equation of motion is

$$\ddot{r}_i = \frac{1}{m} f_i + \alpha \dot{r}_i \tag{6.3.3}$$

and since $\dot{E}_K = 0$, or equivalently

$$\sum_i \dot{r}_i \cdot \ddot{r}_i = 0 \tag{6.3.4}$$

it follows that the value of the Lagrange multiplier α is

$$\alpha = -\frac{\sum_i \dot{r}_i \cdot f_i}{m \sum_i \dot{r}_i^2} \tag{6.3.5}$$

If the thermostat is used together with the PC method, the following function ($m = 1$ is assumed) should be called from *SingleStep* immediately after the force evaluation:

```
void ApplyThermostat ()
{
  real s1, s2, vFac;
  int n;

  s1 = 0.;
  s2 = 0.;
  DO_MOL {
    s1 += VDot (mol[n].rv, mol[n].ra);
    s2 += VLenSq (mol[n].rv);
  }
  vFac = - s1 / s2;
  DO_MOL {
    VVSAdd (mol[n].ra, vFac, mol[n].rv);
  }
}
```

It is also possible to combine the constant-temperature approach with the leapfrog integrator, by embedding[†] the isothermal condition into the integration procedure

---

† This was done in the first edition.

[bro84]. An alternative – one more readily generalized to other problems – is not
to alter the leapfrog integrator, but to modify `ApplyThermostat` so that it uses
estimates of the velocity at the end of the timestep instead of the values available
at the midpoint; this ensures that the contributions to α in (6.3.5) are evaluated at
the same time. The leapfrog version of the thermostat function[♦] is

```
void ApplyThermostat ()
{
  VecR vt;
  real s1, s2, vFac;
  int n;                                                           5

  s1 = 0.;
  s2 = 0.;
  DO_MOL {
    VSAdd (vt, mol[n].rv, 0.5 * deltaT, mol[n].ra);               10
    s1 += VDot (vt, mol[n].ra);
    s2 += VLenSq (vt);
  }
  vFac = - s1 / s2;
  DO_MOL {                                                        15
    VSAdd (vt, mol[n].rv, 0.5 * deltaT, mol[n].ra);
    VVSAdd (mol[n].ra, vFac, vt);
  }
}
```

The temperature constraint is only preserved to the accuracy of the numerical
integration. Any temperature drift must be corrected periodically (at intervals of
`stepAdjustTemp`) by velocity rescaling, although this can now be based on the
instantaneous value of E_K rather than on an average over preceding timesteps as in
§3.6. The function used for this is

```
void AdjustTemp ()
{
  real vFac;
  int n;                                                           5

  vvSum = 0.;
  DO_MOL vvSum += VLenSq (mol[n].rv);
  vFac = velMag / sqrt (vvSum / nMol);
  DO_MOL VScale (mol[n].rv, vFac);
}                                                                 10
```

In Table 6.2 we make a limited comparison between the constant-energy MD
results obtained earlier for an $N_m = 500$ LJ system (with $r_c = 2.2$ and $\rho = 0.8$) and
the corresponding results obtained using isothermal dynamics; leapfrog integration

---

♦ pr_06_2

Table 6.2. Constant-energy and temperature results for LJ fluid.

	constant-E	constant-T		constant-E		constant-T	
T	E	$\langle E \rangle$	$\sigma(E)$	$\langle P \rangle$	$\sigma(P)$	$\langle P \rangle$	$\sigma(P)$
0.8	−3.903	−3.896	0.038	0.998	0.117	1.019	0.195
1.0	−3.411	−3.421	0.044	2.021	0.145	1.999	0.208
1.2	−2.957	−2.947	0.049	2.967	0.165	2.976	0.230
1.4	−2.460	−2.486	0.054	3.880	0.176	3.837	0.247
1.6	−2.029	−2.048	0.062	4.684	0.224	4.636	0.282

is used ($\Delta t = 0.005$). The agreement is satisfactory, as more careful tests will confirm.

Constant pressure and temperature

The idea of using mechanical constraints to fix thermodynamic properties can be extended to include pressure as well [eva84]. The problem is formulated in terms of scaled coordinates, exactly as in the feedback case, and so the unconstrained Lagrangian is

$$\mathcal{L} = \tfrac{1}{2}mV^{2/3}\sum_i \dot{r}_i^2 - \sum_{i<j} u(V^{1/3}r_{ij}) \tag{6.3.6}$$

Here, the kinetic energy is again defined using velocities measured relative to the rate at which the region size changes. The equations for the T and P constraints are

$$N_m E_K = \tfrac{1}{2}mV^{2/3}\sum_i \dot{r}_i^2 \tag{6.3.7}$$

$$3PV = mV^{2/3}\sum_i \dot{r}_i^2 + V^{1/3}\sum_{i<j} r_{ij} \cdot f_{ij} \tag{6.3.8}$$

The equation of motion is

$$\ddot{r}_i = \frac{1}{mV^{1/3}} f_i + (\alpha' - 2\gamma)\dot{r}_i \tag{6.3.9}$$

where

$$\gamma = \dot{V}/3V \tag{6.3.10}$$

is the dilation rate and α' the Lagrange multiplier. From the constant-temperature condition $\dot{E}_K = 0$ we have

$$\sum_i \dot{r}_i \cdot \ddot{r}_i + \gamma \sum_i \dot{r}_i^2 = 0 \tag{6.3.11}$$

and it follows that

$$\alpha \equiv \alpha' - \gamma = -\frac{\sum_i \dot{\boldsymbol{r}}_i \cdot \boldsymbol{f}_i}{mV^{1/3} \sum_i \dot{\boldsymbol{r}}_i^2} \tag{6.3.12}$$

so that (6.3.9) can be expressed in terms of α rather than α'.

The constant-P condition provides the means for computing γ. Starting with

$$\frac{d}{dt}(PV) = P\dot{V} = 3\gamma PV = \frac{1}{3}\sum_{i<j}\frac{d}{dt}(\boldsymbol{r}'_{ij} \cdot \boldsymbol{f}_{ij}) \tag{6.3.13}$$

and noting that for pair potentials (such as LJ) that depend only on distance $\boldsymbol{r}' \cdot \boldsymbol{f} = -r'du(r')/dr'$, we obtain

$$\frac{d}{dt}(\boldsymbol{r}' \cdot \boldsymbol{f}) = -\psi \boldsymbol{r}' \cdot \dot{\boldsymbol{r}}' \tag{6.3.14}$$

where we have defined

$$\psi(r) = \frac{d^2}{dr^2}u(r) + \frac{1}{r}\frac{d}{dr}u(r) \tag{6.3.15}$$

Thus,

$$9\gamma PV = -V^{2/3}\sum_{i<j}\psi_{ij}\left(\boldsymbol{r}_{ij} \cdot \dot{\boldsymbol{r}}_{ij} + \gamma r_{ij}^2\right) \tag{6.3.16}$$

where ψ_{ij} denotes $\psi(r'_{ij})$, and for the LJ (or soft-sphere) potential

$$\psi(r) = 144(4r^{-14} - r^{-8}) \tag{6.3.17}$$

Rearrangement of (6.3.16) yields an expression for γ in terms of known quantities, namely,

$$\gamma = -\frac{V^{2/3}\sum_{i<j}\psi_{ij}\boldsymbol{r}_{ij} \cdot \dot{\boldsymbol{r}}_{ij}}{9PV + V^{2/3}\sum_{i<j}\psi_{ij}r_{ij}^2} \tag{6.3.18}$$

In situations where the use of an interaction cutoff has a significant effect on P the estimated corrections can be included in the evaluation of γ.

The equations of motion need to be supplemented by the dilation equation, where $L = V^{1/3}$,

$$\dot{L} = \gamma L \tag{6.3.19}$$

This equation must be integrated numerically to obtain $L(t)$ at each timestep, given the current value of $\gamma(t)$. The pressure can initially be set to the required value and any subsequent small drift eliminated by solving the nonlinear equation

$$P(V) - P = 0 \tag{6.3.20}$$

to obtain the appropriate V. The solution is obtained using the Newton–Raphson method [pre92]; here this entails iterating the expression

$$L \leftarrow L\left(1 + \frac{P(V) - P}{3P(V) + \sum\limits_{i<j} \psi_{ij}{r'_{ij}}^2/3V}\right) \tag{6.3.21}$$

and recomputing the interactions and pressure at each cycle, until

$$\frac{|P(V) - P|}{P} < \epsilon_P \tag{6.3.22}$$

The equilibrium averages are those of the NPT ensemble [eva84].

Many of the implementation details♠ are based on the feedback case treated previously. The cell method is used, and the required additions to ComputeForces (§3.4) are

```
VecR dv;
real w;
...
dvirSum1 = dvirSum2 = 0.;
...
    w = 144. * rri3 * (4. * rri3 - 1.) * rri;
    VSub (dv, mol[j1].rv, mol[j2].rv);
    dvirSum1 += w * VDot (dr, dv);
    dvirSum2 += w * rr;
```

In the function ApplyThermostat, the second loop contains

```
VSSAdd (mol[n].ra, 1. / varL, mol[n].ra, vFac / varL -
    dilateRate, mol[n].rv);
```

♠ pr_06_3

where `varL` is the value of L. The new variables used in these computations are

```
real dilateRate, dilateRate1, dilateRate2, dvirSum1, dvirSum2,
    extPressure, tolPressure, varL, varLo, varLv, varLv1, varLv2;
int maxEdgeCells, nPressCycle, stepAdjustPress;
```

and the additional inputs

```
  NameR (extPressure),
  NameI (stepAdjustPress),
  NameR (tolPressure),
```

where `tolPressure` is ϵ_P in (6.3.22). A new function is needed to evaluate the dilation rate γ:

```
void ApplyBarostat ()
{
  real vvS;
  int n;
                                                                        5
  vvS = 0.;
  DO_MOL vvS += VLenSq (mol[n].rv);
  dilateRate = - dvirSum1 * varL / (3. * (vvS * Sqr (varL) + virSum) +
    dvirSum2);
}                                                                       10
```

The dilation equation (6.3.19) is solved using the $k = 3$ PC method for first-order differential equations described in §3.5.

```
void PredictorStepBox ()
{
  real c[] = {23., -16., 5.}, div = 12.;

  varLv = dilateRate * varL;                                             5
  varLo = varL;
  varL = varL + (deltaT / div) * (c[0] * varLv + c[1] * varLv1 +
    c[2] * varLv2);
  varLv2 = varLv1;
  varLv1 = varLv;                                                       10
  dilateRate2 = dilateRate1;
  dilateRate1 = dilateRate;
  VSetAll (region, varL);
}
                                                                        15
void CorrectorStepBox ()
{
  real c[] = {5., 8., -1.}, div = 12.;

  varLv = dilateRate * varL;                                            20
```

```
    varL = varLo + (deltaT / div) * (c[0] * varLv + c[1] * varLv1 +
        c[2] * varLv2);
    VSetAll (region, varL);
}
```

The changes that must be made to *SingleStep* to accommodate the extra computations are

```
PredictorStep ();
PredictorStepBox ();
ApplyBoundaryCond ();
UpdateCellSize ();
UnscaleCoords ();                                                    5
ComputeForces ();
ApplyBarostat ();
ApplyThermostat ();
CorrectorStep ();
CorrectorStepBox ();                                                10
ApplyBoundaryCond ();
EvalProps ();
nPressCycle = 0;
if (stepCount % stepAdjustPress == 0) AdjustPressure ();
if (stepCount % stepAdjustPress == 10) AdjustTemp ();              15
```

and in *EvalProps*

```
vvSum *= Sqr (varL);
...
pressure.val = (vvSum + virSum) / (3. * Cube (varL));
```

The reason for separating the pressure and temperature adjustments by several
timesteps is to allow the system to settle down after the volume change; the effect of the delay can be seen in the results.

Pressure adjustments employ a Newton–Raphson procedure to modify the region size, as discussed earlier.

```
void AdjustPressure ()
{
  real rFac, w;
  int maxPressCycle, n;
                                                                    5
  maxPressCycle = 20;
  if (fabs (pressure.val - extPressure) > tolPressure * extPressure) {
    UnscaleCoords ();
    vvSum = vvSum / Sqr (varL);
    for (nPressCycle = 0; nPressCycle < maxPressCycle;              10
      nPressCycle ++) {
```

```
    UpdateCellSize ();
    ComputeForces ();
    w = 3. * Cube (varL);
    pressure.val = (vvSum * Sqr (varL) + virSum) / w;              15
    rFac = 1. + (pressure.val - extPressure) / (3. * pressure.val +
        dvirSum2 / w);
    DO_MOL VScale (mol[n].r, rFac);
    VScale (region, rFac);
    varL *= rFac;                                                  20
    if (fabs (pressure.val - extPressure) <
        tolPressure * extPressure) break;
  }
  ScaleCoords ();
  vvSum *= Sqr (varL);                                            25
  }
}
```

The variable *maxPressCycle* is provided as a safety measure in the unlikely event
of the method failing to converge. The counter *nPressCycle* is globally declared
to allow its inclusion in the output. The previous acceleration values used by the
PC method are not modified following the volume change; since this change ought
to be small the consequences of this omission should be negligible.

The function *AdjustTemp* shown earlier requires the addition after the first loop

```
    vvSum *= Sqr (varL);
```

Initialization (in *SetupJob*) now includes

```
    InitBoxVars ();
    ScaleCoords ();
    ScaleVels ();
```

and the variables associated with the region size are initialized by

```
void InitBoxVars ()
{
  varL = region.x;
  varLv = 0.;
  varLv1 = 0.;                                                     5
  varLv2 = 0.;
  dilateRate1 = 0.;
  dilateRate2 = 0.;
}
```

The variable *maxEdgeCells* is used, as in the earlier feedback case, to allow
for extra cells. Note that *varL* (the current region edge) is used instead of *varV*

Table 6.3. Results from constant-PT run.

timestep	$\langle E \rangle$	$\langle E_K \rangle$	$\langle P \rangle$	L
1 000	1.6334	1.4970	0.7985	11.2765
2 000	1.6327	1.4971	0.8105	10.0277
4 000	2.3421	1.4987	6.7568	8.5371
8 000	2.3222	1.4979	6.5538	8.5432
12 000	2.3170	1.4979	6.5383	8.5484
16 000	2.3152	1.4979	6.5222	8.5846
20 000	2.3155	1.4979	6.5286	8.5835

(the region volume in the feedback case) in converting between real and scaled coordinates.

The demonstration of this method uses a soft-sphere system with $N_m = 500$; the input data are

```
deltaT           0.002
density          0.8
extPressure      6.5
initUcell        5 5 5
stepAdjustPress  2000
stepAvg          500
stepEquil        0
stepLimit        20000
temperature      1.
tolPressure      0.001
```

Edited output of this run is shown in Table 6.3; the results include the current value of the region edge L. The pressure is adjusted every 2000 timesteps, but the drift is sufficiently small that only two cycles of the correction process are required (except on the very first call where 12 are needed). The typical value of $\sigma(P)$ for this run is 0.002.

6.4 Further study

6.1   By examining the relevant partition functions confirm that the equilibrium properties of these methods correspond to the NVT and NPT ensembles.

6.2   If scaled coordinates are not used in the constrained-pressure method, examine the implications for processing the periodic boundaries.

6.3   Make a careful comparison of $E(T)$ and $P(T)$ measurements using feedback and basic MD methods.

6.4 Implement the pressure feedback simulation for the case of variable region shape [nos83].

6.5 A further extension of the feedback approach is to the case of fixed external stress [par81]; investigate applications of this method.

6.6 Study the soft-sphere and LJ melting transitions at constant pressure; when the fluid freezes what is the crystal structure?

7

Nonequilibrium dynamics

7.1 Introduction

In the study of equilibrium behavior, MD is used to probe systems that, at least in principle, are amenable to treatment by statistical mechanics. The fact that statistical mechanics is generally unable to make much headway without resorting to simplification and approximation is merely a practical matter; the concepts and general relationships are extremely important even in the absence of closed-form solutions. When one departs from equilibrium, very little theoretical guidance is available and it is here that MD really begins to fill the role of an experimental tool.

There are many nonequilibrium phenomena worthy of study, but MD applications have so far tended to concentrate on relatively simple systems, and the case studies in this chapter will focus on the simplest of problems. To be more specific, we will demonstrate two very different approaches to questions related to fluid transport. The first approach uses genuine Newtonian dynamics applied to spatially inhomogeneous systems, in which the boundaries play an essential role: simulations of fluids partly constrained by hard walls will be used to determine both shear viscosity and thermal conductivity. The second approach is based on a combination of modified equations of motion and fully homogeneous systems: the same transport coefficients will be measured, but since there are no explicit boundaries the dynamics must be altered in very specific ways to compensate for their absence.

7.2 Homogeneous and inhomogeneous systems

As computational tools, both homogeneous and inhomogeneous nonequilibrium methods have their strengths and weaknesses. Before delving into the case studies, which include a sampling of both approaches, it is appropriate to point out the benefits and limitations of the different methods.

The reason for preferring homogeneous systems is that if physical walls can be eliminated (and replaced by periodic boundaries) all atoms perceive a similar environment. Inhomogeneous systems, on the other hand, must allow for perturbations to the structure and dynamics due to the presence of the walls. Furthermore, inhomogeneous systems may not exist at a uniquely defined temperature or density – essential if any comparison with experiment is to be made – as a consequence of the relatively large force needed to drive the mass or heat flow combined with the small system size. Not all problems offer the homogeneous alternative, although the more familiar transport coefficients can indeed be studied in this way.

The disadvantage of homogeneous nonequilibrium systems in general is the unphysical nature of the dynamics, for not only are the equations of motion modified in such a way that the desired transport coefficient emerges directly from linear response theory (actually a version of the theory extended to handle isothermal systems), they are also altered to mechanically suppress the heat generated by the applied force [eva84, eva90]. The method is therefore best regarded as a computational technique whose results are valid in the limit of zero applied force. The fact that each transport property requires a separate simulation because of the differing dynamical requirements leads to the question: if homogeneous systems are already being used why isn't it better to follow the more straightforward approach with Newtonian trajectories and autocorrelation functions as described in §5.2, where all the transport coefficients can be computed together? Historically, the answer focused on the accuracy of results obtained for comparatively small systems, with non-Newtonian methods having a clear advantage. Whether this advantage still exists now that much more extensive simulation is possible remains to be seen.

7.3 Direct measurement

Viscous flow

Two of the more elementary exercises in fluid dynamics, both with closed-form solutions, are Couette flow and Poiseuille flow. In planar Couette flow the fluid is confined between two parallel walls that slide relative to one another at a constant rate. An example of Poiseuille flow is a fluid forced to flow between two fixed walls. The walls are rough, so that a thin, stationary – relative to the wall – layer of fluid exists close to each wall. In each of these flow problems we can assume the system to be unbounded (or, for MD purposes, periodic) in the remaining two directions.

The viscous nature of the fluid requires sustained work to maintain motion. For Couette flow a force must be applied to keep the walls moving relative to one

another, whereas for Poiseuille flow a pressure head or gravity-like force acts in
the flow direction. This work is converted to heat that must be removed from the
fluid through the walls to limit the temperature rise. Temperature will vary in the
direction perpendicular to the walls, a reflection of the fact that heat generated in
the interior must be transported to the walls. This is true in both experiment and
simulation.

Once walls have been introduced explicitly into the problem the question arises
as to how realistically they need to be modeled. Real walls are complicated and
can only be represented in an average sense because roughness is essentially a
statistical notion; this observation, however, is of little help when trying to develop
a detailed microscopic simulation. All we require are walls that are sufficiently
rough to ensure nonslip flow, but the precise way by which this is achieved is
unlikely to affect the overall flow. One could, for example, use a layer of either
fixed or tethered atoms that mimic the effect of a rough wall [ash75]; by adjusting
the way the wall atoms are arranged the roughness can even be varied to a certain
extent. While this scheme offers a semblance of reality, it presents a problem if the
walls are also required to transfer heat in and out of the fluid; by using a thermostat
applied to the tethered wall atoms this issue can also be resolved, but the question
is whether such an intricate scheme is really necessary.

At the opposite extreme there are 'stochastic' walls [tro84]. Whenever an atom
attempts to cross a wall it is reflected back into the interior; the effects of wall
roughness and temperature are achieved by randomizing the direction of the reflected velocity and scaling its magnitude to match the wall temperature. The approach may appear simplistic and, if not used carefully, could interfere with the
integration of the equations of motion, especially if the region near the wall is at
relatively high density and temperature. Whether such boundary conditions actually work (they do) can only be established by trying them out.

This case study deals with Poiseuille flow; the Couette problem will be discussed
in §7.4, in the context of homogeneous systems. The analytic results for (incom-
pressible) Poiseuille flow can be summarized as follows: Assume that the two fixed
walls lie in the xy plane and that flow is in the x direction. In terms of the normalized cross-stream coordinate z, where $0 \leq z \leq 1$, solving the Navier–Stokes and
heat conduction equations [lan59] leads to polynomial velocity and temperature
profiles

$$v_x(z) = \frac{\rho g L_z^2}{2\eta} \left[\tfrac{1}{4} - (z - \tfrac{1}{2})^2 \right] \tag{7.3.1}$$

$$T(z) = T_w + \frac{\rho^2 g^2 L_z^4}{12\lambda\eta} \left[\tfrac{1}{16} - (z - \tfrac{1}{2})^4 \right] \tag{7.3.2}$$

where L_z is the channel width, T_w the wall temperature and g the external field driving the flow. By dividing the simulation region into slices parallel to the walls and measuring the mean flow velocity and temperature in each slice, the shear viscosity η and thermal conductivity λ can be determined by fitting second- and fourth-degree polynomials to the results.

In order to carry out this simulation[♠] we must first modify the neighbor-list generation function to allow for the fact that the z boundaries are no longer periodic. In *BuildNebrList* (§3.4) the line

```
VCellWrapAll ();
```

is replaced by

```
VCellWrap (x);
VCellWrap (y);
if (m2v.z < 0 || m2v.z >= cells.z) continue;
```

This removes all reference to nonexistent cells beyond the walls during the search for interaction partners.

Flow rate depends directly on the force used to drive the fluid (experimentally the flow is more likely to be due to a pressure difference between the pipe inlet and outlet); the value needed for a given flow rate will have to be determined by experiment because it depends on both ρ and L_z. The variable corresponding to g is declared as

```
real gravField;
```

and its value is included with the input data

```
NameR (gravField),
```

As part of the interaction computation the following function must be called:

```
void ComputeExternalForce ()
{
  int n;

  DO_MOL mol[n].ra.x += gravField;                          5
}
```

The presence of hard walls calls for a change in the usual boundary processing in the z direction. Whenever an atom attempts to cross either of these walls

---

♠ *pr_07_1*

it is reflected back into the interior; the magnitude of the new velocity is set to
a fixed value, corresponding to the wall temperature, and the direction is ran-
domized. In addition, because the atom will have slightly overshot the wall during the current timestep, it is moved back inside (and away from the wall by
a minute amount to avoid any numerical problems). The revised version of the
function is

```
void ApplyBoundaryCond ()
{
  real vSign;
  int n;
                                                                              5
  DO_MOL {
    VWrap (mol[n].r, x);
    VWrap (mol[n].r, y);
    vSign = 0.;
    if (mol[n].r.z >= 0.5 * region.z) vSign = 1.;            10
    else if (mol[n].r.z < -0.5 * region.z) vSign = -1.;
    if (vSign != 0.) {
      mol[n].r.z = 0.49999 * vSign * region.z;
      VRand (&mol[n].rv);
      VScale (mol[n].rv, velMag);                             15
      if (mol[n].rv.z * vSign > 0.) mol[n].rv.z *= -1.;
    }
  }
}
```

Minor changes are required to use this function for the two-dimensional version of
the problem, where it is the y direction that is not periodic.

Analysis of the flow requires the construction of cross-stream v_x and T profiles
based on a series of slices in the xy plane; in the case of T, each profile value
must be computed in a frame of reference moving with the mean flow in that slice.
However, rather than just show the simple profile computation, we will introduce
a more general scheme for computing properties based on a two-dimensional grid
subdivision of the simulation region, with profiles being produced as a byproduct.
This method will prove useful in later work (Chapter 15).

The following function performs the grid averaging, using cells based on the
atomic coordinates; in this example there is no y dependence, so the cell array
size can be set to unity in the y direction. Depending on the value of the argument
opCode, the function initializes the arrays used for collecting the results, accumulates the results for a single measurement, or computes the final averages. The five
quantities collected for each cell (the parameter *NHIST* – used for flexibility – is
equal to 5) are the occupation count, the sums over the squares of the velocities

and the sums of each of the velocity components. The final averaging produces the cell-averaged densities, temperatures and velocities.

```
void GridAverage (int opCode)
{
  VecR invWid, rs, va;
  VecI cc;
  real pSum;                                                                    5
  int c, hSize, j, n;

  hSize = VProd (sizeHistGrid);
  if (opCode == 0) {
    for (j = 0; j < NHIST; j ++) {                                              10
      for (n = 0; n < hSize; n ++) histGrid[j][n] = 0.;
    }
  } else if (opCode == 1) {
    VDiv (invWid, sizeHistGrid, region);
    DO_MOL {                                                                    15
      VSAdd (rs, mol[n].r, 0.5, region);
      VMul (cc, rs, invWid);
      c = VLinear (cc, sizeHistGrid);
      ++ histGrid[0][c];
      histGrid[1][c] += VLenSq (mol[n].rv);                                     20
      histGrid[2][c] += mol[n].rv.x;
      histGrid[3][c] += mol[n].rv.y;
      histGrid[4][c] += mol[n].rv.z;
    }
  } else if (opCode == 2) {                                                     25
    pSum = 0.;
    for (n = 0; n < hSize; n ++) {
      if (histGrid[0][n] > 0.) {
        for (j = 1; j < NHIST; j ++) histGrid[j][n] /= histGrid[0][n];
        VSet (va, histGrid[2][n], histGrid[3][n], histGrid[4][n]);             30
        histGrid[1][n] = (histGrid[1][n] - VLenSq (va)) / NDIM;
        pSum += histGrid[0][n];
      }
    }
    pSum /= hSize;                                                             35
    for (n = 0; n < hSize; n ++) histGrid[0][n] /= pSum;
  }
}
```

The grid computation requires several additional variables, namely,

```
VecI sizeHistGrid;
real **histGrid;
int countGrid, limitGrid, stepGrid;
```

extra input data,

```
NameI (limitGrid),
NameI (sizeHistGrid),
NameI (stepGrid),
```

and array allocation (in *AllocArrays*)

```
AllocMem2 (histGrid, NHIST, VProd (sizeHistGrid), real);
```

The following code is added to *SingleStep*,

```
if (stepCount >= stepEquil &&
    (stepCount - stepEquil) % stepGrid == 0) {
  ++ countGrid;
  GridAverage (1);
  if (countGrid % limitGrid == 0) {                         5
    GridAverage (2);
    EvalProfile ();
    PrintProfile (stdout);
    GridAverage (0);
  }                                                        10
}
```

and to *SetupJob*, for initialization,

```
GridAverage (0);
countGrid = 0;
```

In this case study the grid results will be used to compute profiles, but other kinds of processing could be carried out that utilize the spatial dependence of the data, including graphics (Chapter 15).

The functions that extract the profiles from the grid data and output the results follow.

```
void EvalProfile ()
{
  int k, n;

  for (n = 0; n < sizeHistGrid.z; n ++) {                   5
    profileV[n] = 0.;
    profileT[n] = 0.;
  }
  for (n = 0; n < VProd (sizeHistGrid); n ++) {
    k = n / (sizeHistGrid.x * sizeHistGrid.y);             10
    profileV[k] += histGrid[2] [n];
    profileT[k] += histGrid[1] [n];
  }
```

```
  for (n = 0; n < sizeHistGrid.z; n ++) {
    profileV[n] /= sizeHistGrid.x * sizeHistGrid.y;          15
    profileT[n] /= sizeHistGrid.x * sizeHistGrid.y;
  }
}

void PrintProfile (FILE *fp)                                 20
{
  real zVal;
  int n;

  fprintf (fp, "V profile\n");                               25
  for (n = 0; n < sizeHistGrid.z; n ++) {
    zVal = (n + 0.5) / sizeHistGrid.z;
    fprintf (fp, "%.2f %.3f\n", zVal, profileV[n]);
  }
  ... (ditto for T profile) ...                              30
}
```

The arrays for holding the T and v_x profiles

```
real *profileT, *profileV;
```

are allocated by

```
  AllocMem (profileT, sizeHistGrid.z, real);
  AllocMem (profileV, sizeHistGrid.z, real);
```

A three-dimensional soft-sphere system is used in this study, but the simulation region is not cubic in shape. Since the sheared flow develops in the xz plane the two longer region edges are assigned to the x and z directions; in this way we achieve a relatively large area for examining the flow details while retaining the three-dimensional nature of the system. The input data include

```
deltaT          0.005
density         0.8
gravField       0.1
initUcell       20 5 20
limitGrid       100
sizeHistGrid    1 1 50
stepAvg         2000
stepEquil       1000
stepGrid        50
stepLimit       32000
temperature     1.
```

The parameter `gravField` takes values between 0.1 and 0.4. The initial state is

Table 7.1. Kinetic energy during the runs.

timestep	$g = 0.1$	0.2	0.3	0.4
4 000	1.65	3.60	6.71	11.62
8 000	3.05	9.12	18.32	31.64
12 000	4.03	12.16	22.95	36.57
16 000	4.47	12.87	22.86	35.32
20 000	4.69	12.63	22.37	34.64
24 000	4.72	12.58	22.14	34.60
28 000	4.72	12.23	22.55	34.13
32 000	4.70	12.20	22.53	33.96

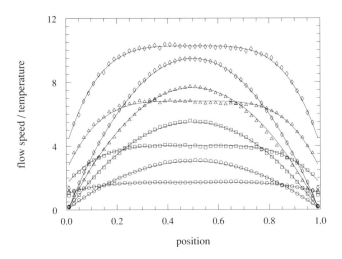

Fig. 7.1. Flow-velocity and temperature profiles for g values 0.1–0.4; polynomial fits are included.

a simple cubic lattice, so that $N_m = 2000$, and at the start of the run there is no overall flow.

A certain amount of time is required for the system to achieve steady flow. Table 7.1 shows how the mean kinetic energy initially increases with time, eventually reaching a steady value after about 16 000 timesteps (there is even a suggestion of overshoot). The flow velocity and temperature measurements made over the last 15 000 timesteps of each run appear in Figure 7.1. Polynomial fits to the data based on (7.3.1) and (7.3.2) are also shown. Estimates of η and λ derived from the fits are listed in Table 7.2.

Table 7.2. Transport coefficients from the fits in Figure 7.1.

g	η	λ
0.1	1.52	6.7
0.2	1.66	7.2
0.3	1.79	8.4
0.4	1.91	9.6

The fits are not forced to comply with the boundary conditions, namely, $v_x = 0$, $T = 1$, because of the limited spatial resolution of the coarse-grained measurements. Other examples of problems that might complicate the analysis include a small amount of slip at the walls and density variations across the flow. The fact that the transport coefficients can depend on ρ, T and even the local shear rate, contributes to the error when trying to fit to analytic results that assume η and λ are constant. Despite all these reservations, the fits obtained here appear remarkably good. On the other hand, there are too many variables involved to determine the reason why η and λ vary with flow rate.

Heat transport

We now turn to another example of the use of MD to mimic a real experiment, in this case heat flow between two parallel walls maintained at different temperatures [ten82]. If heat is transferred only by conduction, from Fourier's law [mcq76] the thermal conductivity is the ratio of the rate of heat (kinetic energy) transfer across the system to the temperature gradient,

$$\lambda = \frac{\Delta E_K^{(in)} + \Delta E_K^{(out)}}{2t_m A \Delta T / L_z} \tag{7.3.3}$$

where t_m is the measurement interval, ΔT the temperature difference, and $A = L_x L_y$ the wall area.

The program♠ is similar to the previous one, differing only in that the external force is absent and the walls are maintained at different temperatures. In the modified version of *ApplyBoundaryCond* shown below, the processing associated with the z boundary has been altered to handle two distinct wall temperatures; it also evaluates the total heat transferred in and out of the system as the impacting atoms

---

♠ pr_07_2

exchange energy with the constant-temperature walls.

```
void ApplyBoundaryCond ()
{
  real vNew, vSign, vvNew, vvOld;
  int n;

  enTransSum = 0.;
  DO_MOL {
    VWrap (mol[n].r, x);
    VWrap (mol[n].r, y);
    vSign = 0.;
    if (mol[n].r.z >= 0.5 * region.z) vSign = 1.;
    else if (mol[n].r.z < -0.5 * region.z) vSign = -1.;
    if (vSign != 0.) {
      mol[n].r.z = 0.49999 * vSign * region.z;
      vvOld = VLenSq (mol[n].rv);
      vNew = sqrt (NDIM * ((vSign < 0.) ? wallTempHi : wallTempLo));
      VRand (&mol[n].rv);
      VScale (mol[n].rv, vNew);
      vvNew = VLenSq (mol[n].rv);
      enTransSum += 0.5 * vSign * (vvNew - vvOld);
      if (mol[n].rv.z * vSign > 0.) mol[n].rv.z *= -1.;
    }
  }
}
```

New variables are

```
real *profileT, enTransSum, wallTempHi, wallTempLo;
Prop thermalCond;
```

and there are additional inputs

```
NameR (wallTempHi),
NameR (wallTempLo),
```

Heat transfer measurements require additions to *EvalProps*,

```
thermalCond.val = 0.5 * enTransSum / (deltaT * region.x *
    region.y * ((wallTempHi - wallTempLo) / region.z));
```

and to *AccumProps*,

```
if (icode == 0) {
  ...
  PropZero (thermalCond);
} else if (icode == 1) {
  ...
```

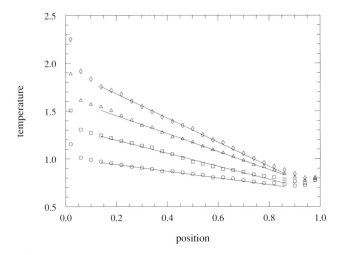

Fig. 7.2. Temperature profiles for ΔT between 0.5 and 2.0; linear fits to the interior values are included.

```
    PropAccum (thermalCond);
  } else if (icode == 2) {
    ...
    PropAvg (thermalCond, stepAvg);
  }
```
                                                                          10

The value of `thermalCond.sum` that should be output by `PrintSummary` is the estimate of λ.

The system used is similar to the previous case, with changed input data

sizeHistGrid	*1 1 25*
stepAvg	*5000*
stepEquil	*0*
stepLimit	*50000*
wallTempHi	*1.5*
wallTempLo	*1.*

The parameter `wallTempHi` ranges between 1.5 and 3.0. The simulation is started with the fluid at uniform temperature; achieving a steady state once again requires a certain amount of time, after which measurement can begin.

The temperature measurements made over the latter 15 000 timesteps of each run appear in Figure 7.2 (the z coordinate is normalized). Linear fits to the data are shown; the fits ignore the three values closest to each wall where the deviations from linearity are strongest. Estimates of λ derived from the fits and the measured heat flow are 4.84, 5.06, 5.92 and 5.75, for $\Delta T = 0.5, 1.0, 1.5$ and 2.0, respectively; these may be compared with the earlier Poiseuille flow values.

As in the case of Poiseuille flow, the fact that the system is inhomogeneous can create problems when trying to interpret the results. Here, a kind of thermal boundary layer exists close to each wall where the fluid temperature varies more rapidly than in the bulk, so that the effective thermal gradient is overestimated. Failure to account for this leads to λ estimates that are some 40% too small. Temperature-dependent density variations may also create problems.

7.4 Modified dynamics

Linear response theory

The question addressed by linear response theory [han86b, eva90] can be stated in the following way. Given a system with Hamiltonian $\mathcal{H}_0$, evaluate the change in some dynamical variable $B(t)$ caused by an external field $F_e(t)$ applied starting at $t = 0$, whose effect on the system can be expressed schematically in terms of atomic coordinates and momenta as

$$\mathcal{H} = \mathcal{H}_0 - A(\mathbf{r}, \mathbf{p})F_e(t) \tag{7.4.1}$$

The actual coupling of F_e to the system may be more general than the form suggested here, with vector or tensor products being involved. The step function is just one possible form of perturbation, and sinusoidal and delta functions are also of interest.

Assuming that the effect of F_e is small enough to permit a linear perturbation treatment, an analysis based on the Liouville equation then leads to

$$\langle B(t)\rangle - \langle B(0)\rangle = \frac{1}{T} \int_0^t \langle B(t - t')\dot{A}(0)\rangle_0 F_e(t')\, dt' \tag{7.4.2}$$

This result is valid in the limit $F_e \to 0$, with $\langle \ldots \rangle_0$ denoting an equilibrium average evaluated in the unperturbed system. The existence of the perturbed Hamiltonian $\mathcal{H}$ is not required for this result to be true [eva84]. If, for example, the equations of motion are

$$\dot{\mathbf{r}}_i = \mathbf{p}_i/m + \mathbf{C}_i(\mathbf{r}, \mathbf{p})F_e(t) \tag{7.4.3}$$

$$\dot{\mathbf{p}}_i = \mathbf{f}_i + \mathbf{D}_i(\mathbf{r}, \mathbf{p})F_e(t) \tag{7.4.4}$$

then if the phase-space distribution function $f(\mathbf{r}, \mathbf{p})$ obeys $df(\mathbf{r}, \mathbf{p})/dt = 0$ (as systems defined by a Hamiltonian always do – the Liouville theorem) and, consequently,

$$\sum_i \left(\nabla_{r_i} \cdot \mathbf{C}_i + \nabla_{p_i} \cdot \mathbf{D}_i\right) = 0 \tag{7.4.5}$$

the result also holds, but with $\dot{A}$ in (7.4.2) replaced by $-J$ defined via

$$
\dot{\mathcal{H}}_0 = \sum_i (\dot{p}_i \cdot p_i/m - \dot{r}_i \cdot f_i)
$$

$$
= -\sum_i (-p_i \cdot D_i/m + f_i \cdot C_i) F_e(t)
$$

$$
\equiv -J(r, p)F_e(t) \tag{7.4.6}
$$

Since the applied force F_e performs mechanical work on the system the temperature rises, and equilibrium is never attained. To eliminate this problem [mor85] a thermostat is included in the dynamics (as in §6.3) by adding a term $\alpha\, p_i$ to (7.4.4); constant kinetic energy is assured if the value of the Lagrange multiplier is[†]

$$
\alpha = -\frac{\sum_i (f_i + D_i F_e) \cdot p_i}{\sum_i p_i^2} \tag{7.4.7}
$$

Transport coefficients can be evaluated by applying the appropriate force F_e and examining the behavior in the limit $F_e \to 0$. If the transport coefficient can be expressed as the integrated autocorrelation function of some dynamical quantity J,

$$
Q = \frac{1}{T} \int_0^\infty \langle J(t) J(0) \rangle \, dt \tag{7.4.8}
$$

then, provided the perturbation is designed so that linear response theory yields an expression that is formally identical to the definition of the transport coefficient (7.4.8) – in other words $B = J$ – we obtain

$$
Q = \lim_{F_e \to 0} \lim_{t \to \infty} \frac{\langle J(t) \rangle}{F_e} \tag{7.4.9}
$$

The order of the limits is important, with the large t results obtained at finite F_e being extrapolated to $F_e = 0$. Since the goal is to use this definition for Q in constant-temperature simulations, but the usual formulation of linear response theory assumes constant-energy (Newtonian) dynamics, the theory must be extended to cover this situation. When this is done [mor85, eva90] the same expressions appear, but with the averages now evaluated at constant temperature. In the remainder of this section we will use this approach to evaluate the two transport coefficients considered previously – shear viscosity and thermal conductivity.

Shear viscosity

We consider the case of Couette flow in which the fluid undergoes sheared flow due to boundary walls that are in relative motion. The equations of motion used in this

---

[†] The momenta appear in the analysis because of Hamilton's equations of motion, but we will dispense with them shortly.

shear viscosity study are based on the constant-temperature dynamics described in
§6.3. A small, but significant, change is required in order to ensure that temperature
is correctly defined in the presence of flow, and this affects the velocity terms in
the Lagrange multiplier used for the thermostat.

The usual microscopic definition of temperature in terms of mean-square velocity assumes that there is no overall motion; any local flow must be subtracted from
the velocities before using them to evaluate temperature (similar situations have
been encountered in earlier case studies). The same holds true for the velocities
used in the thermostat. However, knowing the bulk flow to an accuracy suitable for
use in the equations of motion implies that the problem has already been solved;
this circularity can be removed by assuming the nature of the flow, and only later
checking to see whether consistent results are obtained. A less reliable alternative
is to evaluate local flow by means of coarse-grained averaging and then use the
results in the equations of motion; such an approach is unstable to any fluctuations
in the flow because these variations are interpreted by the equations of motion as
temperature fluctuations that must be suppressed.

In this study we impose the reasonable requirement that the MD flow obeys the
linear velocity profile known from the exact solution of the continuum problem.
Assuming it is the z boundaries that are in motion, then if the relative velocity of
the walls is γL_z, the shear rate dv_x/dz has the constant value γ. The thermostatted
equation of motion is then [eva84]

$$\ddot{r}_i = \frac{1}{m} f_i + \alpha(\dot{r}_i - \gamma r_{iz}\hat{x}) \tag{7.4.10}$$

The value of the Lagrange multiplier α follows from the constant-temperature
constraint,

$$\alpha = -\frac{\sum_i(\dot{r}_i - \gamma r_{iz}\hat{x}) \cdot (f_i/m - \gamma \dot{r}_{iz}\hat{x})}{\sum_i(\dot{r}_i - \gamma r_{iz}\hat{x})^2} \tag{7.4.11}$$

The equation of motion (7.4.10) assumes that the linear velocity profile has already
been established; creating the initial sheared flow is most readily done as part of the
initial conditions, and from the more formal point of view this amounts to applying
an impulse of the correct size and direction to each atom at $t = 0$. The sliding
boundaries, in the form of a special kind of boundary condition (see below), maintain the constant shear rate. The constant-temperature version of linear response
theory for this problem provides an expression for η based on the pressure tensor

$$\eta = -\lim_{\gamma \to 0} \lim_{t \to \infty} \frac{\langle P_{xz} \rangle}{\gamma} \tag{7.4.12}$$

To show that (7.4.10) corresponds to the more general form given by (7.4.3)–(7.4.4) we define the momentum measured relative to the local flow

$$\boldsymbol{p}_i/m = \dot{\boldsymbol{r}}_i - \gamma r_{i\,z}\hat{\boldsymbol{x}} \tag{7.4.13}$$

The first-order equations are then

$$\dot{\boldsymbol{r}}_i = \boldsymbol{p}_i/m + \gamma r_{i\,z}\hat{\boldsymbol{x}} \tag{7.4.14}$$

$$\dot{\boldsymbol{p}}_i = \boldsymbol{f}_i - \gamma p_{i\,z}\hat{\boldsymbol{x}} + \alpha\,\boldsymbol{p}_i \tag{7.4.15}$$

exactly as required (if the thermostat is ignored).

The boundaries are periodic, but of a special form to accommodate the uni-formly sheared flow [lee72]. The idea is to replace sliding walls by sliding replica systems; layers of replicas that are adjacent in the z direction move with relative velocity $\gamma L_z\hat{\boldsymbol{x}}$, an arrangement designed to ensure periodicity at shear rate γ. An atom crossing a z boundary requires special treatment because the x components of position and velocity are both discontinuous – not for the replica system just entered, but relative to the opposite side of the region into which the atom must be inserted. The velocity change whenever a $\pm z$ boundary is crossed is $\mp\gamma L_z\hat{\boldsymbol{x}}$, and the coordinate change is $\mp d_x\hat{\boldsymbol{x}}$, where the total relative displacement of the neighboring replicas – a quantity only meaningful over the range $(-L_x/2, L_x/2)$ – is given by

$$d_x = (\gamma L_z t + L_x/2) \ (\mathrm{mod}\ L_x) - L_x/2 \tag{7.4.16}$$

Note that because the x coordinate changes when a z boundary is crossed, a further correction for periodic wraparound in the x direction may be needed. Interactions that occur between atoms separated by the z boundary require an offset value $-d_x$ to be included in the distance computation.

When using the cell method for the interaction calculation, the range of neighbor cells in the x direction for adjacent cells on opposite sides of the z boundary must extend over four cells, instead of the usual three, to allow for the fact that the cells of the sliding replicas are not usually aligned. If there are M_x cells on an edge, the additional cell offset across the positive z boundary is

$$\Delta m_x = \lfloor M_x\,(1 - d_x/L_x)\rfloor - M_x \tag{7.4.17}$$

Taking these considerations into account, the modified form of the function *ComputeForces*♠ is the following.

```
#define OFFSET_VALS                                          \
   {{0,0,0}, {1,0,0}, {1,1,0}, {0,1,0}, {-1,1,0},            \
    {0,0,1}, {1,0,1}, {1,1,1}, {0,1,1}, {-1,1,1},            \
```

♠ pr_07_3

```
    {-1,0,1}, {-1,-1,1}, {0,-1,1}, {1,-1,1}, {2,-1,1},          \
    {2,0,1}, {2,1,1}}                                                         5

 void ComputeForces ()
 {
   ...
   int cellShiftX, offsetHi;                                                 10
   ...
   cellShiftX = (int) (cells.x * (1. - bdySlide / region.x)) - cells.x;
   pTensorXZ = 0.;
   for (m1z ...
       ...                                                                   15
       m1 = VLinear (m1v, cells) + nMol;
       offsetHi = (m1v.z == cells.z - 1) ? 17 : 14;
       for (offset = 0; offset < offsetHi; offset ++) {
         VAdd (m2v, m1v, vOff[offset]);
         VZero (shift);                                                      20
         if (m1v.z == cells.z - 1 && vOff[offset].z == 1) {
           m2v.x += cellShiftX;
           shift.x = bdySlide;
           if (m2v.x >= cells.x) {
             m2v.x -= cells.x;                                               25
             shift.x += region.x;
           } else if (m2v.x < 0) {
             m2v.x += cells.x;
             shift.x -= region.x;
           }                                                                 30
         } else {
           VCellWrap (x);
         }
         VCellWrap (y);
         VCellWrap (z);                                                      35
         m2 = ...
           ...
         if (rr < rrCut) {
             ...
           pTensorXZ += fcVal * dr.x * dr.z;                                 40
   ...
 }
```

The quantity `bdySlide` corresponding to d_x is computed at the beginning of
`SingleStep`,

```
  bdySlide = shearRate * region.z * timeNow + 0.5 * region.x;
  bdySlide -= (int) (bdySlide / region.x) * region.x + 0.5 * region.x;
```

A modified version of `ApplyBoundaryCond` handles the sliding boundary conditions by treating z boundary crossings in the prescribed manner.

```
void ApplyBoundaryCond ()
{
  int n;

  DO_MOL {                                                              5
    VWrap (mol[n].r, x);
    VWrap (mol[n].r, y);
    if (mol[n].r.z >= 0.5 * region.z) {
      mol[n].r.x -= bdySlide;
      if (mol[n].r.x < 0.5 * region.x) mol[n].r.x += region.x;         10
      mol[n].rv.x -= shearRate * region.z;
      mol[n].r.z -= region.z;
    } else if (mol[n].r.z < -0.5 * region.z) {
      mol[n].r.x += bdySlide;
      if (mol[n].r.x >= 0.5 * region.x) mol[n].r.x -= region.x;        15
      mol[n].rv.x += shearRate * region.z;
      mol[n].r.z += region.z;
    }
  }
}                                                                       20
```

New variables introduced are

```
real bdySlide, pTensorXZ, shearRate, vvSumXZ;
Prop shearVisc;
```

and there is an additional input data item

```
NameR (shearRate),
```

The thermostat plays a key role in this approach, with temperature being evaluated with respect to the local flow to adhere to the correct definition. We assume that local flow is determined by the constant shear rate γ. The changes – PC integration is used – require the x components to be singled out for special treatment (7.4.11),

```
void ApplyThermostat ()
{
  real s1, s2, vFac;
  VecR as, vs;
  int n;                                                                5

  s1 = 0.;
  s2 = 0.;
  DO_MOL {
    vs = mol[n].rv;                                                    10
    vs.x -= shearRate * mol[n].r.z;
    as = mol[n].ra;
    as.x -= shearRate * mol[n].rv.z;
```

```
    s1 += VDot (vs, as);
    s2 += VLenSq (vs);                                                  15
  }
  vFac = - s1 / s2;
  DO_MOL {
    VVSAdd (mol[n].ra, vFac, mol[n].rv);
    mol[n].ra.x -= vFac * shearRate * mol[n].r.z;                       20
  }
}
```

Corresponding modification is needed in `AdjustTemp`. The initial velocities include the uniform shear flow; the addition to `InitVels` is

```
mol[n].rv.x += shearRate * mol[n].r.z;
```

Changes to `EvalProps` for the evaluation of $\eta(\gamma)$ are

```
vvSumXZ = 0.;
DO_MOL {
  v = mol[n].rv;
  v.x -= shearRate * mol[n].r.z;
  VVAdd (vSum, v);                                                       5
  vvSum += VLenSq (v);
  vvSumXZ += (mol[n].rv.x - shearRate * mol[n].r.z) * mol[n].rv.z;
}
...
pTensorXZ = (pTensorXZ + vvSumXZ) * density / nMol;                     10
shearVisc.val = - pTensorXZ / shearRate;
```

The averaging of `shearVisc` is included in `AccumProps`,

```
if (icode == 0) {
  ...
  PropZero (shearVisc);
} else if (icode == 1) {
  ...                                                                    5
  PropAccum (shearVisc);
} else if (icode == 2) {
  ...
  PropAvg (shearVisc, stepAvg);
}                                                                       10
```

and the appropriate output added to `PrintSummary`.

The runs used for the shear viscosity measurements are based on the following input data,

```
deltaT            0.002
density           0.8
```

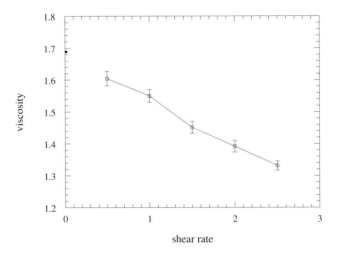

Fig. 7.3. Shear viscosity for various shear rates; the equilibrium value appears at the left edge of the graph.

initUcell	*4 4 4*
shearRate	*0.5*
stepAdjustTemp	*999999*
stepAvg	*1000*
stepEquil	*1000*
stepLimit	*22000*
temperature	*1.*

with the value of *shearRate* ranging between 0.5 and 2.5. An FCC initial state is used, so that $N_m = 256$.

Estimates of $\eta(\gamma)$ are shown in Figure 7.3; the error bars show the standard deviation of the mean $\sigma(\langle \eta \rangle)$, computed from the block averages that are produced every 1000 timesteps. Taking into account the more extensive computation required in the equilibrium case (§5.5) and the larger uncertainty in the final estimate, the advantage of the nonequilibrium approach here is apparent. Further discussion appears in [eva90, fer91, lie92].

Thermal conductivity

The thermal conductivity is another example of a transport coefficient that can be measured by a similar approach, assuming a suitable equation of motion can be synthesized. In this case a fictitious external field F_e of an unusual kind is introduced [eva82, gil83]: it has the effect of driving atoms with a higher than average energy in the direction of F_e, while those with a lower energy are driven in the opposite direction; in other words, F_e generates heat flow and so, at least for small values of the field, produces the effect of an imposed temperature difference.

The additional force acting on each atom is defined[†] as

$$f_i' = e_i F_e + \tfrac{1}{2} \sum_{j(\neq i)} f_{ij}(r_{ij} \cdot F_e) - \frac{1}{2N_m} \sum_{j\neq k} f_{jk}(r_{jk} \cdot F_e) \qquad (7.4.18)$$

where e_i is the excess energy of atom i. Here, f_i' has been chosen so that in terms of the heat current S (5.2.15),

$$\sum_i \dot{r}_i \cdot f_i' = V S \cdot F_e \qquad (7.4.19)$$

The force conserves total momentum because $\sum f_i' = 0$. Since only relative distances occur in f_i', and assuming the force is sufficiently weak that the system remains homogeneous, there is nothing to prevent the use of periodic boundary conditions – exactly the motivation for devising methods of this kind. If $J = S_z$ and $F_e = F_e \hat{z}$, then the constant-temperature version of (7.4.9) leads to the result

$$\lambda = \lim_{F_e \to 0} \lim_{t \to \infty} \frac{\langle S_z \rangle}{F_e T} \qquad (7.4.20)$$

The thermostat is the usual one, based on the total force, so that the equations of motion are simply

$$\ddot{r}_i = f_i + f_i' + \alpha \dot{r}_i \qquad (7.4.21)$$

For computational convenience we introduce a matrix associated with each atom i whose elements are

$$B_{i\,xy} = \sum_{j(\neq i)} f_{ij\,x} r_{ij\,y} \qquad (7.4.22)$$

so that the components of (7.4.18) can be written

$$f_{i\,\mu}' = e_i F_e \delta_{\mu z} + \tfrac{1}{2}\big(B_{i\,\mu z} - \langle B_{\mu z}\rangle\big) F_e \qquad (7.4.23)$$

where $\langle B_{\mu z}\rangle$ is just the mean of $B_{i\,\mu z}$.

The following variables are added to the structure Mol and evaluated exactly as in the equilibrium case (§5.3)

```
VecR rf[3];
real en;
```

---

† The signs differ from [eva82] because of the way r_{ij} is defined.

and the other variables used in this calculation are

```
real heatForce;
Prop thermalCond;
```

The only new input item is the thermal driving force F_e

```
NameR (heatForce),
```

Evaluating the right-hand side of the equation of motion, again assuming the use of PC integration, requires the additional function♠

```
void ComputeThermalForce ()
{
  VecR rfMolSumZ;
  real enMolSum;
  int n;                                                          5

  VZero (rfMolSumZ);
  enMolSum = 0.;
  DO_MOL {
    VVAdd (rfMolSumZ, mol[n].rf[2]);                              10
    mol[n].en += VLenSq (mol[n].rv);
    enMolSum += mol[n].en;
  }
  DO_MOL {
    VVSAdd (mol[n].rf[2], -1. / nMol, rfMolSumZ);                 15
    mol[n].en = 0.5 * (mol[n].en - enMolSum / nMol);
  }
  DO_MOL {
    mol[n].ra.z += mol[n].en * heatForce;
    VVSAdd (mol[n].ra, 0.5 * heatForce, mol[n].rf[2]);            20
  }
}
```

The value of $\lambda(F_e)$ is computed by a modified version of *EvalProps* that includes

```
real thermVecZ;
...
thermVecZ = 0.;
DO_MOL thermVecZ += mol[n].rv.z * mol[n].en +
  0.5 * VDot (mol[n].rf[2], mol[n].rv);                           5
thermalCond.val = thermVecZ * density / (temperature *
  heatForce * nMol);
```

Final averages are processed by *AccumProps* in exactly the same way as the viscosity – simply change the variable names to those used here.

♠ pr_07_4

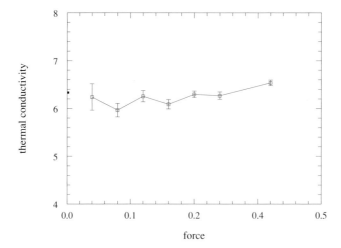

Fig. 7.4. Thermal conductivity for different values of thermal driving force; the equilibrium result is included.

The results of this case study are based on the same input data as previously, except that the shear rate is replaced by the parameter `heatForce`, with values in the range 0.05 to 0.4. Run lengths of 84 000 timesteps are used to reduce the error estimates. The results appear in Figure 7.4 (error bars are computed as before); the benefits of the nonequilibrium approach are not as pronounced here as in the case of shear viscosity, although the fact that these runs are substantially shorter than the run of §5.5 should not be forgotten.

7.5 Further study

7.1   Examine the Couette flow problem when hard sliding walls are included [tro84].

7.2   Compare the transport coefficients obtained by nonequilibrium methods with those from the autocorrelation integrals [erp77]; for a given level of accuracy, which is more efficient, and by how much?

7.3   Consider walls with atomic structure based on atoms that are either fixed or mobile [ash75, tho90, lie92]; how can the effect of a constant-temperature wall be achieved?

7.4   Homogeneous shear MD can produce a spurious 'string' phase [eva86, loo92], and homogeneous heat flow MD can become unstable for large systems [han94]; in both cases the proposed solution is a more carefully designed thermostat – explore this issue.

8

Rigid molecules

8.1 Introduction

The elementary constituents of most substances are structured molecules, rather than the spherically symmetric atoms treated in previous chapters. The emphasis on simple monatomic models is justified for a number of reasons: the dynamics are simpler, thereby making life easier for newcomers; it reflects the historical development of the field, since the original work establishing the viability of the MD approach as a quantitative tool dealt with liquid argon [rah64]; and once the basic techniques have been mastered they can be extended to a variety of more complex situations. In this chapter we discuss the first of these excursions – to molecules constructed from a rigidly linked atomic framework. This approach is suitable for small, relatively compact molecules, where rigidity seems a reasonable assumption, but if this is not true then motion within the molecule must also be taken into account, as we will see in later chapters. There is really no such thing as a rigid molecule, but from the practical point of view it is a very effective simplification of the underlying quantum problem; the model also does not account for chemical processes – no mechanism is provided for molecular formation and dissociation.

The chapter begins with a summary of rigid-body dynamics, but with a slightly unfamiliar emphasis. In treatises on classical mechanics Euler angles play a central part [gol80]; while they provide the most intuitive means for describing the orientation of a rigid body and are helpful for analyzing certain exactly soluble problems, in numerical applications they actually represent a very poor choice. Quaternions, originally a purely theoretical development due to Hamilton, turn out to be a better choice, and the dynamics will be described using such quantities. Linear molecules, with only two rotational degrees of freedom rather than three, are treated separately. A simple but useful model for liquid water is considered in the case studies; the results derived for linear molecules will be used in §13.2. The chapter concludes with a discussion of an alternative approach that employs the molecular rotation

matrices themselves as the orientation variables; this allows the leapfrog method
to be used and the approach is demonstrated on a system of tetrahedral molecules.
The methods shown here also apply when there is no translational motion and the
molecules merely rotate about fixed lattice sites (as in molecular crystals [kle86]);
the force computations are simpler because the neighbors within interaction range
never change.

8.2 Dynamics

Coordinates

Rigid-body motion can be decomposed into two completely independent parts,
translational motion of the center of mass and rotation about the center of mass.
A basic result of classical mechanics is that the former is governed by the total
force acting on the body, whereas the latter depends on the total applied torque.
Thus translation can be treated as before and we need only consider the dynamics
of rotation.

Fully rigid molecules come in two flavors, linear and nonlinear, with each molecule having two or three rotational degrees of freedom respectively. The orientation of a rodlike linear molecule can be specified using two angular coordinates,
whereas the more general case requires three, and it is for the latter that Euler angles are usually introduced as a particularly simple way of describing orientation.
We begin with the nonlinear case [gol80].

The Euler angles are defined in terms of a sequence of rotations of a set of
cartesian coordinate axes about the origin. The first rotation is through an angle
(measured counterclockwise) ϕ about the z axis; this is followed by a rotation θ
about the new x axis; the final rotation is through an angle ψ about the new z axis.
The full rotation matrix R is the product of the individual rotation matrices

$$R = R(\psi)R(\theta)R(\phi) \tag{8.2.1}$$

and, if required, the elements of R can be expressed in terms of the Euler angles.
There are two ways of interpreting the rotation described by R. One is to consider
a vector r' and use R to obtain its components in the rotated coordinate system,
namely,

$$r = Rr' \tag{8.2.2}$$

The other is to rotate the vector, beginning with r and applying the opposite rotations in reverse order by means of the transpose of R, in which case the result is
the rotated vector $r' = R^T r$.

Quaternions

There are other ways of describing rotations, perhaps not as intuitively obvious, but often more convenient for numerical problem solving. Here we consider a particularly useful method – Hamilton's quaternions [gol80].

We begin by specifying the components[†] of a quaternion in terms of the Euler angles,

$$
\begin{aligned}
q_1 &= \sin(\theta/2) \cos\big((\phi - \psi)/2\big) \\
q_2 &= \sin(\theta/2) \sin\big((\phi - \psi)/2\big) \\
q_3 &= \cos(\theta/2) \sin\big((\phi + \psi)/2\big) \\
q_4 &= \cos(\theta/2) \cos\big((\phi + \psi)/2\big)
\end{aligned}
\tag{8.2.3}
$$

The components are normalized,

$$
\sum_m q_m^2 = 1
\tag{8.2.4}
$$

The inverse relations are

$$
\begin{aligned}
\sin\theta &= 2\sqrt{(q_1^2 + q_2^2)(1 - q_1^2 - q_2^2)} \\
\cos\theta &= 1 - 2\,(q_1^2 + q_2^2) \\
\sin\phi &= 2\,(q_1 q_3 + q_2 q_4)/\sin\theta \\
\cos\phi &= 2\,(q_1 q_4 - q_2 q_3)/\sin\theta \\
\sin\psi &= 2\,(q_1 q_3 - q_2 q_4)/\sin\theta \\
\cos\psi &= 2\,(q_1 q_4 + q_2 q_3)/\sin\theta
\end{aligned}
\tag{8.2.5}
$$

These results break down when $\theta = 0$ or π, corresponding to the coincidence of two of the rotation axes; since ϕ and ψ cannot be identified separately in this case we (arbitrarily) set $\psi = 0$ to remove any ambiguity.

An alternative definition is motivated by the fact that any rotation about a fixed point can be expressed in the form

$$
\boldsymbol{r}' = \boldsymbol{r}\cos\zeta + (\boldsymbol{c} \cdot \boldsymbol{r})\boldsymbol{c}(1 - \cos\zeta) + (\boldsymbol{c} \times \boldsymbol{r})\sin\zeta
\tag{8.2.6}
$$

where $\boldsymbol{c}$ is a unit vector specifying the axis of rotation and ζ is the rotation angle. If we define

$$
q_m =
\begin{cases}
c_m \sin(\zeta/2) & m = 1, 2, 3 \\
\cos(\zeta/2) & m = 4
\end{cases}
\tag{8.2.7}
$$

then

$$
\boldsymbol{r}' = (2q_4^2 - 1)\boldsymbol{r} + 2(\boldsymbol{q} \cdot \boldsymbol{r})\boldsymbol{q} + 2q_4\boldsymbol{q} \times \boldsymbol{r}
\tag{8.2.8}
$$

[†] There is a minor change of indices: q_0 is replaced by q_4.

While the definition based on Euler angles (8.2.3) is useful for converting to and from the quaternion representation, the completely equivalent result (8.2.8) leads directly to the rotation matrix

$$\mathbf{R} = 2 \begin{pmatrix} q_1^2 + q_4^2 - \frac{1}{2} & q_1 q_2 + q_3 q_4 & q_1 q_3 - q_2 q_4 \\ q_1 q_2 - q_3 q_4 & q_2^2 + q_4^2 - \frac{1}{2} & q_2 q_3 + q_1 q_4 \\ q_1 q_3 + q_2 q_4 & q_2 q_3 - q_1 q_4 & q_3^2 + q_4^2 - \frac{1}{2} \end{pmatrix} \tag{8.2.9}$$

One of the benefits of quaternions is obvious – no trigonometric functions are required in evaluating $\mathbf{R}$. However, there is a more important advantage that will become apparent shortly.

A more formal treatment of quaternions [gol80] can be summarized as follows. Define a quaternion as the complex sum of a scalar and a vector,

$$\tilde{q} = q_4 + i\mathbf{q} \tag{8.2.10}$$

The product of two quaternions is then

$$\tilde{q}\tilde{q}' = q_4 q_4' - \mathbf{q} \cdot \mathbf{q}' + i(q_4 \mathbf{q}' + \mathbf{q} q_4' + \mathbf{q} \times \mathbf{q}') \tag{8.2.11}$$

itself also a quaternion. The complex conjugate of $\tilde{q}$ is $\tilde{q}^* = q_4 - i\mathbf{q}$, so that normalization implies $\tilde{q}\tilde{q}^* = 1$. The connection with rotation is made by choosing a vector $\mathbf{r}$, defining two quaternions $\tilde{r} = 0 + i\mathbf{r}$ and $\tilde{r}' = 0 + i\mathbf{r}'$, where $\mathbf{r}$ and $\mathbf{r}'$ are related as above, and with a little algebra arriving at the result $\tilde{r}' = \tilde{q}\tilde{r}\tilde{q}^*$. In this way quaternions are seen to provide the correct answers.

There are other ways of describing rotations, and even the quaternions can be expressed in an alternative fashion as complex 2×2 matrices, but our interest is confined to the set of real numbers $\{q_m\}$. The next step is to demonstrate the important role of quaternions in the dynamics of rigid bodies.

Equations of motion for nonlinear molecules

Rigid-body dynamics generally deals with two coordinate frames, one fixed in space, the other attached to the principal axes of the rotating body. The expression for the angular velocity $\boldsymbol{\omega}'$ measured in the body-fixed frame, in terms of Euler angles, is a familiar one [gol80],

$$\begin{pmatrix} \omega_x' \\ \omega_y' \\ \omega_z' \end{pmatrix} = \begin{pmatrix} \sin\theta \sin\psi & \cos\psi & 0 \\ \sin\theta \cos\psi & -\sin\psi & 0 \\ \cos\theta & 0 & 1 \end{pmatrix} \begin{pmatrix} \dot{\phi} \\ \dot{\theta} \\ \dot{\psi} \end{pmatrix} \tag{8.2.12}$$

Texts on mechanics tend to ignore the fact that the matrix in this equation is singular when $\sin\theta = 0$. Since the inverse of the matrix appears in the equations of motion, the numerical treatment is destined to become unstable whenever θ even approaches 0 or π. The simplest and most elegant way to avoid this inconvenience is to abandon Euler angles and use quaternions instead [eva77a, eva77b]; this eliminates the problem of singular matrices.

The angular velocity $\boldsymbol{\omega}'$ is related to $\dot{\tilde{q}}$ by

$$\begin{pmatrix} \omega'_x \\ \omega'_y \\ \omega'_z \\ 0 \end{pmatrix} = 2W \begin{pmatrix} \dot{q}_1 \\ \dot{q}_2 \\ \dot{q}_3 \\ \dot{q}_4 \end{pmatrix} \tag{8.2.13}$$

where

$$W = \begin{pmatrix} q_4 & q_3 & -q_2 & -q_1 \\ -q_3 & q_4 & q_1 & -q_2 \\ q_2 & -q_1 & q_4 & -q_3 \\ q_1 & q_2 & q_3 & q_4 \end{pmatrix} \tag{8.2.14}$$

is an orthogonal matrix. This result [cor60, eva77a] follows from the fact that if

$$\tilde{r}(t) = \tilde{q}(t)\tilde{r}(0)\tilde{q}^*(t) \tag{8.2.15}$$

then the time derivative is

$$\dot{\tilde{r}}(t) = \dot{\tilde{q}}(t)\tilde{r}(0)\tilde{q}^*(t) + \tilde{q}(t)\tilde{r}(0)\dot{\tilde{q}}^*(t) \tag{8.2.16}$$

Now

$$\tilde{r}(0) = \tilde{q}^*(t)\tilde{r}(t)\tilde{q}(t) \tag{8.2.17}$$

and, dropping the explicit t dependence for convenience,

$$\dot{\tilde{q}}\tilde{q}^* = -\tilde{q}\dot{\tilde{q}}^* \tag{8.2.18}$$

so that

$$\dot{\tilde{r}} = \dot{\tilde{q}}\tilde{q}^*\tilde{r} - \tilde{r}\dot{\tilde{q}}\tilde{q}^* \tag{8.2.19}$$

Thus

$$\dot{\boldsymbol{r}} = \boldsymbol{u} \times \boldsymbol{r} - \boldsymbol{r} \times \boldsymbol{u} = 2\boldsymbol{u} \times \boldsymbol{r} \tag{8.2.20}$$

where $\boldsymbol{u}$ is the vector part of the quaternion $\tilde{u} = \dot{\tilde{q}}\tilde{q}^*$; the scalar part is zero. In other words, since $\dot{\boldsymbol{r}} = \boldsymbol{\omega} \times \boldsymbol{r}$,

$$\boldsymbol{\omega} = 2\boldsymbol{u} \tag{8.2.21}$$

The result (8.2.13) follows immediately because

$$\tilde{u}' = \tilde{q}^* \tilde{u} \tilde{q} = \tilde{q}^* \dot{\tilde{q}} \tag{8.2.22}$$

In a space-fixed coordinate frame, torque equals the rate of change of angular momentum,

$$\boldsymbol{n} = \frac{d\boldsymbol{l}}{dt} \tag{8.2.23}$$

Given that the general relation between time derivatives in space- and body-fixed coordinate frames is

$$\left(\frac{d\boldsymbol{l}}{dt}\right)_{\text{space}} = \left(\frac{d\boldsymbol{l}}{dt}\right)_{\text{body}} + \boldsymbol{\omega} \times \boldsymbol{l} \tag{8.2.24}$$

the body-fixed version of (8.2.23) can be written

$$n_x = \dot{l}_x + \omega'_y l_z - \omega'_z l_y \tag{8.2.25}$$

with corresponding expressions for the other two components. In the body-fixed frame, each of the components of $\boldsymbol{l}$ has the simple form $l_x = I_x \omega'_x$, where I_x is one component of the (diagonal) inertia tensor, so that (8.2.25) becomes

$$I_x \dot{\omega}'_x = n_x + (I_y - I_z)\omega'_y \omega'_z \tag{8.2.26}$$

with similar equations for the other components. These are the Euler equations describing the rotation of a rigid body.

The quaternion accelerations are obtained from

$$\dot{\tilde{u}}' = \dot{\tilde{q}}^* \dot{\tilde{q}} + \tilde{q}^* \ddot{\tilde{q}} \tag{8.2.27}$$

Premultiply by $\tilde{q}$ and rearrange to get

$$\ddot{\tilde{q}} = \tilde{q}(\dot{\tilde{u}}' - \dot{\tilde{q}}^* \dot{\tilde{q}}) \tag{8.2.28}$$

or, equivalently,

$$\begin{pmatrix} \ddot{q}_1 \\ \ddot{q}_2 \\ \ddot{q}_3 \\ \ddot{q}_4 \end{pmatrix} = \tfrac{1}{2} \boldsymbol{W}^T \begin{pmatrix} \dot{\omega}'_x \\ \dot{\omega}'_y \\ \dot{\omega}'_z \\ -2 \sum \dot{q}_m^2 \end{pmatrix} \tag{8.2.29}$$

We can eliminate $\dot{\boldsymbol{\omega}}'$ from the right-hand side of (8.2.29) by using the Euler equations, and if the components of $\boldsymbol{\omega}'$ that then appear are replaced by linear combinations of the $\dot{q}_i$ from (8.2.13), the result is a set of equations of motion expressed entirely in terms of quaternions and their derivatives [pow79, rap85]; Euler angles and angular velocities no longer play any part in the computation.

A new C structure type is introduced to hold the quaternion components♠

```
typedef struct {
  real u1, u2, u3, u4;
} Quat;
```

and several useful operations on quaternions are defined,

```
#define QSet(q, s1, s2, s3, s4)                           \
    (q).u1 = s1,                                          \
    (q).u2 = s2,                                          \
    (q).u3 = s3,                                          \
    (q).u4 = s4                                           
#define QZero(q)                                          \
    QSet (q, 0., 0., 0., 0.)
#define QScale(q, s)                                      \
    (q).u1 *= s,                                          \
    ...
#define QSAdd(q1, q2, s3, q3)                             \
    (q1).u1 = (q2).u1 + (s3) * (q3).u1,                   \
    ...
#define QLenSq(q)                                         \
    (Sqr ((q).u1) + Sqr ((q).u2) + Sqr ((q).u3) +         \
    Sqr ((q).u4))
#define QMul(q1, q2, q3)                                  \
    (q1).u1 =   (q2).u4 * (q3).u1 - (q2).u3 * (q3).u2 +   \
                (q2).u2 * (q3).u3 + (q2).u1 * (q3).u4,    \
    (q1).u2 =   (q2).u3 * (q3).u1 + (q2).u4 * (q3).u2 -   \
                (q2).u1 * (q3).u3 + (q2).u2 * (q3).u4,    \
    (q1).u3 = - (q2).u2 * (q3).u1 + (q2).u1 * (q3).u2 +   \
                (q2).u4 * (q3).u3 + (q2).u3 * (q3).u4,    \
    (q1).u4 = - (q2).u1 * (q3).u1 - (q2).u2 * (q3).u2 -   \
                (q2).u3 * (q3).u3 + (q2).u4 * (q3).u4
```

An extended form of the `Mol` structure is used to hold all the molecular state information; the elements include the quaternion components, their first and second derivatives, other values required for the PC integration, and the torque on the molecule with components expressed in the body-fixed coordinate frame,

```
typedef struct {
  VecR r, rv, ra, ra1, ra2, ro, rvo;
  Quat q, qv, qa, qa1, qa2, qo, qvo;
  VecR torq;
} Mol;
```

♠ pr_08_1

The function for evaluating the quaternion accelerations – in other words, the rotational equations of motion – is the following, with $mInert$ a vector whose components are I_x, I_y and I_z.

```
void ComputeAccelsQ ()
{
  Quat qs;
  VecR w;
  int n;                                                                      5

  DO_MOL {
    ComputeAngVel (n, &w);
    QSet (qs,
        (mol[n].torq.x + (mInert.y - mInert.z) * w.y * w.z) / mInert.x,    10
        (mol[n].torq.y + (mInert.z - mInert.x) * w.z * w.x) / mInert.y,
        (mol[n].torq.z + (mInert.x - mInert.y) * w.x * w.y) / mInert.z,
        -2. * QLenSq (mol[n].qv));
    QMul (mol[n].qa, mol[n].q, qs);
    QScale (mol[n].qa, 0.5);                                                15
  }
}
```

The angular velocities w required here are computed by a function that is called separately for each molecule,

```
void ComputeAngVel (int n, VecR *w)
{
  Quat qt, qvt;

  qvt = mol[n].qv;                                                           5
  qvt.u4 *= -1.;
  QMul (qt, qvt, mol[n].q);
  QScale (qt, 2.);
  VSet (*w, qt.u1, qt.u2, qt.u3);
}                                                                            10
```

The interactions between rigid molecules are usually expressed as sums of contributions from pairs of 'interaction sites' on different molecules. It is sufficient to know the center of mass separation of the two molecules and their orientations in order to be able to compute the interactions between the site pairs – the subject is discussed in §8.3. Assuming that the site forces have already been computed, the forces and torques acting on the molecules as a whole are evaluated by the function given below. If k labels the interaction sites, and r_k is the location of a site relative to the center of mass of the molecule to which it belongs, then the total torque acting on the molecule is $\sum_k r_k \times f_k$.

A new C structure associated with the interaction sites

```
typedef struct {
  VecR f, r;
} Site;
```

is used to hold the current coordinates of the site and the results of the force calculation. Rotation matrices are represented by the structure

```
typedef struct {
  real u[9];
} RMat;
```

The variable *sitesMol* is the number of interaction sites in each molecule (all molecules are assumed identical for simplicity). The torques are evaluated in the space-fixed coordinate frame and then transformed to the body-fixed frame as required by the equations of motion.

```
void ComputeTorqs ()
{
  RMat rMat;
  VecR dr, t, torqS;
  int j, n;                                                    5

  DO_MOL {
    VZero (mol[n].ra);
    VZero (torqS);
    for (j = 0; j < sitesMol; j ++) {                          10
      VVAdd (mol[n].ra, site[n * sitesMol + j].f);
      VSub (dr, site[n * sitesMol + j].r, mol[n].r);
      VCross (t, dr, site[n * sitesMol + j].f);
      VVAdd (torqS, t);
    }                                                          15
    BuildRotMatrix (&rMat, &mol[n].q, 0);
    MVMul (mol[n].torq, rMat.u, torqS);
  }
}
```

The rotation matrix for each molecule (or its transpose, as required below) is constructed from the quaternion components by *BuildRotMatrix* and stored in column order as a linear array.

```
void BuildRotMatrix (RMat *rMat, Quat *q, int transpose)
{
  real p[10], tq[4], s;
  int k, k1, k2;
                                                               5
  tq[0] = q->u1;
```

```
  tq[1] = q->u2;
  tq[2] = q->u3;
  tq[3] = q->u4;
  for (k = 0, k2 = 0; k2 < 4; k2 ++) {                              10
    for (k1 = k2; k1 < 4; k1 ++, k ++) p[k] = 2. * tq[k1] * tq[k2];
  }
  rMat->u[0] = p[0] + p[9] - 1.;
  rMat->u[4] = p[4] + p[9] - 1.;
  rMat->u[8] = p[7] + p[9] - 1.;                                    15
  s = transpose ? 1. : -1.;
  rMat->u[1] = p[1] + s * p[8];
  rMat->u[3] = p[1] - s * p[8];
  rMat->u[2] = p[2] - s * p[6];
  rMat->u[6] = p[2] + s * p[6];                                     20
  rMat->u[5] = p[5] + s * p[3];
  rMat->u[7] = p[5] - s * p[3];
}
```

The interaction site coordinates must be computed in preparation for the force calculation. These coordinates are specified when the molecule is in a predefined reference orientation and are kept in the structure

```
typedef struct {
  VecR r;
  int typeF;
} MSite;
```

together with an element *typeF* that is used to distinguish individual sites for the force evaluation. Computation of interaction site coordinates is as follows, where *mSite* is an array of *MSite* structures.

```
void GenSiteCoords ()
{
  RMat rMat;
  VecR t;
  int j, n;                                                         5

  DO_MOL {
    BuildRotMatrix (&rMat, &mol[n].q, 1);
    for (j = 0; j < sitesMol; j ++) {
      MVMul (t, rMat.u, mSite[j].r);                                10
      VAdd (site[sitesMol * n + j].r, mol[n].r, t);
    }
  }
}
```

The operation for multiplying a matrix by a vector, here used in rotating the molecule from its predefined reference orientation to the current state, is defined by

```
#define MVMul(v1, m, v2)                                        \
   (v1).x = (m)[0] * (v2).x + (m)[3] * (v2).y + (m)[6] * (v2).z,  \
   (v1).y = (m)[1] * (v2).x + (m)[4] * (v2).y + (m)[7] * (v2).z,  \
   (v1).z = (m)[2] * (v2).x + (m)[5] * (v2).y + (m)[8] * (v2).z
```

Numerical integration of these second-order equations uses the same PC method as the translational equations. The integration functions, named `PredictorStepQ` and `CorrectorStepQ`, are based on the translational functions of §3.5 and only differ in the name and number of variables processed.

```
#define PQ(t)                                                  \
   PCR4 (mol[n].q, mol[n].q, mol[n].qv, mol[n].qa,              \
      mol[n].qa1, mol[n].qa2, t)
#define PQV(t)                                                 \
   PCV4 (mol[n].q, mol[n].qo, mol[n].qv, mol[n].qa,             \
      mol[n].qa1, mol[n].qa2, t)
#define CQ(t)                                                  \
   PCR4 (mol[n].q, mol[n].qo, mol[n].qvo, mol[n].qa,            \
      mol[n].qa1, mol[n].qa2, t)
#define CQV(t)                                                 \
   PCV4 (mol[n].q, mol[n].qo, mol[n].qv, mol[n].qa,             \
      mol[n].qa1, mol[n].qa2, t)

void PredictorStepQ ()
{
   ...
   DO_MOL {
     mol[n].qo = mol[n].q;
     mol[n].qvo = mol[n].qv;
     PQ (u1);
     PQV (u1);
     ... (ditto for u2, u3, u4) ...
     mol[n].qa2 = mol[n].qa1;
     mol[n].qa1 = mol[n].qa;
   }
}

void CorrectorStepQ ()
{
   ...
   DO_MOL {
     CQ (u1);
     CQV (u1);
     ...
   }
}
```

Normalization of the quaternions must be enforced separately to prevent gradual accumulation of numerical error (the error over a single timestep is very small); the adjustments can be carried out after each integration step.

```
void AdjustQuat ()
{
  real qi;
  int n;

  DO_MOL {
    qi = 1. / sqrt (QLenSq (mol[n].q));
    QScale (mol[n].q, qi);
  }
}
```

The contribution of the rotational motion to the kinetic energy is computed by an addition to *EvalProps*,

```
VecR w;
...
vvqSum = 0.;
DO_MOL {
  ComputeAngVel (n, &w);
  vvqSum += VWLenSq (mInert, w);
}
```

where *VWLenSq* is defined as

```
#define VWLenSq(v1, v2)  VWDot(v1, v2, v2)
#define VWDot(v1, v2, v3)                               \
   ((v1).x * (v2).x * (v3).x + (v1).y * (v2).y * (v3).y +   \
    (v1).z * (v2).z * (v3).z)
```

Tests based on momentum and energy conservation serve as partial checks on the correctness of the calculation. Angular momentum is not conserved however; this is a consequence both of the abrupt changes in angular momentum whenever a molecule crosses a periodic boundary and interaction wraparound. In order to verify angular momentum conservation, an isolated cluster of molecules would have to be simulated in a region that is nominally unbounded, thus eliminating the effects of periodicity.

Equations of motion for linear molecules

Linear rigid bodies are treated in a different way, since there are only two rotational degrees of freedom rather than three. The torque on a linear molecule can be written as a sum over interaction sites,

$$n = \sum_k r_k \times f_k = s \times \sum_k d_k f_k = s \times g \tag{8.2.30}$$

where the orientation is defined by s, the unit vector along the molecular axis, and where d_k is the signed distance along the axis of each interaction site from the center of mass. In the linear case, angular momentum is simply $l = I\omega$, so that the equations of motion are

$$I\dot{\omega} = s \times g \tag{8.2.31}$$

$$\dot{s} = \omega \times s \tag{8.2.32}$$

There is also a two-dimensional version of this problem, in which s is confined to the xy plane and the only nonzero component of ω is ω_z.

We have a choice of either using this pair of first-order equations, or eliminating ω to obtain a single second-order equation

$$\begin{aligned}\ddot{s} &= \dot{\omega} \times s + \omega \times \dot{s} \\ &= I^{-1}(s \times g) \times s + \omega \times (\omega \times s) \\ &= I^{-1}g - \left(I^{-1}(s \cdot g) + \dot{s}^2\right)s \end{aligned} \tag{8.2.33}$$

where we have used the results $\omega \cdot s = 0$ – a consequence of (8.2.31) – and $\dot{s}^2 = \omega^2$. Here, it is important that the initial state be defined consistently, to ensure that (8.2.32) is satisfied. In both cases the length of s must be adjusted at regular intervals (not necessarily at every timestep, although this causes no harm) to avoid any gradual buildup of error.

The PC integration functions (§3.5) for the first-order equations (8.2.31)–(8.2.32) follow (the second-order equation is used in §13.2). Here, $mol[].sv$ and $mol[].sa$ denote ω and $\dot{\omega}$ (for the second-order equation they stand for $\dot{s}$ and $\ddot{s}$); $mol[].svxs$ and related quantities hold the current and previous values of $\omega \times s$ appearing in (8.2.32).

```
#define PC(r, ro, v, v1, v2, t)                              \
   r.t = ro.t + w * (c[0] * v.t + c[1] * v1.t + c[2] * v2.t)

void PredictorStepF ()
{                                                                    5
  real c[] = {23., -16., 5.}, div = 12., w;
  int n;

  w = deltaT / div;
  DO_MOL {                                                          10
    so[n] = mol[n].s;
    svo[n] = mol[n].sv;
    PC (mol[n].s, mol[n].s, mol[n].svxs, mol[n].svxs1,
```

```
      mol[n].svxs2, x);
    PC (mol[n].sv, mol[n].sv, mol[n].sa, mol[n].sa1, mol[n].sa2, x);      15
    ... (ditto for y and z components) ...
    mol[n].sa2 = mol[n].sa1;
    mol[n].sa1 = mol[n].sa;
    mol[n].svxs2 = mol[n].svxs1;
    mol[n].svxs1 = mol[n].svxs;                                           20
    VCross (mol[n].svxs, mol[n].sv, mol[n].s);
  }
}

void CorrectorStepF ()                                                   25
{
  real c[] = {5., 8., -1.}, div = 12., w;
  int n;

  w = deltaT / div;                                                      30
  DO_MOL {
    PC (mol[n].s, so[n], mol[n].svxs, mol[n].svxs1, mol[n].svxs2, x);
    PC (mol[n].sv, svo[n], mol[n].sa, mol[n].sa1, mol[n].sa2, x);
    ... (ditto for y and z components) ...
    VCross (mol[n].svxs, mol[n].sv, mol[n].s);                           35
  }
}
```

The contribution of the rotational motion to the kinetic energy is once again computed by code added to *EvalProps*,

```
vvsSum = 0.;
DO_MOL vvsSum += mInert * VLenSq (mol[n].sv);
vvSum += vvsSum;
```

Temperature control

In the same way that the constant-temperature constraint was applied to simple atoms (§6.3), it can also be applied to nonlinear rigid molecules, but now the constraint must be based on the combined translational and rotational kinetic energy. For each molecule we include a Lagrange multiplier term in the translational equations as before, and a term of the general form $\alpha I_x \omega'_x$ must be added to each Euler equation (8.2.26). Since the total kinetic energy is

$$N_m E_K = \tfrac{1}{2} m \sum_i \dot{r}_i^2 + \tfrac{1}{2} \sum_x I_x \sum_i {\omega'_{i\,x}}^2 \qquad (8.2.34)$$

with $\sum_x$ denoting a sum over the vector components, by setting $\dot{E}_K = 0$ we obtain

$$\alpha = -\frac{\sum_i \dot{r}_i \cdot f_i + \sum_i \omega'_i \cdot n_i}{m \sum_i \dot{r}_i^2 + \sum_x I_x \sum_i {\omega'_i}_x^2} \qquad (8.2.35)$$

where f_i and n_i are the total force and torque on molecule i. When using quaternions, the kth component of the right-hand side of the equation of motion (8.2.29) gains an extra term (we omit the molecule index) $+\alpha \dot{q}_k$.

A similar expression for the Lagrange multiplier also applies in the case of linear molecules. The equation for α is similar to (8.2.35), but involves sums over either $\omega_i \cdot n_i$ and $I\omega_i^2$, or $\dot{s}_i \cdot [g_i - (s_i \cdot g_i + I\dot{s}_i^2)s_i]$ and $I\dot{s}_i^2$, depending on which form of the equation of motion is used.

The version of `ApplyThermostat` needed for nonlinear rigid molecules (assuming $m = 1$) is

```
void ApplyThermostat ()
{
  real s1, s2, vFac;
  VecR w;
  int n;                                                              5

  s1 = 0.;
  s2 = 0.;
  DO_MOL {
    s1 += VDot (mol[n].rv, mol[n].ra);                               10
    s2 += VLenSq (mol[n].rv);
  }
  DO_MOL {
    ComputeAngVel (n, &w);
    s1 += VDot (w, mol[n].torq);                                     15
    s2 += VWLenSq (mInert, w);
  }
  vFac = - s1 / s2;
  DO_MOL {
    VVSAdd (mol[n].ra, vFac, mol[n].rv);                             20
    QSAdd (mol[n].qa, mol[n].qa, vFac, mol[n].qv);
  }
}
```

while the changes for the linear case are

```
  ...
  DO_MOL {
    s1 += mInert * VDot (mol[n].sv, mol[n].sa);
    s2 += mInert * VLenSq (mol[n].sv);
  }                                                                   5
```

```
  ...
  DO_MOL {
    ...
    VVSAdd (mol[n].sa, vFac, mol[n].sv);
  }
```
                                                                    10

Temperature adjustment to correct numerical drift is applied separately to the translational and rotational motion. The addition to *AdjustTemp* for the nonlinear case, assuming a constant-temperature simulation, is

```
  VecR w;
  ...
  vvqSum = 0.;
  DO_MOL {
    ComputeAngVel (n, &w);
    vvqSum += VWLenSq (mInert, w);
  }
  vFac = velMag / sqrt (vvqSum / nMol);
  DO_MOL QScale (mol[n].qv, vFac);
```
                                                                    5

and for linear molecules it is

```
  vvsSum = 0.;
  DO_MOL vvsSum += mInert * VLenSq (mol[n].sv);
  vFac = velMag / sqrt (1.5 * vvsSum / nMol);
  DO_MOL VScale (mol[n].sv, vFac);
```

In both cases the value of *velMag* (see below) determines the correct kinetic energy value.

Initial state

The functions listed below are called from *SetupJob* to handle the initialization of the rotational variables. Here we consider only nonlinear molecules; the linear case will be treated in §13.2. Molecular orientation is randomly assigned (*atan2* is a standard library function), with each angular velocity having a fixed magnitude based on the temperature (through the quantity *velMag*) and a randomly chosen direction. Angular coordinates and velocities are converted to quaternion form, and angular accelerations used by the PC method are set to zero.

```
  void InitAngCoords ()
  {
    VecR e;
    real eAng[3];
    int n;
```
                                                                    5

```
  DO_MOL {
    VRand (&e);
    eAng[0] = atan2 (e.x, e.y);
    eAng[1] = acos (e.z);                                    10
    eAng[2] = 2. * M_PI * RandR ();
    EulerToQuat (&mol[n].q, eAng);
  }
}
                                                            15
void EulerToQuat (Quat *qe, real *eAng)
{
  real a1, a2, a3;

  a1 = 0.5 * eAng[1];                                       20
  a2 = 0.5 * (eAng[0] - eAng[2]);
  a3 = 0.5 * (eAng[0] + eAng[2]);
  QSet (*qe, sin (a1) * cos (a2), sin (a1) * sin (a2),
     cos (a1) * sin (a3), cos (a1) * cos (a3));
}                                                           25

void InitAngVels ()
{
  Quat qe;
  VecR e;                                                   30
  real f;
  int n;

  DO_MOL {
    VRand (&e);                                             35
    QSet (qe, e.x, e.y, e.z, 0.);
    QMul (mol[n].qv, mol[n].q, qe);
    f = 0.5 * velMag / sqrt (VWLenSq (mInert, e));
    QScale (mol[n].qv, f);
  }                                                         40
}

void InitAngAccels ()
{
  int n;                                                    45

  DO_MOL {
    QZero (mol[n].qa);
    QZero (mol[n].qa1);
    QZero (mol[n].qa2);                                     50
  }
}
```

The translational variables are initialized in exactly the same way as for atomic fluids.

8.3 Molecular construction

General features

Now that we have seen how to formulate and solve the dynamical problem we turn
to the details of the molecules themselves. Interactions between rigid molecules are
most readily introduced by specifying the locations of the sites in the molecule at
which the forces act. The total force between two molecules is then simply the sum
of the forces acting between all pairs of interaction sites. The amount of work is
proportional to the square of the number of sites, so this number should be kept as
small as possible. The potential function used for each pair can be defined independently, but molecular symmetry reduces the number of functions needed. Interaction sites may be associated with the positions of the nuclei, but this is not essential
and often just serves as the initial version of a model. There is considerable scope
for fine-tuning the structure and interactions in this engineering-like approach, with
the simulations themselves being used to refine the models; for further details see
[gra84, lev92].

Molecular fluids require substantially more computation per molecule than their
atomic counterparts because of the need to consider all pairs of interaction sites.
Coulomb interactions are usually involved, so the cutoff distance should be as
large as possible, again adding to the computational effort; the specialized methods available for such long-range forces (see Chapter 13) are not used here. The
fact that the interaction range now extends over a substantial fraction of the simulation region can erase the benefits of the cell and neighbor-list methods, so that
the all-pairs approach is often the method of choice for systems that are not too
large.

Model water

The most popular molecular fluid for MD exploration, for obvious reasons, is water. Not only because of its ubiquity and importance, but also for its many unusual
features that defy simple explanation, water has long been associated with MD
simulation [rah71, sti72, sti74] and numerous models have been proposed to help
understand the microscopic mechanisms underlying the behavior.

For our case study we will use one of several available rigid models, the TIP4P
model [jor83]. The model molecule, shown in Figure 8.1, is based on four interaction sites located in a planar configuration, two of which – labeled M and O – are
associated with the oxygen nucleus, and two – labeled H – with the protons; the
site M lies on the symmetry axis, between O and the line joining the H sites. The

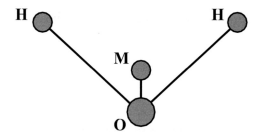

Fig. 8.1. The TIP4P water molecule; site coordinates are given in the text.

distances and angle required to fully specify the site coordinates are

$$r_{\mathrm{OH}} = 0.957 \, \text{Å}$$
$$r_{\mathrm{OM}} = 0.15 \, \text{Å} \tag{8.3.1}$$
$$\angle_{\mathrm{HOH}} = 104.5°$$

The interaction energy between two molecules i and j consists of a double sum over the interaction sites of both molecules; the terms in the sum, indexed by k and l, allow for Coulomb interactions between the electric charges assigned to the sites, as well as contributions of LJ type,

$$u_{ij} = \sum_{k \in i} \sum_{l \in j} \left(\frac{q_k q_l}{r_{kl}} + \frac{A_{kl}}{r_{kl}^{12}} - \frac{C_{kl}}{r_{kl}^{6}} \right) \tag{8.3.2}$$

The charges associated with the sites, while maintaining some resemblance to the actual molecule, are generally regarded as parameters that can be adjusted to fit known molecular properties, such as the multipole moments. The corresponding force is

$$f_{ij} = \sum_{k \in i} \sum_{l \in j} \left(\frac{q_k q_l}{r_{kl}^{3}} + \frac{12 A_{kl}}{r_{kl}^{14}} - \frac{6 C_{kl}}{r_{kl}^{8}} \right) r_{kl} \tag{8.3.3}$$

The charges appearing in the potential function are

$$q_{\mathrm{H}} = 0.52 \, e$$
$$q_{\mathrm{O}} = 0 \tag{8.3.4}$$
$$q_{\mathrm{M}} = -2 q_{\mathrm{H}}$$

where $e = 4.803 \times 10^{-10}$ esu; to convert to units used experimentally note that $e^2 = 331.8$ (kcal/mole) Å. As part of the molecular design process, the negative charge has been shifted away from the O site by a small amount, to the M site introduced specifically for this purpose. The parameters in the LJ part of the potential, which acts only between O sites, are

$$A_{OO} \equiv A = 600 \times 10^3 \,(\text{kcal/mole}) \,\text{\AA}^{12}$$
$$C_{OO} \equiv C = 610 \,(\text{kcal/mole}) \,\text{\AA}^6 \tag{8.3.5}$$

We now switch to reduced MD units appropriate to the problem. Define the length unit σ to be the value of r for which

$$\frac{A}{r^{12}} - \frac{C}{r^6} = 0 \tag{8.3.6}$$

namely

$$\sigma = (A/C)^{1/6} \tag{8.3.7}$$

and the unit of energy to be

$$\epsilon = A/4\sigma^{12} \tag{8.3.8}$$

The mass unit is the mass of the water molecule, 2.987×10^{-23} g. Physical and reduced units are then related by $\sigma = 3.154 \,\text{\AA}$, $\epsilon = 0.155 \,\text{kcal/mole}$ (or 1.08×10^{-14} erg/molecule) and the unit of time is 1.66×10^{-12} s. We also define a reduced unit of charge e in terms of which $q_H = 1$, and for convenience we let

$$b = e^2/\epsilon\sigma \tag{8.3.9}$$

In reduced units, $b = 183.5$.

The coordinates of the interaction sites when the molecule is situated in a reference state in the yz plane with its center of mass at the origin are (in reduced units)

$$r_O = (0,\ 0,\ -0.0206)$$
$$r_M = (0,\ 0,\ 0.0274) \tag{8.3.10}$$
$$r_H = (0,\ \pm 0.240,\ 0.165)$$

Masses denoted by m_O and m_H are associated with the O and H sites, and $m_O = 16 m_H$; in reduced units $m_O + 2m_H = 1$. The principal moments of inertia are

$$I_y = m_O z_O^2 + 2m_H z_H^2 \quad = 0.0034$$
$$I_z = 2m_H y_H^2 \qquad\qquad = 0.0064 \tag{8.3.11}$$

and, of course, $I_x = I_y + I_z$.

Interaction calculations

In terms of the reduced units just introduced, the potential energy and force contributions from the different pairs of interaction sites, namely, LJ between the O sites and Coulomb between all pairs of charges, are

$$\begin{aligned}
\text{OO:} \quad & u = 4(r^{-12} - r^{-6}) & f = 48(r^{-14} - \tfrac{1}{2}r^{-8})\boldsymbol{r} \\
\text{MM:} \quad & u = 4b/r & f = (4b/r^3)\boldsymbol{r} \\
\text{MH:} \quad & u = -2b/r & f = (-2b/r^3)\boldsymbol{r} \\
\text{HH:} \quad & u = b/r & f = (b/r^3)\boldsymbol{r}
\end{aligned} \qquad (8.3.12)$$

The function shown below computes these interactions using an all-pairs approach and assuming periodic boundaries. The different kinds of interaction site are assigned numerical types 1, 2 and 3, corresponding to O, M and H; these values appear in the element *typeF* in the *MSite* structure. The decision as to whether a pair of sites lies within the cutoff range is based on the distance between the centers of mass of the molecules containing the sites, and not on the distance between the sites themselves; not only is this more efficient computationally than testing pairs of sites individually, but it means that there are no partially interacting molecules.

```
void ComputeSiteForces ()
{
  VecR dr, shift;
  real fcVal, rr, rrCut, rri, rri3, uVal;
  int j1, j2, m1, m2, ms1, ms2, n, typeSum;                        5

  rrCut = Sqr (rCut);
  for (n = 0; n < nMol * sitesMol; n ++) VZero (site[n].f);
  uSum = 0.;
  for (m1 = 0; m1 < nMol - 1; m1 ++) {                             10
    for (m2 = m1 + 1; m2 < nMol; m2 ++) {
      VSub (dr, mol[m1].r, mol[m2].r);
      VZero (shift);
      VShiftAll (dr);
      VVAdd (dr, shift);                                          15
      rr = VLenSq (dr);
      if (rr < rrCut) {
        ms1 = m1 * sitesMol;
        ms2 = m2 * sitesMol;
        for (j1 = 0; j1 < sitesMol; j1 ++) {                      20
          for (j2 = 0; j2 < sitesMol; j2 ++) {
            typeSum = mSite[j1].typeF + mSite[j2].typeF;
            if (mSite[j1].typeF == mSite[j2].typeF || typeSum == 5) {
              VSub (dr, site[ms1 + j1].r, site[ms2 + j2].r);
              VVAdd (dr, shift);                                  25
              rr = VLenSq (dr);
              rri = 1. / rr;
```

```
        switch (typeSum) {
          case 2:
            rri3 = Cube (rri);                                    30
            uVal = 4. * rri3 * (rri3 - 1.);
            fcVal = 48. * rri3 * (rri3 - 0.5) * rri;
            break;
          case 4:
            uVal = 4. * bCon * sqrt (rri);                       35
            fcVal = uVal * rri;
            break;
          case 5:
            uVal = -2. * bCon * sqrt (rri);
            fcVal = uVal * rri;                                  40
            break;
          case 6:
            uVal = bCon * sqrt (rri);
            fcVal = uVal * rri;
            break;                                               45
        }
        VVSAdd (site[ms1 + j1].f, fcVal, dr);
        VVSAdd (site[ms2 + j2].f, - fcVal, dr);
        uSum += uVal;
      }                                                          50
    }
   }
  }
 }                                                               55
}
```

Further details

New variables appearing in this simulation are

```
Site *site;
MSite *mSite;
VecR mInert;
real bCon, vvqSum;
int sitesMol;                                                    5
```

and in *SetParams* we set

```
  sitesMol = 4;
```

The additional array allocations in *AllocArrays* are

```
  AllocMem (site, nMol * sitesMol, Site);
  AllocMem (mSite, sitesMol, MSite);
```

The details of the molecule are specified in the function *DefineMol*, called from *SetupJob*,

```
void DefineMol ()
{
  int j;

  for (j = 0; j < sitesMol; j ++) VZero (mSite[j].r);      5
  mSite[0].r.z = -0.0206;
  mSite[1].r.z = 0.0274;
  mSite[2].r.y = 0.240;
  mSite[2].r.z = 0.165;
  mSite[3].r.y = - mSite[2].r.y;                           10
  mSite[3].r.z = mSite[2].r.z;
  VSet (mInert, 0.00980, 0.00340, 0.00640);
  bCon = 183.5;
  mSite[0].typeF = 1;
  mSite[1].typeF = 2;                                      15
  mSite[2].typeF = 3;
  mSite[3].typeF = 3;
}
```

The full sequence of function calls in *SingleStep* used for the interaction computations and integration is

```
PredictorStep ();
PredictorStepQ ();
GenSiteCoords ();
ComputeSiteForces ();
ComputeTorqs ();                    5
ComputeAccelsQ ();
ApplyThermostat ();
CorrectorStep ();
CorrectorStepQ ();
AdjustQuat ();                      10
ApplyBoundaryCond ();
```

A few additional items complete the description: The interaction cutoff is at $r_c = 7.5$ Å, or 2.38 in reduced units; this value of *rCut* can be added to the input data. A density of 1 g/cm^3 is equivalent to a unit cell spacing of 3.103 Å for an initial cubic lattice arrangement, or 0.983 in reduced units. Since the unit of energy corresponds to $\epsilon/k_B = 78.2$ K, a typical temperature of 298 K corresponds to 3.8 in reduced units. The timestep used is $\Delta t = 0.0005$; in real units this equals 8×10^{-16} s.

8.4 Measurements

Types of measurement

A model such as the one described here has a variety of properties that are of experimental relevance, and others that, although not directly measurable in the laboratory, are able to contribute towards understanding the behavior at the microscopic level. We will consider two examples of the former and one of the latter, all in connection with pure water. A particularly important use of water models is in the study of solvation of other kinds of molecules, ranging from simple atoms and ions to complex molecules such as biopolymers; we will not attempt to delve into this extensive subject [bro88, lev92].

The first of the measurements involves the site–site RDFs. Here, rather than simply examining the distribution of center of mass separations, we study RDFs associated with distinct sites on the molecules; together, these RDFs are able to provide clues to local molecular arrangement beyond just the distances themselves. The second measurement deals with rotational diffusion by looking at the rate at which molecules undergo orientational change, an important aspect of certain kinds of spectroscopic study. The final feature examined, the one without a direct experimental counterpart, is the nature of the hydrogen-bond network formed by the fluid.

Other properties, including those of thermodynamic interest, as well as the dielectric constant, can also be measured, although they will not be considered here. A quantity such as the pressure, normally expressed in terms of the virial sum, needs to be redefined for use with rigid molecules. There are in fact two ways of dealing with the virial which, for equilibrium systems, are readily shown to be completely equivalent [cic86b]; it can be expressed either as a sum involving just the intermolecular forces and center of mass separations, ignoring all the internal details, or as a sum over all pairs of interaction sites in each pair of molecules.

Radial distribution functions

When evaluating the RDF we consider three distinct site–site distribution functions that are accessible experimentally – g_{OO}, g_{OH} and g_{HH}. For computational purposes we assign numerical labels to the sites to simplify the task of deciding which site pairs contribute to which function; this is done by adding an element *typeRdf* to *MSite* and initializing the values in *DefineMol*,

```
mSite[0].typeRdf = 1;
mSite[1].typeRdf = -1;
mSite[2].typeRdf = 2;
mSite[3].typeRdf = 2;
```

The array required for the RDFs is

```
real **histRdf;
```

where, unlike §4.3, `histRdf` is now a two-dimensional array that provides for several distinct RDF measurements; the allocation (in `AllocArrays`) is

```
AllocMem2 (histRdf, 3, sizeHistRdf, real);
```

The modified version of `EvalRdf` is as follows.

```
void EvalRdf ()
{
  VecR dr, shift;
  real deltaR, normFac, rr;
  int j1, j2, k, m1, m2, ms1, ms2, n, rdfType, typeSum;        5

  if (countRdf == 0) {
    for (k = 0; k < 3; k ++) {
      for (n = 0; n < sizeHistRdf; n ++) histRdf[k][n] = 0.;
    }
  }                                                            10
  deltaR = rangeRdf / sizeHistRdf;
  for (m1 = 0; m1 < nMol - 1; m1 ++) {
    for (m2 = m1 + 1; m2 < nMol; m2 ++) {
      VSub (dr, mol[m1].r, mol[m2].r);                         15
      VZero (shift);
      VShiftAll (dr);
      VVAdd (dr, shift);
      rr = VLenSq (dr);
      if (rr < Sqr (rangeRdf)) {                               20
        ms1 = m1 * sitesMol;
        ms2 = m2 * sitesMol;
        for (j1 = 0; j1 < sitesMol; j1 ++) {
          for (j2 = 0; j2 < sitesMol; j2 ++) {
            typeSum = mSite[j1].typeRdf + mSite[j2].typeRdf;   25
            if (typeSum >= 2) {
              VSub (dr, site[ms1 + j1].r, site[ms2 + j2].r);
              VVAdd (dr, shift);
              rr = VLenSq (dr);
              if (rr < Sqr (rangeRdf)) {                       30
                n = sqrt (rr) / deltaR;
                if (typeSum == 2) rdfType = 0;
                else if (typeSum == 3) rdfType = 1;
                else rdfType = 2;
                ++ histRdf[rdfType][n];                        35
              }
            }
          }
        }
      }
    }
```

```
      }                                                                        40
    }
  }
  ++ countRdf;
  if (countRdf == limitRdf) {
    normFac = VProd (region) / (2. * M_PI * Cube (deltaR) *                     45
      Sqr (nMol) * countRdf);
    for (k = 0; k < 3; k ++) {
      for (n = 0; n < sizeHistRdf; n ++)
        histRdf[k][n] *= normFac / Sqr (n - 0.5);
    }                                                                          50
    PrintRdf (stdout);
    countRdf = 0;
  }
}
```

It is not necessary to recompute the site coordinates after the corrector step, since
the values computed for use in the interaction calculations are adequate for this
purpose. Because there are two H sites per molecule, and we have not allowed for
this symmetry in the RDF computation, both g_{OH} and g_{HH} must be divided by four.
This can be done by the function *PrintRdf*, which must also be modified to output
three sets of RDF measurements.

  The run used to produce the RDF results involves the following input data:

deltaT	0.0005
density	0.98
initUcell	6 6 6
limitRdf	100
rangeRdf	2.5
rCut	2.38
sizeHistRdf	125
stepAdjustTemp	1000
stepAvg	200
stepEquil	1000
stepLimit	16000
stepRdf	50
temperature	3.8

A cubic initial array is used, so that the system contains $N_m = 216$ molecules.
The value of Δt is an order of magnitude smaller than that used in the soft-sphere
work; this is a result of the higher temperature (in MD units) and the need to allow
for the rotational motion of molecules with a small moment of inertia and hence a
relatively high angular velocity. Constant-temperature MD is used; with the value
of Δt shown, the temperature drift over 1000 timesteps amounts to about 4%, but
if this presents a problem the drift can be reduced by an order of magnitude simply
by halving Δt.

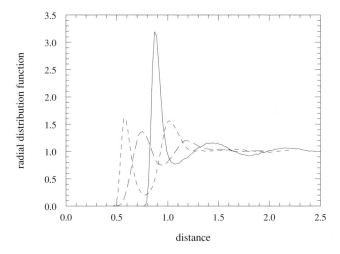

distance

Fig. 8.2. Site–site radial distribution functions for TIP4P water: g_{OO} (solid curve), g_{OH} (short dashes) and g_{HH} (long dashes).

Averaged♠ g_{OO}, g_{OH} and g_{HH} measurements are shown in Figure 8.2. The latter two curves are truncated at distances less than `rangeRdf` because the criterion for limiting the distance between sites is applied to the molecular centers of mass (exactly as in the force computation) and not to the sites themselves. Without going into detail, the results are consistent with the expected tetrahedral, or icelike, structural correlations known to occur in liquid water [jor83]. One example of a measurement demonstrating the loose packed molecular organization of the fluid is the integral of the function $4\pi r^2 g_{OO}(r)$ out to a distance that includes the first peak of g_{OO}; this provides an estimate of the number of molecules that can be regarded as nearest neighbors, and here the value is found to be 4.7.

Rotational diffusion

Rotational diffusion, a measure of the rate at which the direction of the molecular dipole changes, is another quantity of experimental significance. The dipole direction appears as the bottom row of the rotation matrix (8.2.9) and is the unit vector

$$\boldsymbol{\mu} = 2 \begin{pmatrix} q_1 q_3 + q_2 q_4 \\ q_2 q_3 - q_1 q_4 \\ q_3^2 + q_4^2 - \frac{1}{2} \end{pmatrix} \tag{8.4.1}$$

♠ `pr_anrdf`

Rotational diffusion – the mean-square rate of change in orientation – is expressed in terms of the time-dependent dipole autocorrelation function

$$C(t) = \langle \boldsymbol{\mu}_i(t) \cdot \boldsymbol{\mu}_i(0) \rangle \tag{8.4.2}$$

Translational diffusion will also be measured, based on the molecular center of mass coordinates.

The measurement[♠] is organized in the same way as translational diffusion, with extra elements in *TBuf*,

```
VecR *orgD;
real *ddDiffuse;
```

and additions to *EvalDiffusion* (§5.3),

```
VecR e;
...
for (nb = 0; nb < nBuffDiffuse; nb ++) {
  if (tBuf[nb].count == 0) {
    DO_MOL {                                                          5
      ...
      e.x = 2. * (mol[n].q.u1 * mol[n].q.u3 +
        mol[n].q.u2 * mol[n].q.u4);
      e.y = 2. * (mol[n].q.u2 * mol[n].q.u3 -
        mol[n].q.u1 * mol[n].q.u4);                                  10
      e.z = 2. * (Sqr (mol[n].q.u3) + Sqr (mol[n].q.u4) - 0.5);
      tBuf[nb].orgD[n] = e;
    }
  }
  if (tBuf[nb].count >= 0) {                                         15
    ...
    tBuf[nb].ddDiffuse[ni] = 0.;
    DO_MOL {
      e.x = ... (as above) ...
      ...                                                           20
      tBuf[nb].ddDiffuse[ni] += VDot (tBuf[nb].orgD[n], e);
    }
  }
  ...
}                                                                    25
```

Additions to *AccumDiffusion*, in the appropriate places, are

```
for (j = 0; j < nValDiffuse; j ++)
  ddDiffuseAv[j] += tBuf[nb].ddDiffuse[j];
...
```

♠ pr_08_2

```
    fac = 1. / (nMol * limitDiffuseAv);
    for (j = 0; j < nValDiffuse; j ++) ddDiffuseAv[j] *= fac;          5
```

and to *ZeroDiffusion*,

```
    for (j = 0; j < nValDiffuse; j ++) ddDiffuseAv[j] = 0.;
```

In *PrintDiffusion*, the values of *ddDiffuseAv* must be included in the output. Array declaration and allocation (*AllocArrays*) requires

```
real *ddDiffuseAv;

  AllocMem (ddDiffuseAv, nValDiffuse, real);
  for (nb = 0; nb < nBuffDiffuse; nb ++) {
    AllocMem (tBuf[nb].ddDiffuse, nValDiffuse, real);           5
    AllocMem (tBuf[nb].orgD, nMol, VecR);
  }
```

The runs used for these measurements are similar to the one described above, but the system size is reduced to $N_m = 125$ and the following additional input data are required,

```
limitDiffuseAv      10
nBuffDiffuse        20
nValDiffuse         200
stepDiffuse         40
```

The translational diffusion coefficients and the dipole autocorrelations, at $T = 3.8$ and 4.4, are shown in Figure 8.3. The runs of 57 600 timesteps used to produce these results allow averaging over nine sets of data, after skipping the first three.

Hydrogen bonds

The molecular structure of normal ice involves a diamond (or tetrahedral) lattice; short-range correlations reminiscent of this order persist into the liquid state. The forces responsible for this loosely packed arrangement are attributed to hydrogen-bonding, in which each molecule forms four strong, highly directional bonds with its immediate neighbors. One of the basic requirements of any water model is that it should reproduce this behavior. What exactly constitutes a hydrogen bond is not included in the definition of the molecule, since it is a feature whose origin is quantum mechanical, but, for modeling purposes, it is reasonable to assume that the presence of such a bond between two molecules is marked by an interaction energy lying in a particular range, and a molecular alignment that satisfies certain conditions insofar as the distance and angles are concerned. Once all the hydrogen

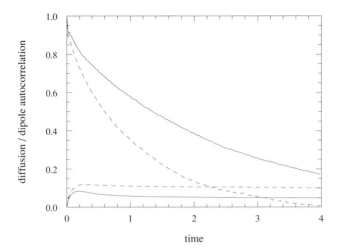

Fig. 8.3.   Water diffusion coefficients and dipole autocorrelation functions at temperatures 3.8 (solid curves) and 4.4 (dashed curves).

bonds have been identified, it is possible to study the properties of the network formed by the bonds [rah73, gei79].

Here we will focus on the pair-energy distribution [jor83], to see whether there is anything special about its form to warrant using it in determining where hydrogen bonds have formed. Since this exercise[♦] turns out to be successful, we then make use of what has been learned to count the numbers of bonds formed by each molecule.

The first step is to separately evaluate the interaction energy of each pair of molecules and construct a histogram of these values. In addition, each pair whose energy lies below a certain threshold is regarded as linked by a hydrogen bond, and bond counts associated with these molecules are incremented. The threshold is determined by a parameter *boundPairEng*. The following alterations and additions to *ComputeSiteForces* are required, where *nBond* is added to the *Mol* structure for use in the counting task.

```
real uSumPair;
int j;
...
DO_MOL mol[n].nBond = 0;
uSum = 0.;
for (m1 = 0; m1 < nMol - 1; m1 ++) {
  for (m2 = m1 + 1; m2 < nMol; m2 ++) {
```
5

```
      ...
      if (rr < rrCut) {
        uSumPair = 0.;                                                    10
        ...
        for (j1 = 0; j1 < sitesMol; j1 ++) {
          for (j2 = 0; j2 < sitesMol; j2 ++) {
            typeSum = ...
            if (mSite[j1].typeF == mSite[j2].typeF || typeSum == 5) {    15
              ...
              uSumPair += uVal;
            }
          }
        }
        uSum += uSumPair;                                                 20
        j = sizeHistPairEng * (uSumPair - minPairEng) /
          (maxPairEng - minPairEng);
        ++ histPairEng[Clamp (j, 0, sizeHistPairEng - 1)];
        if (uSumPair < boundPairEng) {                                   25
          ++ mol[m1].nBond;
          ++ mol[m2].nBond;
        }
      }
    }
  }                                                                      30
}
DO_MOL ++ histBondNum[Min (mol[n].nBond, sizeHistBondNum - 1)];
++ countPairEng;
```

Here, *Clamp* (§18.2) ensures that the array index is in the permitted range.
  The data collected are processed by a function called from *SingleStep*,

```
if (countPairEng == limitPairEng) PrintPairEng (stdout);
```

which computes the average pair-energy distribution over a series of configurations, constructs a histogram of the number of bonds per molecule and outputs the results.

```
void PrintPairEng (FILE *fp)
{
  real eVal, hSum;
  int n;
                                                                          5
  hSum = 0.;
  for (n = 0; n < sizeHistPairEng; n ++) hSum += histPairEng[n];
  for (n = 0; n < sizeHistPairEng; n ++) histPairEng[n] /= hSum;
  hSum = 0.;
  for (n = 0; n < sizeHistBondNum; n ++) hSum += histBondNum[n];         10
  for (n = 0; n < sizeHistBondNum; n ++) histBondNum[n] /= hSum;
  fprintf (fp, "pair energy\n");
  for (n = 0; n < sizeHistPairEng; n ++) {
```

```
    eVal = minPairEng + (n + 0.5) * (maxPairEng - minPairEng) /
       sizeHistPairEng;                                              15
    fprintf (fp, "%8.4f %8.4f\n", eVal, histPairEng[n]);
  }
  fprintf (fp, "bond count\n");
  for (n = 0; n < sizeHistBondNum; n ++)
    fprintf (fp, "%d %8.4f\n", n, histBondNum[n]);                   20
  InitPairEng ();
}
```

Initialization is carried out by the function (called from `SetupJob`)

```
void InitPairEng ()
{
  int n;

  for (n = 0; n < sizeHistPairEng; n ++) histPairEng[n] = 0.;       5
  for (n = 0; n < sizeHistBondNum; n ++) histBondNum[n] = 0.;
  countPairEng = 0;
}
```

The new variables used here are

```
real *histBondNum, *histPairEng, boundPairEng, maxPairEng, minPairEng;
int countPairEng, limitPairEng, sizeHistBondNum, sizeHistPairEng;
```

additional data to be input are

```
    NameR (boundPairEng),
    NameI (limitPairEng),
    NameR (maxPairEng),
    NameR (minPairEng),
    NameI (sizeHistBondNum),                                        5
    NameI (sizeHistPairEng),
```

and the array allocations (`AllocArrays`)

```
    AllocMem (histBondNum, sizeHistBondNum, real);
    AllocMem (histPairEng, sizeHistPairEng, real);
```

To investigate the pair-energy distribution we carry out a run with input data

```
boundPairEng      -8.
deltaT            0.0003
density           0.98
initUcell         6 6 6
limitPairEng      1000
```

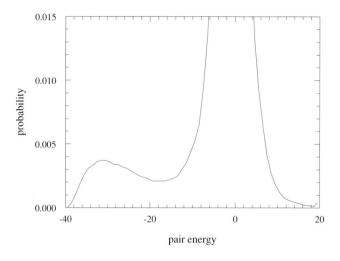

Fig. 8.4. Pair-energy distribution.

maxPairEng	20.
minPairEng	-40.
rCut	2.38
sizeHistBondNum	8
sizeHistPairEng	60
stepAdjustTemp	1000
stepAvg	200
stepEquil	0
stepLimit	5000
temperature	3.8

The value of Δt has been reduced to improve stability. The results shown in Figure 8.4 are obtained by averaging over 1000 timesteps, after skipping the first 4000; the smaller peak corresponds to tightly bound, nearest neighbor molecule pairs.

If we now assume that all pairs of molecules with mutual interaction energy below a certain threshold e_h are hydrogen-bonded, we can actually examine the distribution of hydrogen bonds. By way of example, we use values of *boundPairEng* (corresponding to e_h) of –8, –10 and –12 to obtain the results shown in Figure 8.5. Although a more detailed analysis taking the relative orientation of the molecules into account is required to ensure a consistent picture, the fact that for physically reasonable values of e_h ($e_h = -10$ corresponds to 1.55 kcal/mole) the average number of hydrogen bonds formed by each molecule is close to four is encouraging. Further analysis appears in [jor83].

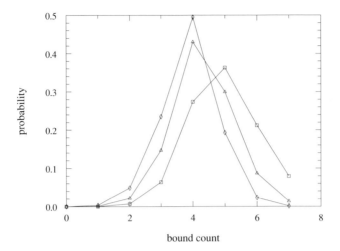

Fig. 8.5.   Distribution of mean number of hydrogen bonds per molecule for threshold energy $e_h = -8$ (squares), -10 (triangles) and -12 (diamonds).

8.5 Rotation matrix representation

Equations of motion

Although the quaternion representation has achieved popularity, there is an alternative method that considers the rotation matrices of the molecules directly [dul97], and in doing so avoids the need for the supplementary normalization calculations required by the quaternions. Furthermore, this approach can be incorporated into a leapfrog integration scheme and does not require a PC integrator. The extra storage needed by the matrix components (nine elements instead of four) is more than compensated for by the reduced quantity of data retained from earlier timesteps. Another potential benefit, although this depends on the relative magnitudes of the rotational and translational velocities (which, in turn, depends on the moment of inertia of the molecules), is the ability to use a larger integration timestep than allowed by the PC method, while maintaining a high degree of energy conservation.

Two ways of formulating the problem are described here. One approach involves expressing the angular velocity and acceleration vectors – ω_i' and α_i' – in the coordinate frame fixed to the principal axes of the molecule. Let R_i denote the rotation matrix of molecule i; the transpose R_i^T would be used, as before, in converting the site coordinates of the molecule to their values in the space-fixed frame. The first stage of the leapfrog integration consists of a half-timestep update of the angular velocities,

$$\omega'_i(t + h/2) = \omega'_i(t) + \frac{h}{2}\alpha'_i(t) \tag{8.5.1}$$

together with a full update of the rotation matrix, expressed in terms of a symmetric product of matrices each describing a small partial rotation about a different axis,

$$\boldsymbol{R}_i(t + h) = \boldsymbol{R}_i(t) \, \boldsymbol{U}_1^T \, \boldsymbol{U}_2^T \, \boldsymbol{U}_3^T \, \boldsymbol{U}_2^T \, \boldsymbol{U}_1^T \tag{8.5.2}$$

where the matrices

$$\boldsymbol{U}_1 = \boldsymbol{U}_x(\omega'_{i\,x}h/2)\,, \quad \boldsymbol{U}_2 = \boldsymbol{U}_y(\omega'_{i\,y}h/2)\,, \quad \boldsymbol{U}_3 = \boldsymbol{U}_z(\omega'_{i\,z}h) \tag{8.5.3}$$

correspond to the rotations around the different axes. Writing the product of the noncommuting rotation matrices in this symmetric form is necessary to ensure the time-reversible nature of the numerical integration.

Next, express the updated angular velocity in the newly rotated frame of the molecule that results from applying these small partial rotations,

$$\omega'_i(t + h/2) \rightarrow \boldsymbol{U}_1 \, \boldsymbol{U}_2 \, \boldsymbol{U}_3 \, \boldsymbol{U}_2 \, \boldsymbol{U}_1 \, \omega'_i(t + h/2) \tag{8.5.4}$$

The translational part of the first stage of the leapfrog integration is also carried out at this point, and the force and torque computations performed as before. Since the torque $\boldsymbol{\tau}_i$ on the molecule is evaluated in the space-fixed frame,

$$\alpha'_i(t + h) = \left(\mathcal{I}_i^c\right)^{-1} \boldsymbol{R}_i^T(t + h) \, \boldsymbol{\tau}_i(t + h) \tag{8.5.5}$$

where $\mathcal{I}_i^c$ is the diagonal moment of inertia matrix in the principal-axes frame of the body. Finally, in the second stage of the leapfrog process, the rotational part consists of

$$\omega'_i(t + h) = \omega'_i(t + h/2) + \frac{h}{2}\alpha'_i(t + h) \tag{8.5.6}$$

The alternative is to work entirely in the space-fixed frame[†]. The first stage of the leapfrog integration is then

$$\omega_i(t + h/2) = \omega_i(t) + \frac{h}{2}\alpha_i(t) \tag{8.5.7}$$

and, since it is more convenient to work with the transpose of $\boldsymbol{R}_i$,

$$\boldsymbol{R}_i^T(t + h) = \boldsymbol{U}_1 \, \boldsymbol{U}_2 \, \boldsymbol{U}_3 \, \boldsymbol{U}_2 \, \boldsymbol{U}_1 \, \boldsymbol{R}_i^T(t) \tag{8.5.8}$$

where the rotation matrices of (8.5.3) are replaced by

$$\boldsymbol{U}_1 = \boldsymbol{U}_x(\omega_{i\,x}h/2)\,, \quad \boldsymbol{U}_2 = \boldsymbol{U}_y(\omega_{i\,y}h/2)\,, \quad \boldsymbol{U}_3 = \boldsymbol{U}_z(\omega_{i\,z}h) \tag{8.5.9}$$

[†] This version is required for dealing with molecules having limited internal degrees of freedom (Chapter 11).

which are evaluated in the space-fixed frame. The required angular acceleration is now

$$\boldsymbol{\alpha}_i(t+h) = \boldsymbol{R}_i(t+h)\left(\boldsymbol{\mathcal{I}}_i^c\right)^{-1}\boldsymbol{R}_i^T(t+h)\,\boldsymbol{\tau}_i(t+h) \tag{8.5.10}$$

and the second stage of the leapfrog integration is

$$\boldsymbol{\omega}_i(t+h) = \boldsymbol{\omega}_i(t+h/2) + \frac{h}{2}\boldsymbol{\alpha}_i(t+h) \tag{8.5.11}$$

The trigonometric functions appearing in the rotation matrices can be approximated to second order in h, but this must be done in a manner that preserves the orthogonality of the matrices, for example,

$$\boldsymbol{U}_x(\theta) \equiv \begin{pmatrix} 1 & 0 & 0 \\ 0 & \cos\theta & -\sin\theta \\ 0 & \sin\theta & \cos\theta \end{pmatrix} \approx \begin{pmatrix} 1 & 0 & 0 \\ 0 & \dfrac{1-\theta^2/4}{1+\theta^2/4} & \dfrac{-\theta}{1+\theta^2/4} \\ 0 & \dfrac{\theta}{1+\theta^2/4} & \dfrac{1-\theta^2/4}{1+\theta^2/4} \end{pmatrix} \tag{8.5.12}$$

Integration

The variables associated with each molecule are contained in the structure

```
typedef struct {
  VecR r, rv, ra;
  VecR wv, wa;
  RMat rMatT;
} Mol;
```
[5]

that differs from the `Mol` structure used in §8.2. Here, the quaternion components, and their first and second derivatives, have been replaced by a rotation matrix (actually its transpose) `rMatT`, together with the angular velocity and acceleration (expressed in the space-fixed frame); acceleration values from earlier timesteps are no longer required.

Since the matrix $\boldsymbol{U}_3 \equiv \boldsymbol{U}_z(\omega_{iz}h)$ that appears in (8.5.8) can be replaced by $\boldsymbol{U}_z(\omega_{iz}h/2)\,\boldsymbol{U}_z(\omega_{iz}h/2)$, evaluating the product of the partial rotation matrices can be carried out by explicitly constructing a pair of matrix products that together make up (8.5.8), namely,

$$\boldsymbol{U}_x(\omega_{ix}h/2)\,\boldsymbol{U}_y(\omega_{iy}h/2)\,\boldsymbol{U}_z(\omega_{iz}h/2) \tag{8.5.13}$$

and the product terms in reversed order (the terms of the two products share common elements), and then multiplying the results. This is accomplished by the following function, in which the first argument points to the matrix and the second to the vector $h\boldsymbol{\omega}/2$.

```
void BuildStepRmatT (RMat *mp, VecR *a)
{
  RMat m1, m2;
  real c[3], s[3], ak, c0c2, c0s2, s0c2, s0s2, t;
  int k;

  for (k = 0; k < 3; k ++) {
    ak = VComp (*a, k);
    t = 0.25 * Sqr (ak);
    c[k] = (1. - t) / (1. + t);
    s[k] = ak / (1. + t);
  }
  c0c2 = c[0] * c[2];
  c0s2 = c[0] * s[2];
  s0c2 = s[0] * c[2];
  s0s2 = s[0] * s[2];
  m1.u[0] = c[1] * c[2];
  m1.u[1] = s0c2 * s[1] + c0s2;
  m1.u[2] = - c0c2 * s[1] + s0s2;
  m1.u[3] = - c[1] * s[2];
  m1.u[4] = - s0s2 * s[1] + c0c2;
  m1.u[5] = c0s2 * s[1] + s0c2;
  m1.u[6] = s[1];
  m1.u[7] = - s[0] * c[1];
  m1.u[8] = c[0] * c[1];
  m2.u[0] = m1.u[0];
  m2.u[1] = - m1.u[3];
  m2.u[2] = - m1.u[6];
  m2.u[3] = s0c2 * s[1] - c0s2;
  m2.u[4] = s0s2 * s[1] + c0c2;
  m2.u[5] = - m1.u[7];
  m2.u[6] = c0c2 * s[1] + s0s2;
  m2.u[7] = c0s2 * s[1] - s0c2;
  m2.u[8] = m1.u[8];
  MulMat (mp->u, m1.u, m2.u, 3);
}
```

The function `MulMat` (§18.4) is used for multiplying matrices.

The leapfrog integration function is an extension of earlier versions; the additional computation (8.5.8) required for the rotation matrices makes use of the functions `BuildStepRmatT` and `MulMat`.

```
void LeapfrogStep (int part)
{
  RMat mc, mt;
  VecR t;
  int n;

  if (part == 1) {
    DO_MOL {
```

```
        VVSAdd (mol[n].wv, 0.5 * deltaT, mol[n].wa);
        VVSAdd (mol[n].rv, 0.5 * deltaT, mol[n].ra);          10
      }
      DO_MOL {
        VSCopy (t, 0.5 * deltaT, mol[n].wv);
        BuildStepRmatT (&mc, &t);
        MulMat (mt.u, mc.u, mol[n].rMatT.u, 3);               15
        mol[n].rMatT = mt;
      }
      DO_MOL VVSAdd (mol[n].r, deltaT, mol[n].rv);
    } else {
      DO_MOL {                                                 20
        VVSAdd (mol[n].wv, 0.5 * deltaT, mol[n].wa);
        VVSAdd (mol[n].rv, 0.5 * deltaT, mol[n].ra);
      }
    }
  }                                                            25
}
```

Interaction calculations

Instead of revisiting the water model considered previously, the case study used to
demonstrate the rotation matrix approach (the method also appears in Chapter 11)
deals with a fluid of rigid molecules constructed from tetrahedral assemblies of soft
spheres♠. The simulations differ from the water study in a number of respects. The
moment of inertia components of the water molecule are quite small, resulting in
an integration timestep that is limited by the relatively high angular velocity; those
of the tetrahedral molecule are considerably larger, leading to similar contributions
to the site velocities from the translational and angular velocities, so that a larger
timestep can be used. The interactions here are short ranged, so the force calculations can utilize neighbor lists rather than the all-pair approach used earlier. One
additional new feature is the introduction of hard, reflecting container walls, as an
alternative to the periodic boundaries used in most other case studies.

The neighbor list should only include pairs of interaction sites belonging to distinct molecules; the function *BuildNebrList* in §3.4 must therefore be modified
by replacing the quantity *nMol* with *nSite*, and references to coordinates *mol[].r*
are replaced by *site[].r*. The test that ensures each pair of sites is examined only
once must be augmented with a check that the sites involved belong to different
molecules,

```
      if ((m1 != m2 || j2 < j1) && j1 / sitesMol != j2 / sitesMol)
```

The change resulting from the replacement of periodic boundaries by hard walls

♠ pr_08_4, pr_08_5

involves ensuring that the second member of the cell pair lies within the region; this is accomplished by inserting

```
if (m2v.x < 0 || m2v.x >= cells.x ||
    m2v.y < 0 || m2v.y >= cells.y || m2v.z >= cells.z) continue;
```

prior to the innermost pair of loops over cell contents and removing all references to periodic wraparound.

The function for evaluating site forces, `ComputeSiteForces`, is derived from the function *ComputeForces* used for the simplest soft-sphere case, in which the interaction site forces `site[].f` are evaluated for those pairs in the neighbor list that lie within interaction range; site coordinates `site[].r` are used and the reference to periodic wraparound removed. The site coordinates themselves are generated from the available rotation matrices, so the function *GenSiteCoords* (§8.2) now contains the replacement line

```
MVMul (t, mol[n].rMatT.u, mSite[j].r);
```

and the code for generating the rotation matrices is omitted.

Wall interactions

When any of the molecular interaction sites approach a container wall they are subjected to a repulsive force in the direction perpendicular to the wall. The interaction is based on the soft-sphere potential, but only the distance component normal to the wall enters the calculation. Each wall contributes independently, so that a site near a container edge or corner can experience the sum of two or three separate wall interactions. While an extension to the neighbor-list procedure could have been used to select the sites affected, here we simply examine all sites, although the amount of work is reduced by first testing whether the center of mass of the molecule is sufficiently close to the wall for such interactions to be possible.

A pair of definitions are introduced for convenience,

```
#define NearWall(t)                                      \
    fabs (fabs (mol[n].r.t) - 0.5 * region.t) <          \
        farSiteDist + 0.5 * rCut

#define WallForce(t)                                      \      5
    { dr = ((site[j].r.t >= 0.) ? site[j].r.t :          \
        - site[j].r.t) - 0.5 * (region.t + rCut);        \
      if (dr > - rCut) {                                  \
        if (site[j].r.t < 0.) dr = - dr;                 \
        rri = 1. / Sqr (dr);                              \      10
```

```
      rri3 = Cube (rri);                                        \
      site[j].f.t += 48. * rri3 * (rri3 - 0.5) * rri * dr;      \
      uSum += 4. * rri3 * (rri3 - 1.) + 1.;                     \
    }                                                           \
  }                                                             15
```

where *farSiteDist* is the distance of the furthest site in the molecule from its
center of mass. The function that evaluates wall forces is then simply

```
void ComputeWallForces ()
{
  real dr, rri, rri3;
  int j, n;
                                                                5
  DO_MOL {
    if (NearWall (x)) {
      for (j = n * sitesMol; j < (n + 1) * sitesMol; j ++)
        WallForce (x);
    }                                                          10
    ... (ditto for y and z components) ...
  }
}
```

The wall forces ensure that the molecules remain within the container. Their introduction here is just for demonstration purposes; under normal circumstances
periodic boundaries are preferable to hard walls (since the effect of walls on the
behavior can extend a considerable distance into the bulk).

Other details

Once the total force acting on each of the sites has been determined the forces and
torques acting on the molecules – the latter expressed in the space-fixed frame –
can be evaluated by a suitably modified version of *ComputeTorqs*,

```
void ComputeTorqs ()
{
  VecR dr, t, torqS, waB;
  int j, n;
                                                                5
  DO_MOL {
    VZero (mol[n].ra);
    VZero (torqS);
    for (j = 0; j < sitesMol; j ++) {
      VVAdd (mol[n].ra, site[n * sitesMol + j].f);             10
      VSub (dr, site[n * sitesMol + j].r, mol[n].r);
      VCross (t, dr, site[n * sitesMol + j].f);
      VVAdd (torqS, t);
```

```
    }
    MVMulT (waB, mol[n].rMatT.u, torqS);                              15
    VDiv (waB, waB, mInert);
    MVMul (mol[n].wa, mol[n].rMatT.u, waB);
  }
}
```

where *MVMulT* corresponds to *MVMul* with a transposed matrix.

The initialization procedure is

```
void SetupJob ()
{
  AllocArrays ();
  DefineMol ();
  stepCount = 0;                                                      5
  InitCoordsWalls (farSiteDist);
  InitVels ();
  InitAccels ();
  InitAngCoords ();
  InitAngVels ();                                                     10
  InitAngAccels ();
  AccumProps (0);
  nebrNow = 1;
}
```

in which *AllocArrays* includes allocation of the arrays needed for both the rigid-body simulation and the neighbor-list method, and *SingleStep* includes

```
LeapfrogStep (1);
GenSiteCoords ();
if (nebrNow) {
  nebrNow = 0;
  dispHi = 0.;                                                        5
  BuildNebrList ();
}
ComputeSiteForces ();
ComputeWallForces ();
ComputeTorqs ();                                                      10
LeapfrogStep (2);
```

The characteristics of the tetrahedral molecules considered here, namely their site coordinates and moments of inertia, are specified in the following function, with the parameter *siteSep* allowing for an adjustable molecule size. To allow generalization[†] the center of mass position *rCm* is computed from the site positions; here, the same mass value is associated with each site of the tetrahedron and the

---

† Linear molecules are allowed, provided a nonzero moment of inertia is associated with the longitudinal axis.

molecules have unit total mass.

```
void DefineMol ()
{
  VecR rCm;
  int j;

  VSet (mSite[0].r, 0., 0.5 / sqrt (3.), sqrt (2.) / sqrt (3.));
  VSet (mSite[1].r, 0., 1.5 / sqrt (3.), 0.);
  VSet (mSite[2].r, 0.5, 0., 0.);
  VSet (mSite[3].r, - 0.5, 0., 0.);
  for (j = 0; j < sitesMol; j ++) VScale (mSite[j].r, siteSep);
  VZero (rCm);
  for (j = 0; j < sitesMol; j ++) VVAdd (rCm, mSite[j].r);
  for (j = 0; j < sitesMol; j ++)
    VVSAdd (mSite[j].r, -1. / sitesMol, rCm);
  VZero (mInert);
  for (j = 0; j < sitesMol; j ++) {
    mInert.x += Sqr (mSite[j].r.y) + Sqr (mSite[j].r.z);
    mInert.y += Sqr (mSite[j].r.z) + Sqr (mSite[j].r.x);
    mInert.z += Sqr (mSite[j].r.x) + Sqr (mSite[j].r.y);
  }
  VScale (mInert, 1. / sitesMol);
}
```

Generating the initial state uses a function based on *InitCoords* that ensures molecules are not placed too close to the walls,

```
void InitCoordsWalls (real border)
{
  VecR ... regionI;

  VAddCon (regionI, region, - 2. * border);
```

where *VAddCon* (§18.2) adds the same value to each of the vector components, and the reduced region size *regionI* is used instead of *region* for initializing the coordinates. The value of *border* is set equal to *farSiteDist* to ensure that all sites lie within the container. Function *SetParams* sets various quantities, including *farSiteDist*,

```
siteSep = 0.8;
farSiteDist = siteSep / (2. * sqrt (2./3.));
sitesMol = 4;
nSite = nMol * sitesMol;
nebrTabMax = nebrTabFac * nSite;
```

and the required new variables are

```
real farSiteDist, siteSep;
int nSite;
```

For the initial angular coordinates, now expressed in terms of the rotation matrices, a minor change to the function *InitAngCoords* is required,

```
Quat qe;
...
BuildRotMatrix (&mol[n].rMatT, &qe, 1);
```

while angular velocities are set using

```
void InitAngVels ()
{
  VecR e, wvB;
  real f;
  int n;                                                    5

  DO_MOL {
    VRand (&e);
    f = velMag / sqrt (VWLenSq (mInert, e));
    VSCopy (wvB, f, e);                                     10
    MVMul (mol[n].wv, mol[n].rMatT.u, wvB);
  }
}
```

and the angular accelerations are initialized to zero by *InitAngAccels*.

Computation of the energy, together with a fairly generous estimate of the maximum site displacement that is used in deciding when to rebuild the neighbor list, is carried out as follows.

```
void EvalProps ()
{
  VecR wvB;
  real vv, vvMax, vvrMax, vvwMax;
  int n;                                                    5

  VZero (vSum);
  vvSum = 0.;
  vvrMax = 0.;
  vvwMax = 0.;                                              10
  DO_MOL {
    VVAdd (vSum, mol[n].rv);
    vv = VLenSq (mol[n].rv);
    vvSum += vv;
    vvrMax = Max (vvrMax, vv);                              15
```

```
  MVMulT (wvB, mol[n].rMatT.u, mol[n].wv);
  vvSum += VWLenSq (mInert, wvB);
  vv = VLenSq (wvB);
  vvwMax = Max (vvwMax, vv);
}                                                                              20
vvMax = Sqr (sqrt (vvrMax) + farSiteDist * sqrt (vvwMax));
...
}
```

In order to use the constant-temperature constraint together with rotation matrices and leapfrog integration, the following form of the thermostat function – based on the method in §6.3 – is required.

```
void ApplyThermostat ()
{
  RMat mc, mt;
  VecR vt, waB, wvB;
  real s1, s2, vFac;                                                          5
  int n;

  s1 = 0.;
  s2 = 0.;
  DO_MOL {                                                                    10
    VSAdd (vt, mol[n].rv, 0.5 * deltaT, mol[n].ra);
    s1 += VDot (vt, mol[n].ra);
    s2 += VLenSq (vt);
    VSAdd (vt, mol[n].wv, 0.5 * deltaT, mol[n].wa);
    MVMulT (wvB, mol[n].rMatT.u, vt);                                         15
    MVMulT (waB, mol[n].rMatT.u, mol[n].wa);
    s1 += VWDot (mInert, wvB, waB);
    s2 += VWLenSq (mInert, wvB);
  }
  vFac = - s1 / s2;                                                           20
  DO_MOL {
    VSAdd (vt, mol[n].rv, 0.5 * deltaT, mol[n].ra);
    VVSAdd (mol[n].ra, vFac, vt);
    VSAdd (vt, mol[n].wv, 0.5 * deltaT, mol[n].wa);
    VVSAdd (mol[n].wa, vFac, vt);                                             25
  }
}
```

Finally, when adjusting the temperature, the contribution from the rotational motion must also be rescaled, so the relevant code in *AdjustTemp* is changed to

```
VecR wvB;

vvSum = 0.;
DO_MOL {
  MVMulT (wvB, mol[n].rMatT.u, mol[n].wv);                                     5
```

Table 8.1. Energy measurements for fluid of tetrahedral molecules.

timestep	$\langle E \rangle$	$\sigma(E)$	$\langle E_K \rangle$	$\sigma(E_K)$
1 000	3.3961	0.0004	2.9858	0.0283
2 000	3.3969	0.0002	2.9717	0.0298
3 000	3.3970	0.0002	2.9820	0.0270
4 000	3.3970	0.0002	2.9883	0.0284
5 000	3.3970	0.0002	2.9722	0.0234
10 000	3.3974	0.0002	2.9773	0.0268

```
    vvSum += VWLenSq (mInert, wvB);
  }
  vFac = velMag / sqrt (vvSum / nMol);
  DO_MOL VScale (mol[n].wv, vFac);
```

Measurements

The measurements shown here focus on how well the rotation matrix approach succeeds in conserving energy. During the initial equilibration period the temperature is reset to the desired value every 100 timesteps, but then the system is allowed to run without any further adjustment. The run includes the following data and the results are shown in Table 8.1. The energy conservation in this case is of similar quality to the soft-sphere fluid of §3.7.

```
deltaT          0.005
density         0.15
initUcell       8 8 8
stepAvg         200
stepEquil       1000
stepLimit       10000
temperature     1.
```

8.6 Further study

8.1   Explore the relation between quaternions and other representations of orientation [gol80].

8.2   The leapfrog method can also be used with quaternions [fin93]; examine its effectiveness.

8.3   Study the relative orientation of neighboring water molecules.

8.4   Study the nature of the hydrogen-bond network formed and the bond lifetimes [gei79].

8.5   Compute the dielectric constant [spr91, smi94]; how sensitive is this to the choice of model and how significant is the effect of truncating the long-range Coulomb forces?

8.6   Because of the unusual properties of water, various models have been used in MD studies to account for the experimental observations; investigate the factors contributing to the design of different models.

8.7   The rigid molecule used for water ignores important polarizability effects; how can the model be extended [spr88] to incorporate such behavior?

8.8   Constant-pressure techniques can also be applied to rigid bodies; investigate [nos83].

8.9   A subtle, but important, property of water is the density maximum that occurs while still in the liquid state (the reason why ice floats); how successful has MD been in studying this phenomenon [bil94]?

8.10 Compare the efficiency of the quaternion and rotation matrix methods as applied to water.

8.11 Study other examples of rigid molecular models – both linear and nonlinear – for real fluids [lev92].

9

Flexible molecules

9.1 Introduction

The rigid molecule approach described in Chapter 8 is limited in its applicability, because it is really only appropriate for small, compact molecules. Here we consider the opposite extreme, namely, completely flexible molecules of a type used in certain kinds of polymer studies. No new principles are involved, since the intramolecular forces that maintain structural integrity by holding the molecule together, as well as providing any other necessary internal interactions, are treated in the same way as intermolecular forces. Later, in Chapters 10 and 11, we will consider more complex models, in which molecules exhibit a certain amount of flexibility but are also subject to various structural constraints that restrict the internal motion. The first case study in this chapter deals with the configurational properties of a single chain molecule in solution. The second deals with a model of a surfactant in solution, in which very short, interacting chain molecules are just one of the components of a three-component fluid; this very simple system is capable of producing coherent structures on length scales greatly exceeding the molecular size, as the results will demonstrate.

9.2 Description of molecule

Polymer chains

Owing to the central role played by polymers in a variety of fields, biochemistry and materials science are just two examples, model polymer systems have been the subject of extensive study, both by MD and by other methods such as Monte Carlo [bin95]. Of the many kinds of polymer topology that occur, chains have received the most attention, but other types, including stars [grc94] and membranes [abr89], have not been neglected. Chain properties can be divided into two categories, equilibrium and dynamical; much of the equilibrium behavior – especially in the case of long chains – actually falls under the heading of critical phenomena, and here MD is unable to compete with lattice-based methods because of their far less demanding nature, but when it comes to transport phenomena MD is, once again, the only viable method.

Polymer chain models can be studied for different reasons. At one extreme is the attempt to reproduce the behavior of a real polymer, an example being the alkane model we will meet in Chapter 10, or complex biopolymers such as proteins [ber86a, bro88, bro90a, dua98]. Here we concentrate on a much simpler model that aims at capturing some of the more general aspects of chain behavior, rather than all the myriad quantitative details. One can regard this model as analogous to the soft-sphere fluid, but while for simple fluids there is just one basic model, for polymers there are a number of different systems that might be regarded as basic. The simplest is a single chain in the vacuum, used for studying the configurational properties of an isolated polymer. This is followed by a chain in an inert soft-sphere solvent, the purpose of the solvent being to introduce a certain amount of hydrodynamic coupling into the motion of the chain [pie92, smi92, dun93]. Then there are multiple-chain fluids [kre92]; here the chain density is an important parameter, because it determines how much of the dynamics is due to the chain interacting with itself and how much is due to interactions between chains. In each instance, the details of the interactions between chain atoms, as well as the nature of the solvent, if present, must be addressed.

A problem that must be faced when studying polymers is the range of timescales over which configurational change occurs. At one extreme are the localized changes in internal arrangement that involve only short segments of the chain; at the other are large-scale conformational changes and chain diffusion, processes that are seriously impeded by effects such as mutual obstruction and entanglement. This means that some of the more interesting rheological properties of polymer liquids, and the challenging problems of protein folding, appear to be beyond the limits of what can be simulated by MD. But a great deal can still be done within the timescales that are currently accessible.

Chain structure

The goal of the simplest models is to represent the excluded volume of the individual monomers out of which the polymer is constructed and the bonds that link them into chains. The monomers can be simple atoms modeled using a soft-sphere potential, while bonds with limited length variation can be produced by means of an attractive interaction between chain neighbors. Single or multiple chains can be included and a soft-sphere solvent is readily added. Chains constructed in this way are totally flexible, within the limits set by the repulsive potential; a controlled degree of stiffness can be introduced by means of an interaction regulating the separation of next-nearest neighbors, although we will not do this here. More specific structural requirements are best addressed using the methods described in subsequent chapters.

In the model treated here, all pairs of atoms interact via the familiar soft-sphere repulsive force, which we will call f_{ss}; in addition, there is an attractive interaction between each pair of adjacent bonded atoms of form

$$
f_{bb}(r) = \begin{cases} f_{ss}\big((1 - r_m/r)r\big) & r_m - r_c < r < r_m \\ 0 & \text{otherwise} \end{cases} \tag{9.2.1}
$$

In (9.2.1), the direction of the soft-sphere force has been reversed, and its origin shifted to produce a force that limits the separation of bonded atoms; in practice, the bond-length variation can be restricted to a (not too) narrow range by a suitable choice of r_m ($> r_c$). The energy and length scales characterizing the potential, ϵ and σ, are left unchanged.

9.3 Implementation details

Interactions

The evaluation♦ of the forces between nonbonded atoms belonging to the same chain, as well as between atoms in different chains and between solvent–chain and solvent–solvent atom pairs, are all handled by the soft-sphere functions of §3.4, with just one minor alteration. If we assume that neighbor lists are used, the change affects the condition for selecting atom pairs in *BuildNebrList*. The modified form is

```
if ((m1 != m2 || j2 < j1) && (mol[j1].inChain == -1 ||
    mol[j1].inChain != mol[j2].inChain || abs (j1 - j2) > 1))
```

so that bonded atom pairs are excluded – they will be treated separately. The element *inChain* that has been added to the *Mol* structure indicates whether the particular atom belongs to a polymer chain, and if so – denoted by a value ≥ 0 – the identity of the chain.

An additional function is required for evaluating the forces between bonded atoms. The total number of chains is given by *nChain*, the number of atoms per chain – assuming all chains to have the same length – by chainLen, and r_m is represented by the variable bondLim.

```
void ComputeChainBondForces ()
{
  VecR dr;
  real fcVal, rr, rrCut, rri, rri3, uVal, w;
  int i, j1, j2, n;                                              5

  rrCut = Sqr (rCut);
  for (n = 0; n < nChain; n ++) {
    for (i = 0; i < chainLen - 1; i ++) {
      j1 = n * chainLen + i;                                     10
      j2 = j1 + 1;
      VSub (dr, mol[j1].r, mol[j2].r);
      VWrapAll (dr);
      rr = VLenSq (dr);
      if (rr < rrCut) {                                          15
        ... (same as ComputeForces) ...
      }
      w = 1. - bondLim / sqrt (rr);
      if (w > 0.) ErrExit (ERR_BOND_SNAPPED);
      rr *= Sqr (w);                                             20
      if (rr < rrCut) {
        rri = 1. / rr;
        rri3 = Cube (rri);
        fcVal = 48. * w * rri3 * (rri3 - 0.5) * rri;
        ... (same as ComputeForces) ...                         25
      }
    }
  }
}
```

In computing the attractive part of the bond interaction a safety check is included to ensure that the bond has not 'snapped', either because of numerical error or due to incorrectly formulated initial conditions (see below).

Initial state

When preparing the initial state it is essential that the atoms of each chain be positioned so that the bond lengths are all within their permitted ranges, and that no significant overlap occurs between atoms belonging to either the same or different chains. Neither of these issues presents any difficulty in this particular case study, especially if the density is not too high, but questions of how to pack molecules correctly into a reasonably low energy state while avoiding overlap between molecules can arise in other situations [mck92]. Solvent atoms pose less of a problem because they can be added after the chains are in place.

Possible initial chain states include fully stretched and planar zigzag configurations; another option is the linear helix which is even more compact than the zigzag form, a useful feature when chain packing becomes problematic at higher densities. The following function arranges the atoms of each chain in a zigzag state, with the major axis of the chain aligned in the x direction. The chains themselves are organized as a BCC lattice, and after they have been positioned the coordinates are corrected to allow for any periodic wraparound. We also show how the solvent is added; the very simple but inefficient approach demonstrated here attempts to place solvent atoms at the sites of a simple cubic lattice by checking whether the proposed location overlaps any of the chain atoms already in position, and if overlap is found to occur the tentative solvent atom is discarded (for large systems a method based on the use of cells would be preferable).

```
void InitCoords ()
{
  VecR c, dr, gap;
  real by, bz;
  int i, j, m, n, nx, ny, nz;                                          5

  by = rCut * cos (M_PI / 4.);
  bz = rCut * sin (M_PI / 4.);
  n = 0;
  VDiv (gap, region, initUchain);                                      10
  for (nz = 0; nz < initUchain.z; nz ++) {
    for (ny = 0; ny < initUchain.y; ny ++) {
      for (nx = 0; nx < initUchain.x; nx ++) {
        VSet (c, nx + 0.25, ny + 0.25, nz + 0.25);
        VMul (c, c, gap);                                              15
        VVSAdd (c, -0.5, region);
        for (j = 0; j < 2; j ++) {
          for (m = 0; m < chainLen; m ++) {
            VSet (mol[n].r, 0., (m % 2) * by, m * bz);
            VVSAdd (mol[n].r, 0.5 * j, gap);                           20
            VVAdd (mol[n].r, c);
            ++ n;
          }
        }
      }
    }                                                                  25
  }
  nMol = n;
  ApplyBoundaryCond ();
  VDiv (gap, region, initUcell);                                       30
  for (nz = 0; nz < initUcell.z; nz ++) {
    for (ny = 0; ny < initUcell.y; ny ++) {
      for (nx = 0; nx < initUcell.x; nx ++) {
        VSet (c, nx + 0.5, ny + 0.5, nz + 0.5);
        VMul (c, c, gap);                                              35
```

```
        VVSAdd (c, -0.5, region);
        for (i = 0; i < nChain * chainLen; i ++) {
          VSub (dr, mol[i].r, c);
          if (VLenSq (dr) < Sqr (rCut)) break;
        }                                                    40
        if (i == nChain * chainLen) {
          mol[n].r = c;
          ++ n;
        }
      }
    }
  }                                                          45
  nMol = n;
}
```

The variables introduced here are

```
VecI initUchain;
real bondLim;
int chainLen, nChain;
```

and there is additional input data

```
    NameR (bondLim),
    NameI (chainLen),
    NameI (initUchain),
```

The number of chains, assuming the BCC arrangement, is computed in *SetParams*,

```
    nChain = 2 * VProd (initUchain);
    if (nChain == 2) nChain = 1;
```

where the values in `initUchain` (an integer vector) specify the number of unit cells in the lattice used for arranging the chains. To enable the study of just a single chain, we have assumed that if a single unit cell is specified the intention is to have just one chain; to accommodate this case a change is needed in `InitCoords`,

```
  if (nChain == 1) {
    for (m = 0; m < chainLen; m ++) {
      VSet (mol[n].r, 0., (m % 2) * by, m * bz);
      VVSAdd (mol[n].r, -0.25, region);
      ++ n;                                                   5
    }
  } else {
    ... (as before) ...
  }
```

The maximum possible number of atoms, subject to later reduction because of overlap between solvent and chain atoms, is set in `SetParams`,

```
nMol = VProd (initUcell) + nChain * chainLen;
```

where `initUcell` now specifies the number of unit cells that contain solvent atoms (ignoring overlap).

The final stage of the initialization process involves explicit assignment of atoms to chains for use in the interaction calculations; since the chains are constructed consecutively this is a trivial task.

```
void AssignToChain ()
{
  int i, j, n;

  n = 0;
  for (i = 0; i < nChain; i ++) {
    for (j = 0; j < chainLen; j ++) {
      mol[n].inChain = i;
      ++ n;
    }
  }
  for (n = nChain * chainLen; n < nMol; n ++) mol[n].inChain = -1;
}
```

9.4 Properties

Chain conformation

Three spatial properties of polymer chains are frequently studied because of their experimental relevance. The first is the mean-square end-to-end distance $\langle R^2 \rangle$ from which it is possible to learn whether, on average, the chain is in an open or compact configuration; the distribution of R^2 values, or at least the moments of the distribution, can be used to determine the importance of effects such as excluded volume. Then there is the mean-square radius of gyration $\langle S^2 \rangle$ that provides information on the entire mass distribution of the chain and plays a central role in interpreting light scattering and viscosity measurements. Lastly, since the actual mean spatial distribution of the chain mass – essentially its 'shape' – need not be spherical, details of the moments of the mass (or monomer) distribution can be informative.

For a chain of n_s monomers

$$\langle R^2 \rangle = \langle |r_{n_s} - r_1|^2 \rangle \tag{9.4.1}$$

and, if all monomers have the same mass,

$$\langle S^2 \rangle = \frac{1}{n_s} \left\langle \sum_{i=1}^{n_s} |r_i - \bar{r}|^2 \right\rangle \tag{9.4.2}$$

where $\bar{r}$ is the center of mass. Elements of the tensor describing the mass distribution have the form

$$G_{xy} = \frac{1}{n_s} \sum_{i=1}^{n_s} (r_{ix} - \bar{r}_x)(r_{iy} - \bar{r}_y) \tag{9.4.3}$$

The three eigenvalues of G are denoted by g_1, g_2 and g_3; their sum is just $\langle S^2 \rangle$, but it is their ratios that are of interest because if they are not equal to unity it means that the distribution is nonspherical[†]. Rearrangement of (9.4.3) leads to an alternative expression that is used in the computations, namely,

$$G_{xy} = \frac{1}{n_s} \sum_{i=1}^{n_s} r_{ix} r_{iy} - \frac{1}{n_s^2} \left[\sum_{i=1}^{n_s} r_{ix} \right] \left[\sum_{i=1}^{n_s} r_{iy} \right] \tag{9.4.4}$$

The function shown below accumulates these chain properties over a sequence of configurations; it is called from *SingleStep* by

```
if (stepCount >= stepEquil &&
    (stepCount - stepEquil) % stepChainProps == 0) EvalChainProps ();
```

New variables, input data items and the initialization of this calculation (each in the appropriate place) are

```
real bbDistSq, bondLim, eeDistSq, gMomRatio1, gMomRatio2, radGyrSq;
int countChainProps, limitChainProps, stepChainProps;

  NameI (limitChainProps),
  NameI (stepChainProps),                                           5

  countChainProps = 0;
```

The following function measures and averages the end-to-end distance, the radius of gyration, the eigenvalue ratios and the actual bond lengths. Evaluating the eigenvalues $\{g_i\}$ requires diagonalizing a 3×3 matrix; to do this, simply expand the determinant $\det|G - gI|$ to obtain the cubic characteristic equation, so that $\{g_i\}$ are just the solutions of this equation obtained by a call to *SolveCubic* (§18.4). The organization of this function adheres to a pattern that should be familiar by

---

[†] The familiar inertia tensor [gol80] has components $S^2 \delta_{xy} - G_{xy}$.

now; the output function is trivial.

```
void EvalChainProps ()
{
  VecR c, cs, dr, shift;
  real a[3], g[6], gVal[3];
  int i, j, k, n, n1;                                                    5

  if (countChainProps == 0) {
    bbDistSq = 0.;
    eeDistSq = 0.;
    radGyrSq = 0.;                                                      10
    gMomRatio1 = 0.;
    gMomRatio2 = 0.;
  }
  n = 0;
  for (i = 0; i < nChain; i ++) {                                       15
    VZero (shift);
    VZero (cs);
    for (k = 0; k < 6; k ++) g[k] = 0.;
    n1 = n;
    for (j = 0; j < chainLen; j ++) {                                   20
      if (j > 0) {
        VSub (dr, mol[j].r, mol[j - 1].r);
        VShiftWrap (dr, x);
        VShiftWrap (dr, y);
        VShiftWrap (dr, z);                                             25
        bbDistSq += VLenSq (dr);
      }
      VAdd (c, mol[n].r, shift);
      VVAdd (cs, c);
      g[0] += Sqr (c.x);                                                30
      g[1] += Sqr (c.y);
      g[2] += Sqr (c.z);
      g[3] += c.x * c.y;
      g[4] += c.z * c.x;
      g[5] += c.y * c.z;                                                35
      ++ n;
    }
    VVSub (c, mol[n1].r);
    eeDistSq += VLenSq (c);
    VScale (cs, 1. / chainLen);                                         40
    for (k = 0; k < 6; k ++) g[k] /= chainLen;
    g[0] -= Sqr (cs.x);
    g[3] -= cs.x * cs.y;
    ... (similarly for other elements) ...
    a[0] = - g[0] - g[1] - g[2];                                        45
    a[1] = g[0] * g[1] + g[1] * g[2] + g[2] * g[0] -
        Sqr (g[3]) - Sqr (g[4]) - Sqr (g[5]);
    a[2] = g[0] * Sqr (g[5]) + g[1] * Sqr (g[4]) + g[2] * Sqr (g[3]) -
        2. * g[3] * g[4] * g[5] - g[0] * g[1] * g[2];
```

```
    SolveCubic (g, a);                                               50
    gVal[0] = Max3 (g[0], g[1], g[2]);
    gVal[2] = Min3 (g[0], g[1], g[2]);
    gVal[1] = g[0] + g[1] + g[2] - gVal[0] - gVal[2];
    radGyrSq += gVal[0] + gVal[1] + gVal[2];
    gMomRatio1 += gVal[1] / gVal[0];                                 55
    gMomRatio2 += gVal[2] / gVal[0];
  }
  ++ countChainProps;
  if (countChainProps == limitChainProps) {
    bbDistSq /= nChain * (chainLen - 1) * limitChainProps;           60
    eeDistSq /= nChain * limitChainProps;
    ... (ditto for radGyrSq, gMomRatio1, gMomRatio2) ...
    PrintChainProps (stdout);
    countChainProps = 0;
  }                                                                  65
}
```

Here, `Min3` and `Max3` (§18.2) extend `Min` and `Max` to three arguments, and

```
#define VShiftWrap(v, t)                                        \
  if (v.t >= 0.5 * region.t) {                                  \
    shift.t -= region.t;                                        \
    v.t -= region.t;                                            \
  } else if (v.t < -0.5 * region.t) {                           \     5
    shift.t += region.t;                                        \
    v.t += region.t;                                            \
  }
```

Measurements

The results shown here are for a single chain in a soft-sphere solvent. We consider chains consisting of $n_s = 8$, 16 and 24 monomers. The input data for a chain with $n_s = 8$ include

```
bondLim            2.1
chainLen           8
density            0.5
initUcell          10 10 10
initUchain         1 1 1
limitChainProps    100
stepAdjustTemp     1000
stepAvg            20000
stepChainProps     20
stepEquil          0
stepLimit          500000
temperature        2.
```

Table 9.1. Chain measurements.

n_s	$\langle l \rangle$	$\langle R^2 \rangle$	$\sigma(R^2)$	$\langle S^2 \rangle$	$\sigma(S^2)$	$\langle g_2/g_1 \rangle$	$\langle g_3/g_1 \rangle$
8	1.0531	13.08	0.61	2.105	0.050	0.2433	0.0780
16	1.0534	31.92	2.91	5.040	0.261	0.2700	0.0894
24	1.0533	59.07	10.30	9.015	0.937	0.2411	0.0751

Table 9.2. Block-averaged error estimates.

	$n_s = 8$		$n_s = 16$		$n_s = 24$	
b	$\sigma(\langle R^2 \rangle)$	$\sigma(\langle S^2 \rangle)$	$\sigma(\langle R^2 \rangle)$	$\sigma(\langle S^2 \rangle)$	$\sigma(\langle R^2 \rangle)$	$\sigma(\langle S^2 \rangle)$
1	0.245	0.021	1.016	0.091	2.149	0.185
2	0.269	0.022	1.207	0.106	2.813	0.238
4	0.268	0.020	1.268	0.114	3.210	0.279
8	0.258	0.019	1.257	0.112	3.120	0.282
16	0.255	0.020	1.133	0.112	3.267	0.323
32	0.270	0.021	1.346	0.132	2.920	0.260
64	0.257	0.023	0.640	0.075	4.496	0.372

Since a simple cubic lattice is used for the initial positions of the solvent atoms, the maximum number of solvent atoms is 1000 (the values in `initUcell` also help determine the region size), although overlap with chain monomers may reduce this number very slightly. For $n_s = 16$ the values in `initUcell` are increased to 12, and for $n_s = 24$ to 16 – the region must be large enough to hold the chain in its initial state and avoid unwanted wraparound effects. The empirically determined value of `bondLim` ensures that bond length variation is confined to a fairly narrow range. Constant-temperature dynamics and leapfrog integration are used.

Results[♠] obtained from runs of 5×10^5 timesteps are listed in Table 9.1 for the three chain lengths studied. The mean bond lengths $\langle l \rangle$ are practically the same in each case; the value of $\sigma(l)$ is typically 2×10^{-4}, so that bond length is seen to be tightly controlled. The values of $\langle R^2 \rangle$ and $\langle S^2 \rangle$ are comparable to published results, although the values do depend on solvent density [smi92]. The eigenvalue ratios $\langle g_2/g_1 \rangle$ and $\langle g_3/g_1 \rangle$ provide clear evidence that the mean shape of the chain is far from spherical, indeed the shape is more like a flattened cigar.

In order to obtain error estimates for $\langle R^2 \rangle$ and $\langle S^2 \rangle$ we resort to the block averaging described in §4.2. The results of this analysis over a series of block sizes b are shown in Table 9.2. The quality of the estimates is seen to decrease as the chain length grows, suggesting a need for even longer runs.

♠ `pr_anchprops`

9.5 Modeling structure formation

One of the more fascinating processes associated with polymers in solution is the formation of large-scale spatial structures in certain types of three-component fluids. Two of the fluid components are typically water and an oil-like liquid that is insoluble in water. The third component is a relatively small fraction of amphiphilic chain molecules, or surfactants; the term amphiphilic means that one end of the chain is hydrophilic, in other words, it has an affinity for water, and the other end is hydrophobic with a preference for oil-rich surroundings. What occurs in such systems is that the amphiphilic chains form surfaces separating the water and oil; in the case of low oil concentration, the oil will be packaged by the chains into droplets, or micelles, that are themselves water soluble; at higher oil concentrations, layers, bilayer vesicles, or a variety of other structures can form, again corresponding to oil-rich regions separated from the water by surfaces formed out of chain molecules [mye88, gel94]. There are many processes, both natural and industrial, where this kind of supramolecular self-assembly occurs; the nature of the constituent molecules and their interactions determine the morphology of the structures that can develop.

Since phenomena of this kind occur for many different combinations of molecules, it seems reasonable that the underlying behavior ought to be understandable using simplified models that ignore much of the specific molecular detail [kar96]. In particular, these models should be able to reveal how the collective behavior of aggregates – whose sizes greatly exceed those of the individual molecules – are related to the molecular properties. Due to the computationally intensive nature of the MD approach it is important to simplify the model as much as possible; this requires isolating the molecular characteristics that dominate the behavior.

A particularly successful class of model is based on representing both the water and oil molecules, labeled W and O, as simple spherical atoms. The forces between pairs of like atoms, $f_{WW}(r)$ and $f_{OO}(r)$, involve the LJ potential (which is attractive except upon close approach), whereas the interaction between unlike pairs, $f_{WO}(r)$, is a soft-sphere repulsion used simply to prevent molecular overlap. This choice ensures the immiscibility of the W and O species and, in order to reduce the number of parameters, $f_{WW}(r)$ and $f_{OO}(r)$ are assumed identical. The surfactant molecules are constructed from short, completely flexible chains, in which one or more W atoms form the hydrophilic head group, and one or more O atoms make up the hydrophobic tail; the atoms in the chain are linked by the springlike force $f_{bb}(r)$ defined in (9.2.1). A series of studies based on this kind of approach – although with a linear form for $f_{bb}(r)$ – are described in [smi91, ess94, kar94].

Even such a highly simplified representation provides ample scope for exploring a range of phenomena by suitably modifying the model parameters. Here, the case study focuses principally on micelle growth and on how the micelle sizes vary as a function of time. Measuring the properties of these structures calls for cluster analysis (§4.5) in order to mechanize the task of micelle identification. Minor extensions of the model (not discussed here) include varying the relative monomer sizes (for example, using larger W atoms for the chain head groups), the relative interaction strengths, and the chain lengths and structure; another possibility is reducing the flexibility of the surfactants by introducing constraints (Chapter 10).

In general, as the simulated systems become more complicated, visualization begins to play a particularly important role. The ability to follow the evolution of the system as a whole, as well as the motions of individual molecules, can prove extremely valuable, both while developing the simulations and then in analyzing those aspects of the behavior that are not readily expressed in quantitative form. This is particularly true in the present case and the case study will include 'snapshots' of the structures that develop in the course of the simulations; such images can be generated from the molecular coordinates, but details of the algorithms and three-dimensional graphics software required lie outside the scope of this book. The role of computer graphics extends beyond mere static images, since sets of snapshots produced over a period of time can be used to generate animated sequences showing the evolution of the system, from which time-dependent aspects of the behavior can be deduced. Such a highly visual approach complements more conventional quantitative methods of analyzing simulation results.

9.6 Surfactant models

Interactions

The program♠ used for simulating surfactant solutions is partly based on the earlier program for a single chain in solution (§9.3). Now, however, there are multiple chains and two species of atoms – representing oil and water molecules – in the solution. The different kinds of interactions are the following:

- LJ between pairs of like atoms (OO and WW) and soft-sphere repulsion between unlike (OW) pairs, irrespective of whether both atoms belong to the solution, or one atom belongs to the solution and the other is a monomer in a chain, or both atoms are monomers in different chains;
- soft-sphere repulsion between nonlinked monomers in the same chain, irrespective of type;
- bonding forces between linked monomers in the same chain.

♠ pr_09_2

The last of these are processed by *ComputeChainBondForces* (§9.3); the others are handled using a slight modification of the neighbor-list method.

The neighbor list is constructed using cells large enough to include all the attractive OO and WW pairs; while this implies the inclusion of repulsive OW pairs that may lie outside their cutoff range, it is a more concise approach than the alternative of maintaining multiple neighbor lists for the different types of interactions. Associated with each entry in the neighbor list is an element from an array *intType* specifying the kind of interaction required by the atom pair, with the values 1 and 2 denoting soft-sphere and LJ, respectively; this is determined in advance when constructing the neighbor list, rather than during the subsequent force evaluations. The *Mol* structure is

```
typedef struct {
  VecR r, rv, ra;
  int inChain, typeA;
} Mol;
```

where *typeA* values of 1 and 2 distinguish O from W atoms and *inChain* is the chain to which the atom belongs (or −1 if the atom is not a chain monomer).

The modifications to *BuildNebrList* are

```
  real rrNebrA;
  int iType, sameChain;

  rrNebrA = Sqr (rCutA + rNebrShell);
  ...                                                                5
    if (m1 != m2 || j2 < j1) {
      VSub (dr, mol[j1].r, mol[j2].r);
      VVSub (dr, shift);
      sameChain = (mol[j1].inChain == mol[j2].inChain &&
        mol[j1].inChain >= 0);                                       10
      iType = 0;
      if (mol[j1].typeA == mol[j2].typeA && ! sameChain) {
        if (VLenSq (dr) < rrNebrA) iType = 2;
      } else if (! sameChain || abs (j1 - j2) > 1) {
        if (VLenSq (dr) < rrNebr) iType = 1;                         15
      }
      if (iType > 0) {
        ...
        nebrTab[2 * nebrTabLen] = j1;
        nebrTab[2 * nebrTabLen + 1] = j2;
        intType[nebrTabLen] = iType;                                 20
        ...
```

where `rCut` is the usual soft-sphere interaction range and `rCutA` that of the LJ interaction. The corresponding modifications to the neighbor-list version of the function `ComputeForces` are

```
  real rrCutA;

  rrCutA = Sqr (rCutA);
  ...
    if (rr < rrCut || intType[n] == 2 && rr < rrCutA) {          5
    rri = 1. / rr;
    rri3 = Cube (rri);
    fcVal = 48. * rri3 * (rri3 - 0.5) * rri;
    uVal = 4. * rri3 * (rri3 - 1.);
    if (intType[n] == 1) uVal += 1.;                             10
    ...
```

Other parts of the calculation, namely, the integration, treatment of periodic boundaries and temperature adjustment, are handled in the usual way. The additional array associated with the neighbor list requires

```
int *intType;

  AllocMem (intType, nebrTabMax, int);
```

Initial state

Assigning the initial coordinates can be carried out in a variety of ways. The following, chosen for its simplicity, demonstrates just one possibility and can be extended to accommodate differing requirements.

As in other cases, atoms are initially positioned on the sites of a lattice, here the simple cubic; since the nominal chain bond length can differ from the lattice spacing, only one of the chain monomers is actually placed on a lattice site and the other monomers are then suitably spaced along the x axis. The chains are positioned first; random locations are chosen in a way that ensures the chains do not overlap and the head–tail direction is randomly chosen. A temporary array `initSiteOcc` is used to flag sites that become occupied in the course of this process[†]. Note that this technique will fail at sufficiently high chain concentration (for a given chain length) if there are no gaps of sufficient size between the chains already in place to allow insertion of any remaining chains. Atom types and chain assignment are also determined; the way atoms are numbered ensures that atoms belonging to a single

---

† Even though only one of the chain atoms is placed on a lattice site, we assume for simplicity that a number of lattice sites equal to the chain length are filled, a correct assumption if the bond length is reasonably close to, but does not exceed, the lattice spacing.

chain are indexed sequentially from head to tail and that chain atoms have smaller indices than solvent and solute atoms. Finally, the remaining lattice sites are filled with O or W atoms, chosen randomly according to the required concentration.

New variables introduced here are

```
real solConc;
int chainHead;
```

for specifying the relative concentration of the O species and the number of mono-mers in the hydrophilic chain head; other quantities are taken from §9.4. Coordinate initialization is as follows; other initialization is carried out as before.

```
void InitCoords ()
{
  VecR c, gap;
  VecI cc;
  int *initSiteOcc, dir, j, n, nc, nn, nx, ny, nz;              5

  AllocMem (initSiteOcc, nMol, int);
  DO_MOL initSiteOcc[n] = 0;
  for (nc = 0; nc < nChain; nc ++) {
    while (1) {                                                 10
      VSet (cc, RandR () * (initUcell.x - chainLen),
        RandR () * initUcell.y, RandR () * initUcell.z);
      n = VLinear (cc, initUcell);
      for (j = 0; j < chainLen; j ++) {
        if (initSiteOcc[n + j]) break;                         15
      }
      if (j == chainLen) {
        for (j = 0; j < chainLen; j ++) initSiteOcc[n + j] = 1;
        break;
      }                                                        20
    }
  }
  VDiv (gap, region, initUcell);
  nc = 0;
  n = 0;                                                       25
  nn = 0;
  for (nz = 0; nz < initUcell.z; nz ++) {
    for (ny = 0; ny < initUcell.y; ny ++) {
      for (nx = 0; nx < initUcell.x; nx ++) {
        VSet (c, nx + 0.5, ny + 0.5, nz + 0.5);                30
        VMul (c, c, gap);
        VVSAdd (c, -0.5, region);
        if (initSiteOcc[nn]) {
          dir = (RandR () < 0.5) ? 0 : 1;
          for (j = 0; j < chainLen; j ++) {                    35
            mol[n].r = c;
            mol[n].r.x += (dir ? j : chainLen - 1 - j) * rCut;
```

```
            mol[n].typeA = (j >= chainHead) ? 1 : 2;
            mol[n].inChain = nc;
            ++ n;                                                          40
          }
          nx += chainLen - 1;
          ++ nc;
          nn += chainLen;
        } else ++ nn;                                                      45
      }
    }
  }
  nn = 0;
  for (nz = 0; nz < initUcell.z; nz ++) {                                  50
    for (ny = 0; ny < initUcell.y; ny ++) {
      for (nx = 0; nx < initUcell.x; nx ++) {
        VSet (c, nx + 0.5, ny + 0.5, nz + 0.5);
        VMul (c, c, gap);
        VVSAdd (c, -0.5, region);                                         55
        if (initSiteOcc[nn]) {
          nx += chainLen - 1;
          nn += chainLen;
        } else {
          mol[n].r = c;                                                    60
          mol[n].typeA = (RandR () < solConc) ? 1 : 2;
          mol[n].inChain = -1;
          ++ n;
          ++ nn;
        }                                                                 65
      }
    }
  }
  free (initSiteOcc);
}                                                                          70
```

Cluster properties

In order to analyze micelle growth we will use the cluster analysis technique introduced in §4.5 to measure the properties of clusters of O atoms; no distinction is made between the O atoms of the solvent and those that are chain monomers. If two O atoms are separated by less than the distance `rClust` they are regarded as belonging to the same cluster. The results of such an analysis are, of course, only relevant to the micelle problem if the O atoms in solution gather together in compact groups surrounded by chains whose tails of O monomers tend to be on the inside; establishing that this is indeed what occurs is not part of this study, although computer generated images show that this is precisely the outcome.

The only change required to the functions used in the earlier cluster analysis (§4.5) is in `BuildClusters`, where the condition for adding a bonded pair is changed to deal only with OO (type 1) atom pairs

```
if ((m1 != m2 || j2 < j1) && mol[j1].typeA == 1 &&
    mol[j2].typeA == 1)
```

Since, for convenience, the same cell array used for neighbor-list construction is also used here, *rClust* must not be allowed to exceed the neighbor shell size, a requirement that is normally satisfied. The input data includes

```
NameR (bondLim),
NameI (chainHead),
NameI (chainLen),
NameI (nChain),
NameR (rClust),
NameR (rCutA),
NameR (solConc),
```

9.7 Surfactant behavior

Micelle growth

In order to ensure adequate space for the molecules to organize themselves into structures that are considerably bigger than the molecules themselves, the surfactant simulations require larger systems than those considered in earlier case studies. The results shown here are for a fairly modest system of 27 000 atoms, with an O concentration of 0.05. The system includes 1600 chains of four monomers, two of which are O atoms and two W; this represents a relatively high surfactant concentration, but it is necessary because the system is really quite small. Periodic boundaries are used and the run is based on data that includes the following:

```
bondLim             2.1
chainHead           2
chainLen            4
density             0.8
initUcell           30 30 30
nChain              1600
rClust              1.8
rCutA               2.
solConc             0.05
stepAdjustTemp      1000
stepAvg             200
stepLimit           200000
stepSnap            2000
temperature         0.7
```

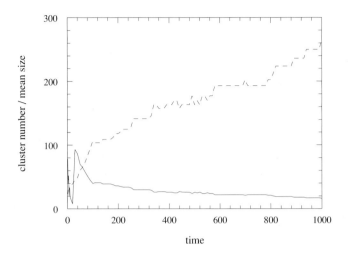

Fig. 9.1. Number of micelle clusters (solid curve) and their mean size (dashed curve) for low solute concentration.

As the run progresses, the system forms a number of compact O regions surrounded by chains that are typically oriented with outward pointing W ends. More detailed aspects of the behavior can depend on the relative concentrations and on other factors such as chain length and interaction strength. Figure 9.1 shows the number of micelles, as reported by the cluster analysis algorithm, together with the mean cluster size, both as functions of time. These results appear consistent with visual observation of the structures. The initial cluster growth from the uniform solution is followed by a certain amount of cluster merging, although the decreasing merge rate makes it difficult to determine whether the final state has been reached in this run; the behavior of the larger structures will obviously be influenced by the size of the system.

Structure formation

While quantitative measurements, such as the cluster analysis, provide some idea of what is happening, there is no better way of examining the behavior than by actually looking at the spatially organized structures as they form.

Recording the coordinates and other relevant data for producing images can employ the function *PutConfig* described in §18.6. The parameters `chainHead`, `chainLen` and `nChain` must be added to the output. Additional information concerning atom types and chain membership must also be included in the configuration snapshot file; such details are needed for creating the images. After writing out the coordinate data, the following code records the atom types, whether each atom belongs to a chain and if so which one,

```
DO_MOL rI[n] = ((mol[n].inChain + 1) << 2) + mol[n].typeA;
WriteFN (rI, nMol);
```

where *WriteFN* (defined in §18.2) writes out the entire array.

These details really need be specified only once at the start of the snapshot file; here we have chosen to have all data blocks identical for simplicity (at the expense of disk storage). Note that since high precision is not required for the graphics, the coordinate data can be scaled to, for example, 10-bit integers for storage, so that a coordinate x,y,z triple can be packed into a single 32-bit integer word; the resulting additional savings in disk storage are likely to be worthwhile when large systems and long runs are contemplated.

The program used to draw the configurations reads this file and displays an animated sequence of frames depicting the evolution of the system. Since these full-color images are based on the complete three-dimensional coordinates, they are able to provide depth perception and can be rotated and zoomed interactively. The details recorded in the file are adequate for the linear chains employed here, assuming the monomers belonging to the chains are indexed as specified earlier; by storing additional information about chain membership and connectivity more general situations can be accommodated. As mentioned previously, details of the actual visualization software are beyond the scope of the discussion, since specialized computer graphics techniques are required; monochrome renditions of the screen images are reproduced here.

The figures show the final states of two systems after 2×10^5 timesteps. The first, in Figure 9.2, is the system considered previously, in which a number of relatively compact micelle clusters develop. The second, in Figure 9.3, shows a different system, in which there are equal concentrations of the O and W species. This results in a state in which each of the species occupies one or more extended regions separated by surfactant chain layers; the layers are far from planar and their form is also affected by the periodic boundaries.

Two kinds of visualization are used for the images. The first is a perspective rendering of stick figures representing the actual surfactant molecules (in the case of low solute concentration, the O atoms are included as points). The second shows the continuous surfaces separating the domains of each species. Such surfaces are produced by evaluating the O and W concentrations over a discrete grid and then, with the aid of interpolation, locating the isosurface on which the concentrations

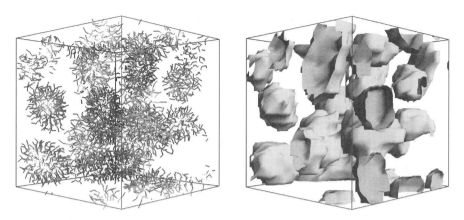

Fig. 9.2. Formation of micelles at low solute concentration: two views of the same system, one showing surfactant chains and solute O atoms (the solvent W atoms are omitted for clarity), the other, the corresponding surfaces delineating the micelles.

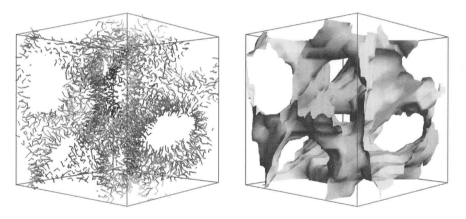

Fig. 9.3. Surfactant layers separating O and W atoms that are present in equal concentrations: one view shows the chains, the other the surfaces between the domains.

are equal (this is analogous to a contour plot, but in three dimensions); the surface roughly follows the chain midpoint locations[†].

These are, of course, many other structures that might develop, depending on the model parameters. Considerable variation can also occur over repeated runs with different initial conditions. The use of even larger systems would provide extra space for structural development.

[†] In principle, the micelle surfaces should be closed; the fact that they do not appear to be so is due to the periodic boundaries – the surfaces are in fact continuous across opposite faces of the simulation cell.

9.8 Further study

9.1   Implement block averaging and determine the simulation lengths needed to ensure convergence of the chain properties.

9.2   The length dependence of $\langle R^2 \rangle$ and $\langle S^2 \rangle$ has been studied extensively for chains on lattices [kre88], and while MD cannot reach the extremely long chains that lattice-based Monte Carlo methods can handle, the results for shorter chains are still of interest; investigate.

9.3   Study the rate at which the chain structure relaxes by examining the time-dependent autocorrelation function of a quantity such as $\langle R^2 \rangle$; relaxation rates are very sensitive to chain length and solvent density [smi92].

9.4   How does the presence of a solvent alter the chain dynamics [dun93]?

9.5   Model a pure polymer liquid; here, reptation is considered to be an important mechanism for molecular motion [kre92]. A suitable initial state must be constructed for this problem.

9.6   Explore how the morphology of the surfactant structures depends on the solute and chain concentrations and on the chain length.

9.7   Devise other ways of constructing initial states for surfactant simulations.

9.8   Generalize the surfactant model to allow for chains whose monomers vary in size from head to tail.

10

Geometrically constrained molecules

10.1 Introduction

Some internal degrees of freedom are important to molecular motion, while others can be regarded as frozen. Classical mechanics allows geometrical relations between coordinates to be included as holonomic constraints. We have already encountered constraints in connection with non-Newtonian modifications of the dynamical equations (Chapter 6); here the constraints occur in a Newtonian context, so that there is little doubt as to the physical nature of the trajectories.

In this chapter we focus on a class of model where constraints play an important role, namely, the polymer models used for studying alkane chains and more complex molecules, in which a combination of geometrical constraints and internal motion is required. The treatment of constraints is not the only new feature of such models; the interactions responsible for bond bending and torsion are essentially three- and four-body potentials, and some rather intricate vector algebra is required to determine the forces. The particular alkane model described here incorporates one further simplification, namely, the use of the often encountered 'united atom' approximation – the hydrogen atoms attached to each carbon atom in the backbone are absorbed into the backbone atoms and are thereby eliminated from the problem.

10.2 Geometric constraints

Role of constraints

The notion of a constraint acting at the molecular level is merely an attempt at simplification; the justification for assuming that certain bond lengths and angles are constant is that, at the prevailing temperature, there is insufficient energy to excite the associated vibrational degrees of freedom (or modes) out of their quantum ground states. Or, adopting a classical perspective, the potential function responsible for limiting the variation of the bond length or angle must involve a very deep and narrow well; the natural frequency associated with such a potential will be much higher than those of other kinds of internal motion and is therefore likely to demand an intolerably small integration timestep. To avoid this situation it is customary to eliminate such degrees of freedom entirely by the simple expedient of replacing them with constraints.

The only unanswered question is whether a completely frozen mode is an accurate way of representing a mode that is really only 'stiff', in the sense that its vibration frequency is much greater than that of other modes and coupling with the rest of the system is weak; there is no completely satisfactory answer since constraints and stiff potentials are both attempts to describe what is fundamentally a quantum problem. The distinction between stiff and frozen modes is important in statistical mechanics, and configurational averages depend on the choice [hel79]; the same is true for dynamical properties [van82].

Problem formulation

Consider a molecule whose structure is subject to one or more geometrical constraints; fixing the distance between any two atoms introduces a constraint of the form

$$|\mathbf{r}_i - \mathbf{r}_j|^2 = b_{ij}^2 \tag{10.2.1}$$

thereby eliminating one degree of freedom. If i and j are bonded neighbors within a molecule, then this constraint amounts to fixing the bond length; if they are next-nearest neighbors, and the two intervening bonds also have constant length, then it is the bond angle that is fixed. While these are examples of replacing stiff interactions between pairs and triplets of atoms, there are other types of structural constraint, such as those used for maintaining the planarity of a molecule; constraints must be formulated with care to ensure the correct selection is made [cic82]. Assuming there are a total of n_c distance constraints imposed on a particular molecule, then if the kth constraint acts between atoms i_k and j_k, the constraints can be summarized by the set of equations

$$\sigma_k \equiv r_{i_k j_k}^2 - b_{i_k j_k}^2 = 0, \qquad k = 1, \dots n_c \tag{10.2.2}$$

For simplicity, the indexing used here considers just a single molecule, but this is readily extended. Note that, because constraints remove degrees of freedom that would otherwise contribute to the temperature, allowance must be made when relating temperature to kinetic energy.

The equations of motion follow directly from the Lagrangian formulation described in §3.2. The result (now allowing for different masses) is

$$m_i \ddot{r}_i = f_i + g_i \qquad (10.2.3)$$

where f_i is the usual force term, m_i the mass of the ith atom, or group of atoms combined into a single monomer, and the additional forcelike term g_i that expresses the effect of the constraints on atom i can be written

$$g_i = -\sum_{k \in \mathcal{C}_i} \lambda_k \nabla_i \sigma_k \qquad (10.2.4)$$

Here, $\mathcal{C}_i$ denotes the set of constraints that directly involve r_i, and the $\{\lambda_k\}$ are the Lagrange multipliers introduced into the problem (the reversed sign in (10.2.4) follows custom [ryc77]). The force f_i includes all non-constraint interactions within the molecule, as well as the intermolecular forces acting on individual atoms (or monomers). There are three scalar equations of motion for each atom, as well as n_c constraint equations for the molecule as a whole, exactly the number needed to evaluate the Lagrange multipliers and integrate the equations of motion.

Solving the problem can be carried out in various ways. A particularly simple method is to advance the system over a single timestep by integrating the unconstrained equations of motion – ignoring g_i – and then adjusting all the coordinates, in practice by only a small amount, so that the constraints are again satisfied in the new state [ryc77]. This adjustment is carried out by means of an iterative relaxation procedure that modifies each pair of constrained coordinates in turn until all constraints are satisfied to the required accuracy. The alternative is to solve the full problem, by first computing the Lagrange multipliers from the time-differentiated constraint equations and then using these values in solving the equations of motion [edb86]. But, unlike the relaxation approach, which restores the constraints to their correct values, here the constraints are subject to numerical integration error. In practice, the error is small and can be corrected by, for example, including an occasional series of relaxation cycles. Both methods will be described below, but first the subject of how to label the atoms and constraints systematically must be addressed.

Atom and constraint indexing

For the linear chain molecules discussed here the indexing problem has a simple solution. For more complex molecular structures, which can involve both tree and ring topologies, the problem is a little more difficult [mor91]. We concentrate

on the case of a simple chain subjected to bond-length constraints and, optionally, to bond-angle constraints as well. Once the constraints have been identified
the remainder of the processing need not be concerned with the topology of the
molecule.

Consider a polymer chain consisting of n_s monomers – atoms for short. If only
the bond lengths are constrained there will be a total of $n_c = n_s - 1$ constraints,
with constraint k relating the coordinates of atoms k and $k + 1$. If, on the other
hand, the chain is subject to both length and angle constraints, there will be $n_s - 1$
of the former and $n_s - 2$ of the latter, so $n_c = 2n_s - 3$. Each of the constraints
acting on atom i then involves one of the four atoms $j = i \pm 1$, $i \pm 2$; length and
angle constraints can be indexed in alternating fashion, leading to the simple result
that the kth constraint acts between atoms $\lfloor (k + 1)/2 \rfloor$ and $\lfloor (k + 4)/2 \rfloor$.

10.3 Solving the constraint problem

Matrix method

Of the two methods, solving the equations of motion together with the constraints
seems to be the more appealing approach from a strictly aesthetic point of view.
This entails expressing the constraint equations in matrix form and then solving
the resulting linear algebra problem using standard numerical techniques. The constraints will of course be subject to numerical error, but if this turns out to be sufficiently small the results can be corrected from time to time using the relaxation
method discussed later in this section; such corrections can also be carried out by,
for example, using standard optimization methods to minimize a penalty function
that measures constraint deviations [edb86].

The constraint forces can be rewritten in the form

$$\mathbf{g}_i = -2 \sum_{k \in \mathcal{C}_i} \lambda_k \mathbf{r}_{i_k j_k} = \sum_{k=1}^{n_c} M_{ik} \lambda_k \mathbf{s}_k \qquad (10.3.1)$$

where

$$\mathbf{s}_k = \mathbf{r}_{\min(i_k,\ j_k)} - \mathbf{r}_{\max(i_k,\ j_k)} \qquad (10.3.2)$$

and the elements of the matrix $\mathbf{M}$, which has n_s rows and n_c columns, are

$$M_{pk} = \begin{cases} +2 & k \in \mathcal{C}_p, \quad j_k < i_k \\ -2 & k \in \mathcal{C}_p, \quad j_k > i_k \\ 0 & k \notin \mathcal{C}_p \end{cases} \qquad (10.3.3)$$

Since s_k^2 is constant, it follows that

$$\ddot{\mathbf{s}}_k \cdot \mathbf{s}_k + \dot{s}_k^2 = 0 \qquad (10.3.4)$$

The acceleration $\ddot{s}_k$ appearing in (10.3.4) can be replaced by the actual equation of motion obtained from (10.2.3). If the indices of the atoms associated with the kth constraint are arranged so that $i_k < j_k$, and we define a new matrix

$$L_{kk'} = \left(M_{i_k k'}/m_{i_k} - M_{j_k k'}/m_{j_k} \right) s_k \cdot s_{k'} \tag{10.3.5}$$

then the result of this replacement is

$$\sum_{k'=1}^{n_c} L_{kk'}\lambda_{k'} = -\left(f_{i_k}/m_{i_k} - f_{j_k}/m_{j_k} \right) \cdot s_k - \dot{s}_k^2, \qquad k = 1,\ldots n_c \tag{10.3.6}$$

The matrix L is of size $n_c \times n_c$ and the only unknowns in (10.3.6) are the $\lambda_{k'}$.

If only bond lengths are constrained there will be exactly two nonzero elements in the ith row of M, corresponding to the two constraints that involve atom i,

$$M_{i,i-1} = +2, \qquad M_{ii} = -2 \tag{10.3.7}$$

If both lengths and angles are constrained there are four nonzero elements per row, namely,

$$M_{i,2i-4} = M_{i,2i-3} = +2, \qquad M_{i,2i-1} = M_{i,2i} = -2 \tag{10.3.8}$$

As an example, the equation of motion of an atom that is subject to both kinds of constraints and not located at the chain ends is

$$\ddot{r}_i = f_i + 2\lambda_{2i-4}r_{i-2,i} + 2\lambda_{2i-3}r_{i-1,i} - 2\lambda_{2i-1}r_{i,i+1} - 2\lambda_{2i}r_{i,i+2} \tag{10.3.9}$$

where we assume that all masses are the same and use MD units. The corresponding equations for the two atoms at either end of the chain omit the terms referring to nonexistent neighbors.

The function that constructs M for the case of bond-length constraints follows; the matrix is stored columnwise as a linear array $mMat^{\dagger}$, $chainLen$ corresponds to n_s and $nCons$ to n_c.

```
void BuildConstraintMatrix ()
{
   int i, m;

   for (i = 0; i < chainLen * nCons; i ++) mMat[i] = 0;            5
   for (i = 0; i < chainLen; i ++) {
      m = i - 1;
      if (m >= 0) mMat[m * chainLen + i] = 2;
      ++ m;
      if (m < nCons) mMat[m * chainLen + i] = -2;                 10
   }
   for (m = 0; m < nCons; m ++) {
```

† `mMat[m * chainLen + i]` corresponds to $M_{i+1,m+1}$; +1 is required because m and i start from zero.

```
      cons[m].distSq = Sqr (bondLen);
      cons[m].site1 = m;
      cons[m].site2 = m + 1;                                        15
  }
}
```

An array of structures `cons` of type

```
typedef struct {
  VecR vec;
  real bLenSq, distSq;
  int site1, site2;
} Cons;                                                             5
```

appears in this function; `vec` holds the constraint vector s_k, `bLenSq` is the current bond length $r^2_{i_k j_k}$, `distSq` corresponds to $b^2_{i_k j_k}$, and `site1` and `site2` denote the atoms i_k and j_k involved in the constraint. The version of the function for constructing M in the case that angle constraints are also present[♠] is

```
void BuildConstraintMatrix ()
{
  int i, m;

  for (i = 0; i < chainLen * nCons; i ++) mMat[i] = 0;             5
  for (i = 0; i < chainLen; i ++) {
    m = 2 * i - 3;
    if (m >= 0) mMat[m * chainLen + i] = 2;
    ++ m;
    if (m >= 0) mMat[m * chainLen + i] = 2;                        10
    m += 2;
    if (m < nCons) mMat[m * chainLen + i] = -2;
    ++ m;
    if (m < nCons) mMat[m * chainLen + i] = -2;
  }                                                                 15
  for (m = 0; m < nCons; m ++) {
    cons[m].distSq = Sqr (bondLen);
    if (m % 2 == 1) cons[m].distSq *= 2. * (1. - cos (bondAng));
    cons[m].site1 = m / 2;
    cons[m].site2 = (m + 3) / 2;                                   20
  }
}
```

Evaluating the Lagrange multipliers and including the constraint forces in the equations of motion is the task of the function that follows. In the course of the processing the matrix L, represented by `lMat`, is constructed and the linear equations (10.3.6) solved using a standard method such as LU decomposition [pre92];

♠ pr_10_1

the solution is evaluated by the function $SolveLineq$ described in §18.4. Both the right-hand side of (10.3.6) and, subsequently, the solution (the set of $\lambda_{k'}$) are stored in $consVec$.

```
void ComputeConstraints ()
{
  VecR da, dv;
  real w;
  int i, m, m1, m2, mDif, n, nn;                              5

  for (n = 0; n < nChain; n ++) {
    nn = n * chainLen;
    for (m = 0; m < nCons; m ++) {
      VSub (cons[m].vec, mol[nn + cons[m].site1].r,         10
          mol[nn + cons[m].site2].r);
      VWrapAll (cons[m].vec);
    }
    m = 0;
    for (m1 = 0; m1 < nCons; m1 ++) {                        15
      for (m2 = 0; m2 < nCons; m2 ++) {
        lMat[m] = 0.;
        mDif = mMat[m1 * chainLen + cons[m2].site1] -
          mMat[m1 * chainLen + cons[m2].site2];
        if (mDif != 0)                                       20
          lMat[m] = mDif * VDot (cons[m1].vec, cons[m2].vec);
        ++ m;
      }
    }
    for (m = 0; m < nCons; m ++) {                           25
      VSub (dv, mol[nn + cons[m].site1].rv,
          mol[nn + cons[m].site2].rv);
      VSub (da, mol[nn + cons[m].site1].ra,
          mol[nn + cons[m].site2].ra);
      consVec[m] = - VDot (da, cons[m].vec) - VLenSq (dv);   30
    }
    SolveLineq (lMat, consVec, nCons);
    for (m = 0; m < nCons; m ++) {
      for (i = 0; i < chainLen; i ++) {
        w = mMat[m * chainLen + i];                          35
        if (w != 0.)
          VVSAdd (mol[nn + i].ra, w * consVec[m], cons[m].vec);
      }
    }
  }                                                          40
}
```

Any residual drift in the constraints can be removed when necessary, as discussed later, but the drift should be sufficiently small that such adjustments are infrequent.

A list of the new quantities appearing in the calculations, including some used later, is

```
Cons *cons;
real *consVec, *lMat, bondAng, bondLen, consDevA, consDevL, consPrec;
int *mMat, chainLen, nChain, nCons, nCycleR, nCycleV, stepRestore;
```

and the related array allocations (in *AllocArrays*) are

```
AllocMem (cons, nCons, Cons);
AllocMem (consVec, nCons, real);
AllocMem (lMat, Sqr (nCons), real);
AllocMem (mMat, chainLen * nCons, int);
```

The deviations of the supposedly constrained bond lengths from the correct values are easily monitored (here only averages are computed).

```
void AnlzConstraintDevs ()
{
  VecR dr1;
  real sumL;
  int i, n, ni;                                              5

  sumL = 0.;
  for (n = 0; n < nChain; n ++) {
    for (i = 0; i < chainLen - 1; i ++) {
      ni = n * chainLen + i;                                10
      VSub (dr1, mol[ni + 1].r, mol[ni].r);
      VWrapAll (dr1);
      cons[i].bLenSq = VLenSq (dr1);
      sumL += cons[i].bLenSq;
    }                                                        15
  }
  consDevL = sqrt (sumL / (nChain * (chainLen - 1))) - bondLen;
}
```

If bond angles are also constrained add

```
  VecR dr2;
  real sumA;
  ...
  sumA = 0.;
  ...                                                        5
    for (i = 1; i < chainLen - 1; i ++) {
      ni = n * chainLen + i;
      VSub (dr1, mol[ni + 1].r, mol[ni].r);
      VWrapAll (dr1);
      VSub (dr2, mol[ni - 1].r, mol[ni].r);                 10
      VWrapAll (dr2);
```

```
        sumA += Sqr (VDot (dr1, dr2)) / (cons[i - 1].bLenSq *
            cons[i].bLenSq);
    }
    ...
    consDevA = sqrt (sumA / (nChain * (chainLen - 2))) -
        cos (M_PI - bondAng);
```

15

Relaxation method

This approach to dealing with constraints – the so-called 'shake' method – [ryc77] begins by advancing the system over a single timestep while ignoring the constraints. If the simple Verlet integration method (3.5.2) is used, we obtain a set of uncorrected coordinates

$$r_i'(t + h) = 2\,r_i(t) - r_i(t - h) + (h^2/m_i)\,f_i(t) \tag{10.3.10}$$

We now want to adjust all the r_i' to obtain corrected coordinates r_i that satisfy the constraints. This can be done by adding in the missing constraint force term (10.2.4); since

$$\nabla_i \sigma_k = 2\,r_{ij}(t) \tag{10.3.11}$$

this leads to

$$r_i(t + h) = r_i'(t + h) - 2(h^2/m_i) \sum_{k \in \mathcal{C}_i} \lambda_k r_{ij}(t) \tag{10.3.12}$$

At this point we change the meaning of λ_k. It will no longer be regarded as a Lagrange multiplier, but rather as an additional variable whose value is determined by having the constraint satisfied to full numerical accuracy and not subject to the truncation error of the integration method. This will ensure that, despite the numerical error experienced by the atomic trajectories, the constrained bond lengths and angles always maintain their correct values.

Implementation of the iterative method begins by setting $r_i' = r_i'(t + h)$ and then, for each constraint, applying corrections along the direction of $r_{ij}(t)$,

$$\begin{aligned} r_i'' &= r_i' - 2(h^2/m_i)\gamma\,r_{ij}(t) \\ r_j'' &= r_j' + 2(h^2/m_j)\gamma\,r_{ij}(t) \end{aligned} \tag{10.3.13}$$

The correction factor γ is determined from the solution of

$$r_{ij}''^{\,2} = b_{ij}^2 \tag{10.3.14}$$

namely,

$$|r_{ij}' - 2h^2(1/m_i + 1/m_j)\gamma\,r_{ij}|^2 = b_{ij}^2 \tag{10.3.15}$$

which, to lowest order in h^2, is

$$\gamma = \frac{r_{ij}'^2 - b_{ij}^2}{4h^2(1/m_i + 1/m_j)r_{ij}' \cdot r_{ij}} \tag{10.3.16}$$

The estimated coordinates r_i' and r_j' are then updated by using this value of γ in (10.3.13). The process is repeated, cycling through each of the constraints in turn, until all the constraints satisfy

$$|r_{ij}'^2 - b_{ij}^2| < \epsilon_r b_{ij}^2 \tag{10.3.17}$$

where ϵ_r is the specified tolerance[†].

As shown here, the method is tied to a specific integration method, but a very similar result can be used for restoring the constraints in general. Simply write

$$\begin{aligned} r_i' &= r_i - \gamma r_{ij} \\ r_j' &= r_j + \gamma r_{ij} \end{aligned} \tag{10.3.18}$$

where γ is now just a small number, and then solve the equations

$$r_{ij}'^2 = b_{ij}^2 \tag{10.3.19}$$

iteratively as before; here we assume that all atoms have the same mass, otherwise the inverse mass terms must be included to avoid moving the center of mass. If terms quadratic in γ are neglected, the solution is reminiscent of (10.3.16),

$$\gamma = \frac{r_{ij}^2 - b_{ij}^2}{4\,r_{ij}^2} \tag{10.3.20}$$

and b_{ij}^2 can replace r_{ij}^2 in the denominator.

The velocities can be corrected in a similar manner, thereby ensuring that the atoms have zero relative velocity along the direction of their mutual constraint; corrections of this kind can also be incorporated in the original 'shake' method [and83]. Each such restriction is equivalent to

$$\dot{\sigma}_k = 2\,\dot{r}_{ij} \cdot r_{ij} = 0 \tag{10.3.21}$$

Following the same approach as before, the velocities are adjusted by iterating the equations

$$\begin{aligned} \dot{r}_i' &= \dot{r}_i - \gamma r_{ij} \\ \dot{r}_j' &= \dot{r}_j + \gamma r_{ij} \end{aligned} \tag{10.3.22}$$

where the value of γ is now chosen to ensure that $\dot{r}_{ij}' \cdot r_{ij} = 0$,

$$\gamma = \frac{\dot{r}_{ij} \cdot r_{ij}}{2\,r_{ij}^2} \tag{10.3.23}$$

[†] The value of λ_k is just the sum of all the γ corrections for that constraint, but it is not needed in the calculation.

Since r_{ij} already satisfies the constraint, b_{ij}^2 can be used in the denominator. The process is repeated until all corrections fall below a specified tolerance.

The function for restoring the coordinates and velocities to their constrained values in this more general case is

```
void RestoreConstraints ()
{
  VecR dr, dv;
  real cDev, cDevR, cDevV, g, ga;
  int changed, m, m1, m2, maxCycle, n;                              5

  maxCycle = 200;
  cDevR = cDevV = 0.;
  for (n = 0; n < nChain; n ++) {
    nCycleR = 0;                                                    10
    changed = 1;
    while (nCycleR < maxCycle && changed) {
      ++ nCycleR;
      changed = 0;
      cDev = 0.;                                                    15
      for (m = 0; m < nCons; m ++) {
        m1 = n * chainLen + cons[m].site1;
        m2 = n * chainLen + cons[m].site2;
        VSub (dr, mol[m1].r, mol[m2].r);
        VWrapAll (dr);                                              20
        g = (VLenSq (dr) - cons[m].distSq) / (4. * cons[m].distSq);
        ga = fabs (g);
        cDev = Max (cDev, ga);
        if (ga > consPrec) {
          changed = 1;                                              25
          VVSAdd (mol[m1].r, - g, dr);
          VVSAdd (mol[m2].r, g, dr);
        }
      }
    }                                                               30
    cDevR = Max (cDevR, cDev);
    nCycleV = 0;
    changed = 1;
    while (nCycleV < maxCycle && changed) {
      ++ nCycleV;                                                   35
      changed = 0;
      cDev = 0.;
      for (m = 0; m < nCons; m ++) {
        m1 = n * chainLen + cons[m].site1;
        m2 = n * chainLen + cons[m].site2;                          40
        VSub (dr, mol[m1].r, mol[m2].r);
        VWrapAll (dr);
        VSub (dv, mol[m1].rv, mol[m2].rv);
        g = VDot (dv, dr) / (2. * cons[m].distSq);
        ga = fabs (g);                                             45
        cDev = Max (cDev, ga);
```

```
        if (ga > consPrec) {
          changed = 1;
          VVSAdd (mol[m1].rv, - g, dr);
          VVSAdd (mol[m2].rv, g, dr);                    50
        }
      }
    }
    cDevV = Max (cDevV, cDev);
  }                                                      55
}
```

Here, `consPrec` is the tolerance used in establishing convergence. The limit `maxCycle` is introduced as a safety measure; the number of iterations should generally not be much greater than ten or so (for small molecules), otherwise one might be inclined to suspect the reliability of the whole approach.

10.4 Internal forces

Bond-torsion force

The torsional force associated with twisting around a bond is another example of an effective interaction. This particular motion provides the means for local changes in spatial arrangement of the polymer chain; simultaneous twisting around two bonds, for example, can be enough to provide a crankshaft form of motion. The force associated with the twist, or torsional degree of freedom, is defined in terms of the relative coordinates of four consecutive atoms, here, for convenience, labeled 1...4; this force depends on the angle of rotation around the bond between atoms 2 and 3 – the dihedral angle. The dihedral angle is defined as the angle between the planes formed by atoms 1,2,3 and 2,3,4 measured in the plane normal to the 2–3 bond; it is zero when all four atoms are coplanar and atoms 1 and 4 are on opposite sides of the bond. Only these four atoms are directly affected by the torsion due to this bond, and the purpose of the following analysis is to determine the force on each.

Labeling the atoms, bonds and angles of a linear polymer in a systematic manner is trivial for linear chains; for molecules with other topologies the problem is more complex, requiring an algorithm (not addressed here) that systematically traverses the graph describing the connectivity of the molecule. Here bond i joins atoms $i - 1$ and i, and is denoted by the vector

$$b_i = r_i - r_{i-1} \tag{10.4.1}$$

As shown in Figure 10.1, the angle between bonds $i - 1$ and i is given by

$$\cos \alpha_i = \frac{b_{i-1} \cdot b_i}{|b_{i-1}||b_i|} \tag{10.4.2}$$

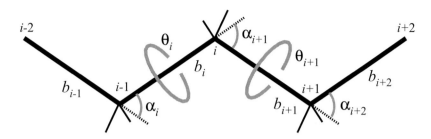

Fig. 10.1. The sites, bonds, bond angles and dihedral angles in a portion of an alkane chain.

so that $\alpha_i = 0$ if the bonds are parallel; by convention, the 'bond angle' refers to $\pi - \alpha_i$. The dihedral angle associated with bond i is obtained from

$$\cos \theta_i = -\frac{(\boldsymbol{b}_{i-1} \times \boldsymbol{b}_i) \cdot (\boldsymbol{b}_i \times \boldsymbol{b}_{i+1})}{|\boldsymbol{b}_{i-1} \times \boldsymbol{b}_i||\boldsymbol{b}_i \times \boldsymbol{b}_{i+1}|} \qquad (10.4.3)$$

There are two parts to the torsional force calculation; the functional dependence on the dihedral angle θ_i, and the vector algebra used to derive expressions for the forces on each of the four affected atoms. We begin with the second part [pea79, dun92]. If we define

$$c_{ij} = \boldsymbol{b}_i \cdot \boldsymbol{b}_j \qquad (10.4.4)$$

then we can express the bond and dihedral angles as

$$\cos \alpha_i = c_{i-1,i}/(c_{i-1,i-1}c_{ii})^{1/2} \qquad (10.4.5)$$

$$\cos \theta_i = p_i/q_i^{1/2} \qquad (10.4.6)$$

where, for conciseness, we have introduced the quantities

$$p_i = c_{i-1,i+1}c_{ii} - c_{i-1,i}c_{i,i+1} \qquad (10.4.7)$$

$$q_i = \left(c_{i-1,i-1}c_{ii} - c_{i-1,i}^2\right)\left(c_{ii}c_{i+1,i+1} - c_{i,i+1}^2\right) \qquad (10.4.8)$$

The torque caused by a rotation about bond i produces forces on the four atoms $j = i - 2, \ldots i + 1$ equal to

$$-\nabla_{\boldsymbol{r}_j} u(\theta_i) = -\frac{du(\theta)}{d(\cos \theta)}\bigg|_{\theta = \theta_i} \boldsymbol{f}_j^{(i)} \qquad (10.4.9)$$

where $u(\theta)$ is the torsion potential and

$$\boldsymbol{f}_j^{(i)} = \nabla_{\boldsymbol{r}_j} \cos \theta_i \qquad (10.4.10)$$

It is clear that the sums of the forces and torques acting on the four atoms are zero, therefore

$$\sum_{j=i-2}^{i+1} f_j^{(i)} = 0 \tag{10.4.11}$$

$$\sum_{j=i-2}^{i+1} r_j \times f_j^{(i)} = 0 \tag{10.4.12}$$

so that

$$(\boldsymbol{b}_{i-1} + \boldsymbol{b}_i) \times \boldsymbol{f}_{i-2}^{(i)} + \boldsymbol{b}_i \times \boldsymbol{f}_{i-1}^{(i)} - \boldsymbol{b}_{i+1} \times \boldsymbol{f}_{i+1}^{(i)} = 0 \tag{10.4.13}$$

Since (10.4.11) and (10.4.12) provide two relations between the four $\boldsymbol{f}_j^{(i)}$ we can write

$$\boldsymbol{f}_{i-1}^{(i)} = \beta_1 \boldsymbol{f}_{i-2}^{(i)} + \beta_2 \boldsymbol{f}_{i+1}^{(i)} \tag{10.4.14}$$

so that (10.4.13) becomes

$$(\boldsymbol{b}_{i-1} + \boldsymbol{b}_i + \beta_1 \boldsymbol{b}_i) \times \boldsymbol{f}_{i-2}^{(i)} + (\beta_2 \boldsymbol{b}_i - \boldsymbol{b}_{i+1}) \times \boldsymbol{f}_{i+1}^{(i)} = 0 \tag{10.4.15}$$

and because both $\boldsymbol{f}_{i-2}^{(i)}$ and $\boldsymbol{f}_{i+1}^{(i)}$ are normal to $\boldsymbol{b}_i$ it follows that

$$\begin{aligned} \beta_1 &= -1 - c_{i-1,i}/c_{ii} \\ \beta_2 &= c_{i,i+1}/c_{ii} \end{aligned} \tag{10.4.16}$$

Hence,

$$\boldsymbol{f}_{i-1}^{(i)} = -(1 + c_{i-1,i}/c_{ii}) \, \boldsymbol{f}_{i-2}^{(i)} + (c_{i,i+1}/c_{ii}) \, \boldsymbol{f}_{i+1}^{(i)} \tag{10.4.17}$$

and, since the four $\boldsymbol{f}_j^{(i)}$ sum to zero,

$$\boldsymbol{f}_i^{(i)} = (c_{i-1,i}/c_{ii}) \, \boldsymbol{f}_{i-2}^{(i)} - (1 + c_{i,i+1}/c_{ii}) \, \boldsymbol{f}_{i+1}^{(i)} \tag{10.4.18}$$

We next evaluate $\boldsymbol{f}_{i-2}^{(i)}$ and $\boldsymbol{f}_{i+1}^{(i)}$ by expanding (10.4.10),

$$\boldsymbol{f}_j^{(i)} = q_i^{-3/2} \left(q_i \nabla_{r_j} p_i - p_i \nabla_{r_j} q_i / 2 \right) \tag{10.4.19}$$

In order to complete the evaluation we need the derivatives of all the scalar products $\boldsymbol{b}_\alpha \cdot \boldsymbol{b}_\beta$ with respect to $\boldsymbol{r}_j$; the full list (apart from results due to the symmetry

$c_{\alpha\beta} = c_{\beta\alpha}$) is

$$\nabla_{r_j} c_{\alpha\beta} = \begin{cases} 2\boldsymbol{b}_j & \alpha = \beta = j \\ -2\boldsymbol{b}_{j+1} & \alpha = \beta = j+1 \\ \boldsymbol{b}_{j+1} - \boldsymbol{b}_j & \alpha = j, \quad \beta = j+1 \\ \boldsymbol{b}_\beta & \alpha = j, \quad \beta \neq j, j+1 \\ -\boldsymbol{b}_\beta & \alpha = j+1, \quad \beta \neq j, j+1 \\ 0 & \text{otherwise} \end{cases} \tag{10.4.20}$$

A certain amount of algebra using the above results leads to

$$f_{i-2}^{(i)} = \frac{c_{ii}}{q_i^{1/2}(c_{i-1,i-1}c_{ii} - c_{i-1,i}^2)} \left[t_1 \boldsymbol{b}_{i-1} + t_2 \boldsymbol{b}_i + t_3 \boldsymbol{b}_{i+1} \right] \tag{10.4.21}$$

$$f_{i+1}^{(i)} = \frac{c_{ii}}{q_i^{1/2}(c_{ii}c_{i+1,i+1} - c_{i,i+1}^2)} \left[t_4 \boldsymbol{b}_{i-1} + t_5 \boldsymbol{b}_i + t_6 \boldsymbol{b}_{i+1} \right] \tag{10.4.22}$$

where

$$\begin{aligned} t_1 &= c_{i-1,i+1}c_{ii} - c_{i-1,i}c_{i,i+1} \\ t_2 &= c_{i-1,i-1}c_{i,i+1} - c_{i-1,i}c_{i-1,i+1} \\ t_3 &= c_{i-1,i}^2 - c_{i-1,i-1}c_{ii} \\ t_4 &= c_{ii}c_{i+1,i+1} - c_{i,i+1}^2 \\ t_5 &= c_{i-1,i+1}c_{i,i+1} - c_{i-1,i}c_{i+1,i+1} \\ t_6 &= -t_1 \end{aligned} \tag{10.4.23}$$

Both force vectors are normal to $\boldsymbol{b}_i$, although this may be less than obvious from an expression such as (10.4.21). In certain cases considerable simplification is possible; if, for example, all $|\boldsymbol{b}_i| = b$ and $\alpha_i = \alpha$, then

$$\cos\theta_i = \frac{\boldsymbol{b}_{i-1} \cdot \boldsymbol{b}_{i+1}/b^2 - \cos^2\alpha}{1 - \cos^2\alpha} \tag{10.4.24}$$

The force expressions are simplified, but the reduction in computational effort is probably not large enough to justify separate functions for individual cases.

The torsional potential function is typically expressed in polynomial form [ryc78],

$$u(\theta) = \sum_{j \geq 0} w_j \cos^j \theta \tag{10.4.25}$$

so that the derivative used for the forces in (10.4.9) is

$$-\frac{du(\theta)}{d(\cos\theta)} = -\sum_{j \geq 1} j w_j \cos^{j-1} \theta \tag{10.4.26}$$

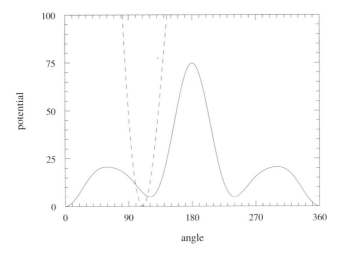

Fig. 10.2.   Bond torsion (solid curve) and bond angle (dashed curve) potentialfunctions.

For the alkane model, the potential (whose coefficients, in energy units appropriate to the problem, are incorporated into the function listed below) is shown in
Figure 10.2. The deepest minimum is at the 'trans' angle $\theta = 0$, two secondary
minima at the 'gauche' angles ($\pm 2\pi/3$), barriers at $\theta = \pm\pi/3$ and a maximum at
π. Other similar functions are also in use for this model [cla90].

The torsional contribution to the interactions is computed by the following
function.

```
void ComputeChainTorsionForces ()
{
  VecR dr1, dr2, dr3, w1, w2;
  real c, c11, c12, c13, c22, c23, c33, ca, cb1, cb2, cd,
      cr1, cr2, f, t1, t2, t3, t4, t5, t6,                          5
      g[6] = {1.000, 1.310, -1.414, -0.330, 2.828, -3.394},
      tCon = 15.50;
  int i, n, nn;

  for (n = 0; n < nChain; n ++) {                                   10
    for (i = 0; i < chainLen - 3; i ++) {
      nn = n * chainLen + i;
      VSub (dr1, mol[nn + 1].r, mol[nn].r);
      VWrapAll (dr1);
      VSub (dr2, mol[nn + 2].r, mol[nn + 1].r);                     15
      VWrapAll (dr2);
      VSub (dr3, mol[nn + 3].r, mol[nn + 2].r);
      VWrapAll (dr3);
```

```
c11 = VLenSq (dr1);
c12 = VDot (dr1, dr2);                                              20
c13 = VDot (dr1, dr3);
c22 = VLenSq (dr2);
c23 = VDot (dr2, dr3);
c33 = VLenSq (dr3);
ca = c13 * c22 - c12 * c23;                                        25
cb1 = c11 * c22 - c12 * c12;
cb2 = c22 * c33 - c23 * c23;
cd = sqrt (cb1 * cb2);
c = ca / cd;
f = - tCon * (g[1] + (2. * g[2] + (3. * g[3] +                     30
    (4. * g[4] + 5. * g[5] * c) * c) * c) * c);
t1 = ca;
t2 = c11 * c23 - c12 * c13;
t3 = - cb1;
t4 = cb2;                                                          35
t5 = c13 * c23 - c12 * c33;
t6 = - ca;
cr1 = c12 / c22;
cr2 = c23 / c22;
VSSAdd (w1, t1, dr1, t2, dr2);                                     40
VVSAdd (w1, t3, dr3);
VScale (w1, f * c22 / (cd * cb1));
VSSAdd (w2, t4, dr1, t5, dr2);
VVSAdd (w2, t6, dr3);
VScale (w2, f * c22 / (cd * cb2));                                 45
VVAdd (mol[nn].ra, w1);
VVSAdd (mol[nn + 1].ra, - (1. + cr1), w1);
VVSAdd (mol[nn + 1].ra, cr2, w2);
VVSAdd (mol[nn + 2].ra, cr1, w1);
VVSAdd (mol[nn + 2].ra, - (1. + cr2), w2);                         50
VVAdd (mol[nn + 3].ra, w2);
uSum += tCon * (g[0] + (g[1] + (g[2] + (g[3] +
    (g[4] + g[5] * c) * c) * c) * c) * c);
   }
  }                                                                55
}
```

Bond angle force

While bond lengths are generally held fixed by constraints, there is no clear preference for bond angles, and both constraints and potentials are in use [cla90, ryc90]. Here we treat the case where interactions ensure that bond angles have only limited variation (the angle and torsion forces are assumed fully independent). The notation is the same as before.

A change in the angle α_i produces forces on the three atoms $j = i - 2, i - 1, i$ given by

$$-\nabla_{r_j} u(\alpha_i) = -\left.\frac{du(\alpha)}{d(\cos\alpha)}\right|_{\alpha=\alpha_i} f_j^{(i)} \qquad (10.4.27)$$

where $u(\alpha)$ is the angle potential and

$$f_j^{(i)} = \nabla_{r_j} \cos\alpha_i \qquad (10.4.28)$$

The sum of the three forces is zero. More of the above algebra leads to

$$f_{i-2}^{(i)} = (c_{i-1,i-1}c_{ii})^{-1/2}\left[(c_{i-1,i}/c_{i-1,i-1})b_{i-1} - b_i\right] \qquad (10.4.29)$$

$$f_i^{(i)} = (c_{i-1,i-1}c_{ii})^{-1/2}\left[b_{i-1} - (c_{i-1,i}/c_{ii})b_i\right] \qquad (10.4.30)$$

The potential associated with bond angle variation for the alkane model is

$$u(\alpha) = (w/2)(\cos\alpha - \cos\alpha_0)^2 \qquad (10.4.31)$$

where w is a constant [cla90] and $\cos\alpha_0 = 1/3$; a plot of the potential function is included in Figure 10.2.

The function that carries out the force and energy computations for this interaction follows.

```
void ComputeChainAngleForces ()
{
  VecR dr1, dr2, w1, w2;
  real c, c11, c12, c22, cCon, cd, f, aCon = 868.6;
  int i, n, nn;                                              5

  cCon = cos (M_PI - bondAng);
  for (n = 0; n < nChain; n ++) {
    for (i = 0; i < chainLen - 2; i ++) {
      nn = n * chainLen + i;                                 10
      VSub (dr1, mol[nn + 1].r, mol[nn].r);
      VWrapAll (dr1);
      VSub (dr2, mol[nn + 2].r, mol[nn + 1].r);
      VWrapAll (dr2);
      c11 = VLenSq (dr1);                                    15
      c12 = VDot (dr1, dr2);
      c22 = VLenSq (dr2);
      cd = sqrt (c11 * c22);
      c = c12 / cd;
      f = - aCon * (c - cCon);                               20
      VSSAdd (w1, c12 / c11, dr1, -1., dr2);
      VScale (w1, f / cd);
      VSSAdd (w2, 1., dr1, - c12 / c22, dr2);
      VScale (w2, f / cd);
      VVAdd (mol[nn].ra, w1);                                25
      VVSub (mol[nn + 1].ra, w1);
```

```
      VVSub (mol[nn + 1].ra, w2);
      VVAdd (mol[nn + 2].ra, w2);
      uSum += 0.5 * aCon * Sqr (c - cCon);
    }
  }
}
```
30

Other interactions

So far we have discussed just two of the interactions in the model, namely, the bond torsion and bond angle forces. Pairs of atoms in each molecule that are neither directly linked by a constraint, nor jointly involved in these three- and four-body forces, interact with the usual LJ potential (the butane molecule studied later on is sufficiently small that there are no pairs in this category). Atoms in different molecules interact with the same force, and solvent atoms can also be included with similar or distinct interaction parameters, depending on what is being modeled; here, for faster computation, the LJ interactions are replaced by soft spheres [tox88].

If the neighbor-list method is used for computing the interactions between pairs of atoms not involved in constraints or bond forces, the only change required in *BuildNebrList* is the elimination of such pairs. This is done by modifying the condition used to select atom pairs for the list,

```
    if ((m1 != m2 || j2 < j1) && (mol[j1].inChain == -1 ||
        mol[j1].inChain != mol[j2].inChain || abs (j1 - j2) > 3))
```

The additional test checks whether both atoms belong to the same molecule (the element *inChain* is used in the same way as for flexible chains, see §9.3) and if this is true then how far apart they are. The interaction functions called from *SingleStep* are (the first of them only if relevant)

```
    ComputeChainAngleForces ();
    ComputeChainTorsionForces ();
    ComputeConstraints ();
```

Adjustment of minor constraint deviations is carried out at suitable intervals by

```
    nCycleR = 0;
    nCycleV = 0;
    if (stepCount % stepRestore == 0) {
      RestoreConstraints ();
      ApplyBoundaryCond ();
    }
```
5

To obtain reports on how well the constraints are preserved, add the following to *SingleStep* prior to any call to *RestoreConstraints*,

```
if (stepCount % stepAvg == 0) AnlzConstraintDevs ();
```

and to *PrintSummary* add

```
fprintf (fp, "constraint devs: %.3e %.3e cycles: %d %d\n",
    consDevL, consDevA, nCycleR, nCycleV);
```

The function *RestoreConstraints* should also be called at the beginning of the run to correct the randomly assigned initial velocities.

10.5 Implementation details

Initial state and parameters

The initial state uses the same BCC lattice arrangement and planar zigzag (or trans) conformation used previously for flexible chains (§9.3) but with distances and angles (in *InitCoords*) modified,

```
by = bondLen * cos (bondAng / 2.);
bz = bondLen * sin (bondAng / 2.);
```

New input data items are

```
NameR (bondAng),
NameR (bondLen),
NameI (chainLen),
NameR (consPrec),
NameI (initUchain),
NameI (stepRestore),
```

and initialization (*SetupJob*) requires

```
AssignToChain ();
BuildConstraintMatrix ();
```

No solvent is used here, so the values of *initUchain* replace *initUcell* when determining the region size. In the case of a simulation involving both multiple chains and solvent there will be two independent densities – for chains and for solvent atoms – that must be specified.

The effect of the constraints must be taken into account when choosing initial velocities that correspond to a given temperature. The total number of degrees of freedom per chain is reduced from $3n_s$ to $2n_s + 1$ in the case of bond-length constraints, and to $n_s + 3$ if bond angles are also constrained. Allowance for this, and the loss of three more degrees of freedom because of momentum conservation, are incorporated in the quantity *velMag* evaluated in *SetParams* (the value of *nCons* shown here is for the case of length and angle constraints),

```
nCons = 2 * chainLen - 3;
velMag = sqrt ((NDIM * (1. - 1. / nMol) - (real) nCons / chainLen) *
    temperature);
```

Temperature adjustment early in the run uses the function *InitAdjustTemp*.

The reduced length and energy units [ryc78] are $\sigma = 3.92\,\text{Å}$ and $\epsilon/k_B = 72\,\text{K}$. All atoms (or monomers) are assumed to have mass 2.411×10^{-23} g, and this is defined as the unit of mass in MD units. The unit of time is then 1.93×10^{-12} s. The bond length of 1.53 Å used in the model corresponds to 0.390; the bond angle is 109.47°. At the density of liquid butane (0.675 g/cm³) there are 0.422 molecules per unit volume, again in MD units.

Structural properties

Properties of the chain fluid as a whole can be studied using the atomic RDF, as in §4.3. An extra test is needed in *EvalRdf* to eliminate the very sharp peaks at the fixed distance between nearest neighbors, and at either the fixed or narrowly spread next-nearest neighbor distance,

```
if (mol[j1].inChain == mol[j2].inChain &&
    mol[j1].inChain != -1 && abs (j1 - j2) < 3) continue;
```

The first example♠ of a measurement specific to this chain model constructs a normalized histogram of the dihedral angle distribution averaged over all chains and over all angles in each chain. If there is some reason to believe that the distribution depends on where the bond is located in the chain (not for the example studied here), then the results for each bond would have to be maintained separately. Usage of this function follows the familiar pattern established in earlier case studies; *Sgn* is defined in §18.2.

```
void AccumDihedAngDistn (int icode)
{
  VecR dr1, dr2, dr3, w;
  real c11, c12, c13, c22, cosAngSq, dihedAng, t;
  int i, j, n, nn;                                          5
```

♠ pr_10_2

```
   if (icode == 0) {
     for (j = 0; j < sizeHistDihedAng; j ++) histDihedAng[j] = 0.;
   } else if (icode == 1) {
     for (n = 0; n < nChain; n ++) {                                  10
       for (i = 0; i < chainLen - 3; i ++) {
         nn = n * chainLen + i;
         VSub (dr1, mol[nn + 1].r, mol[nn].r);
         VWrapAll (dr1);
         VSub (dr2, mol[nn + 2].r, mol[nn + 1].r);                    15
         VWrapAll (dr2);
         VSub (dr3, mol[nn + 3].r, mol[nn + 2].r);
         VWrapAll (dr3);
         c11 = VLenSq (dr1);
         c12 = VDot (dr1, dr2);                                       20
         c13 = VDot (dr1, dr3);
         c22 = VLenSq (dr2);
         cosAngSq = Sqr (c12) / (c11 * c22);
         t = (c13 / Sqr (bondLen) - cosAngSq) / (1. - cosAngSq);
         if (fabs (t) > 1.) t = Sgn (1., t);                         25
         dihedAng = acos (t);
         VCross (w, dr2, dr3);
         if (VDot (dr1, w) < 0.) dihedAng = 2. * M_PI - dihedAng;
         j = dihedAng * sizeHistDihedAng / (2. * M_PI);
         ++ histDihedAng[j];                                         30
       }
     }
   } else if (icode == 2) {
     t = 0.;
     for (j = 0; j < sizeHistDihedAng; j ++) t += histDihedAng[j];   35
     for (j = 0; j < sizeHistDihedAng; j ++) histDihedAng[j] /= t;
   }
 }
```

The function is called from `SingleStep` by

```
   if (stepCount >= stepEquil && (stepCount - stepEquil) %
       stepChainProps == 0) AccumDihedAngDistn (1);
```

New variables, input data items and array allocation are

```
real *histDihedAng;
int sizeHistDihedAng, stepChainProps;

  NameI (sizeHistDihedAng),
  NameI (stepChainProps),                                            5

  AllocMem (histDihedAng, sizeHistDihedAng, real);
```

The results are output as part of *PrintSummary* (§2.3),

```
real hVal;
int n;
...
fprintf (fp, "dihed ang\n");
for (n = 0; n < sizeHistDihedAng; n ++) {                          5
  hVal = (n + 0.5) * 360. / sizeHistDihedAng;
  fprintf (fp, "%5.1f %.4f\n", hVal, histDihedAng[n]);
}
```

which is called (in *SingleStep*) as part of the sequence

```
AccumDihedAngDistn (2);
PrintSummary (stdout);
AccumDihedAngDistn (0);
```

to normalize the accumulated results prior to their output and zero them afterwards.

The next example[♠] considers the bond angle distribution, and is obviously only relevant when bond angles are controlled by a potential instead of constraints. The computation is very similar to the preceding one (with the appropriate additional variables and other details).

```
void AccumBondAngDistn (int icode)
{
  VecR dr1, dr2;
  real bondAng, c11, c12, c22, t;
  int i, n, j, nn;                                                 5

  if (icode == 0) {
    for (j = 0; j < sizeHistBondAng; j ++) histBondAng[j] = 0.;
  } else if (icode == 1) {
    for (n = 0; n < nChain; n ++) {                                10
      for (i = 0; i < chainLen - 2; i ++) {
        nn = n * chainLen + i;
        VSub (dr1, mol[nn + 1].r, mol[nn].r);
        VWrapAll (dr1);
        VSub (dr2, mol[nn + 2].r, mol[nn + 1].r);                  15
        VWrapAll (dr2);
        c11 = VLenSq (dr1);
        c22 = VLenSq (dr2);
        c12 = VDot (dr1, dr2);
        bondAng = M_PI - acos (c12 / sqrt (c11 * c22));            20
        j = bondAng * sizeHistBondAng / M_PI;
        ++ histBondAng[j];
      }
    }
```

♠ pr_10_3

```
    } else if (icode == 2) {                                              25
      t = 0.;
      for (j = 0; j < sizeHistBondAng; j ++) t += histBondAng[j];
      for (j = 0; j < sizeHistBondAng; j ++) histBondAng[j] /= t;
    }
  }                                                                       30
```

The final example♠ considers the time dependence of the dihedral angle autocorrelation function. The quantity measured is

$$C(t) = \langle \cos(\theta_i(t) - \theta_i(0)) \rangle \tag{10.5.1}$$

and again no distinction is made between different bonds, although it is quite likely that for longer chains the time dependence will vary with the position in the chain. We also omit any mention of overlapped data buffers (§5.3) that could be used to improve the quality of the results.

```
void EvalDihedAngCorr ()
{
  VecR dr1, dr2, dr3, w;
  real c11, c12, c13, c22, cosAngSq, dihedAng, t;
  int i, j, n, nn;                                                        5

  dihedAngCorr[countDihedAngCorr] = 0.;
  j = 0;
  for (n = 0; n < nChain; n ++) {
    for (i = 0; i < chainLen - 3; i ++) {                                 10
      nn = ...
      ... (same as AccumDihedAngDistn) ...
      if (VDot (dr1, w) < 0.) ...
      if (countDihedAngCorr == 0) dihedAngOrg[j] = dihedAng;
      dihedAngCorr[countDihedAngCorr] +=                                  15
          cos (dihedAng - dihedAngOrg[j]);
      ++ j;
    }
  }
  ++ countDihedAngCorr;                                                   20
  if (countDihedAngCorr == limitDihedAngCorr) {
    for (n = 0; n < limitDihedAngCorr; n ++)
        dihedAngCorr[n] /= nDihedAng;
    PrintDihedAngCorr (stdout);
    countDihedAngCorr = 0;                                                25
  }
}
```

♠ pr_10_4

New variables and input data are

```
real *dihedAngCorr, *dihedAngOrg;
int countDihedAngCorr, limitDihedAngCorr, nDihedAng, stepDihedAngCorr;

  NameI (limitDihedAngCorr),
  NameI (stepDihedAngCorr),                                          5
```

and there is a quantity computed in *SetParams*,

```
  nDihedAng = nChain * (chainLen - 3);
```

The array allocations, initialization and the function call for the processing (each in the appropriate place) are

```
  AllocMem (dihedAngCorr, limitDihedAngCorr, real);
  AllocMem (dihedAngOrg, nDihedAng, real);

  countDihedAngCorr = 0;
                                                                     5
  if (stepCount >= stepEquil && (stepCount - stepEquil) %
      stepDihedAngCorr == 0) EvalDihedAngCorr ();
```

10.6 Measurements

Constraint preservation

The first part of the case study is an examination of the behavior of the constraint algorithm itself. The runs include the following input data:

bondAng	1.91063
bondLen	0.39
chainLen	4
consPrec	1.0e-05
deltaT	0.002
density	0.422
initUchain	3 3 3
stepAvg	1000
stepEquil	1000
stepInitlzTemp	100
stepRestore	200
temperature	4.17

Since a BCC lattice is used for the initial state the total number of chains is 54.

The above data are for a model butane liquid in which both bond-length and bond-angle constraints are applied. Constant-energy MD is used together with PC integration. No energy adjustments are made after the correct temperature is reached, but the energy drift over a run of 70 000 timesteps is just 4%.

If we examine the degree to which the constraints are maintained over the first few thousand timesteps we find that if constraints are restored using the relaxation method every 200 timesteps (*stepRestore*), then the deviations measured by *AnlzConstraintDevs* are typically 2×10^{-4} for *consDevL* and 10^{-3} for *consDevA*. If restoration occurs every 100 timesteps then both deviations are reduced by a factor of three. Typical numbers of restoration cycles, *nCycleR* and *nCycleV*, needed each time are in the approximate range 3–15.

The alternative is to replace the bond angle constraint by a potential. Because of the very stiff nature of this interaction the timestep must be reduced by a factor of four to 0.0005 in order to achieve the same degree of energy conservation. If constraints are restored every 1600 timesteps (equivalent to 400 of the larger timesteps in the preceding test), then the deviation measured by *consDevL* is 6×10^{-5}; more frequent restoration at intervals of 800 or 400 timesteps leads to deviations of size 2×10^{-5} and 6×10^{-6} (close to the tolerance level) respectively. Typically, 2–4 restoration cycles are required in this case.

Properties

The RDF obtained from the butane simulation is shown in Figure 10.3 for the case of both bond-length and bond-angle constraints. Additional input data needed for this computation are

limitRdf	100
rangeRdf	2.2
sizeHistRdf	110
stepLimit	21000
stepRdf	50

Only the final set of averaged RDF values is considered. The computation excludes the contributions from nearest and next-nearest neighbor pairs within each chain whose separations are fixed by the constraints.

We now turn to the distribution of dihedral angles (θ) for both kinds of constraint – bond length only and combined length and angle – and for the former the bond angle (α) distribution as well. Total run lengths are (a relatively short) 70 000 timesteps for length and angle constraints and four times this value for length constraints. In the former case we omit the first two sets of output and average over the remaining 15 sets; in the latter the first eight sets are skipped, leaving 64 sets for

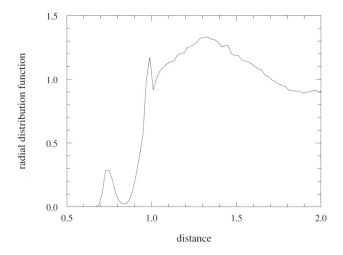

Fig. 10.3. Radial distribution function for liquid butane; intramolecular pairs with fixed separations are excluded.

computing the average distributions. The extra input data are

```
sizeHistBondAng    36
sizeHistDihedAng   36
```

The results of this analysis appear in Figure 10.4. Combined length and angle constraints produce a slightly sharper distribution at zero dihedral angle, although it is not clear from the data shown whether the deviations are statistically significant (the omitted error bars might also account for this difference). The bond angle distribution is relatively narrow, as might be expected from the stiff potential involved.

In Figure 10.5 we show the behavior of the dihedral angle autocorrelation function for the case of length and angle constraints. Additional input data needed here are

```
limitDihedAngCorr 100
stepDihedAngCorr  100
```

Only a single series of measurements covering just 10 000 timesteps are included here; the large fluctuations in the results are due to the very limited sample size. The large-time limit is determined by the average dihedral angle, but the results have not been adjusted to allow for this.

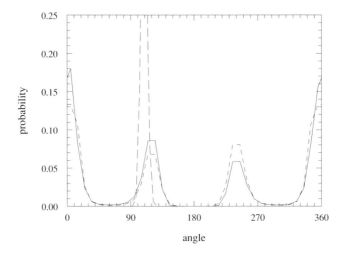

Fig. 10.4.   Butane dihedral angle distributions subject to length and angle constraints (solid curve), and to length constraints alone (short dashes); for the latter, the narrow bond angle distribution is also shown (long dashes).

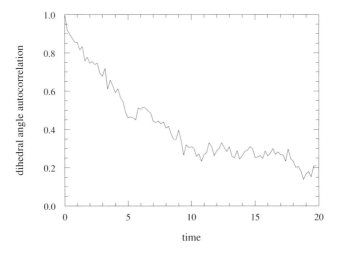

Fig. 10.5.   Dihedral angle autocorrelation function.

10.7 Further study

10.1   Compare the computational requirements of the matrix method for constraints with the 'shake' method using just relaxation; how does the result depend on chain length?

10.2   Include a thermostat that acts on the centers of mass of the molecules [edb86]; how can one apply the thermostat to the intramolecular motion as well?

10.3   Pressure can be studied after establishing that the constraints do not contribute to the virial [cic86b]; measure the pressure for butane (or some other molecule involving constraints) and compare with previous work.

10.4   Consider how to describe branched and ring polymers systematically in order to construct their constraint matrices [mor91].

10.5   In addition to constraints that preserve distances and angles there is a need for constraints that will, for example, maintain the planarity of a molecule; how can this problem be handled without introducing more than a minimal number of constraints [cic82]?

10.6   Study the rate at which crossings of the dihedral potential barrier occur [bro90b].

10.7   There is almost unlimited scope for studying more complex molecules, with proteins and other biopolymers providing the most exciting challenges; explore the capabilities and limitations of the MD approach in this field.

11

Internal coordinates

11.1 Introduction

In earlier chapters, polymer chains were represented as series of atoms coupled by customized springs (Chapter 9), or atoms coupled by rigid links whose length and angle constraints are handled by computations that supplement the timestep integration (Chapter 10). It is also possible to formulate the problem so that the only internal coordinates of the molecule are those actually corresponding to the physical degrees of freedom. Though the formalism involved, which is based on techniques used in robot dynamics, is more complex than the previous methods, the elegance of the approach and the fact that it provides an effective solution to the problem cannot be denied.

11.2 Chain coordinates

Consider a linear polymer chain of monomers. While in principle, each monomer (assumed to be a rigid object) contributes six mechanical degrees of freedom – abbreviated DOFs – to the chain, we use the argument of §10.2 to justify freezing the DOFs associated with variations in bond length and bond angle. Thus, apart from the first monomer which has six DOFs, each additional monomer contributes just a single DOF to the chain. Each such DOF corresponds to torsional motion, or twist, around the appropriate bond axis and is represented by a dihedral angle[†].

If each torsional DOF is regarded as a mechanical joint with a single rotational DOF that is associated with the site at one end of the link, then the system corresponds to a standard problem in the field of robotic manipulators for which techniques are available that express the dynamical equations of motion in a particularly effective manner [jai91, rod92]. Applications of the method to MD simulation

---

† The terms site and monomer (which, in the case considered here, consists of a single atom) are synonymous, as are link and bond.

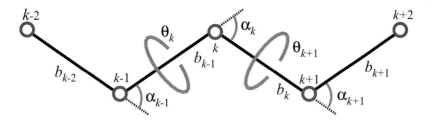

Fig. 11.1. The elements in a section of the linked chain.

appear in [jai93, ber98]. The description of the approach presented in this chapter deals with a linear chain having a single torsional DOF per joint [rap02a], but the treatment is readily generalized (for example, variable bond angles can be introduced, either by allowing two DOFs per joint, or by decomposing each joint into a pair of coincident joints with just one DOF each). The formalism can also be extended to deal with treelike structures; closed loops can be handled, but with extra effort needed to maintain ring closure.

The chain configuration is defined by the site positions r_k; the bond vectors between adjacent sites are denoted by $\boldsymbol{b}_k$, where, as shown[†] in Figure 11.1,

$$r_{k+1} = r_k + \boldsymbol{b}_k \tag{11.2.1}$$

The internal configuration of the chain can be specified by a set of bond rotation matrices $\boldsymbol{R}_k$. The transformation between coordinate frames attached to bonds $k-1$ and k ($k \geq 1$) involves two rotations: a rotation through the bond angle α_k – where $\cos \alpha_k = \hat{\boldsymbol{b}}_{k-1} \cdot \hat{\boldsymbol{b}}_k$ – about the axis $\hat{x}_{k-1}$, followed by a rotation through the dihedral angle θ_k about the joint axis $\hat{z}_{k-1}$. The (transposed) rotation matrix corresponding to the two operations is

$$
\boldsymbol{R}^T_{k-1,k} = \begin{pmatrix} \cos \theta_k & -\sin \theta_k & 0 \\ \sin \theta_k & \cos \theta_k & 0 \\ 0 & 0 & 1 \end{pmatrix} \begin{pmatrix} 1 & 0 & 0 \\ 0 & \cos \alpha_k & -\sin \alpha_k \\ 0 & \sin \alpha_k & \cos \alpha_k \end{pmatrix}
$$

$$
= \begin{pmatrix} \cos \theta_k & -\sin \theta_k \cos \alpha_k & \sin \theta_k \sin \alpha_k \\ \sin \theta_k & \cos \theta_k \cos \alpha_k & -\cos \theta_k \sin \alpha_k \\ 0 & \sin \alpha_k & \cos \alpha_k \end{pmatrix} \tag{11.2.2}
$$

The complete rotation matrix for the kth bond is

$$\boldsymbol{R}^T_k = \boldsymbol{R}^T_0 \boldsymbol{R}^T_{0,1} \cdots \boldsymbol{R}^T_{k-1,k} \tag{11.2.3}$$

---

† Note that the indices of $\boldsymbol{b}_k$ and α_k used in §10.4 are shifted by unity.

where R_0^T is the rotation matrix of the initial bond; in terms of this matrix, (11.2.1) can be written

$$r_{k+1} = r_k + |b_k| R_k^T \hat{z} \tag{11.2.4}$$

It is assumed that all bond lengths $|b_k|$ and angles α_k are constant; thus the only internal DOFs are those associated with the dihedral angles θ_k. Define $\hat{h}_k$ to be the rotation axis of the joint between bonds $k - 1$ and k that is fixed in the frame of bond $k - 1$; here $\hat{h}_k \equiv \hat{z}_{k-1}$. Insofar as indexing is concerned, there are n_r internal rotational joints labeled $1, \ldots n_r$, while the $n_b = n_r + 1$ bonds are labeled $0, \ldots n_r$ and the $n_r + 2$ sites $0, \ldots n_r + 1$. In order to complete the description of the chain state, an additional joint with three translational and three rotational DOFs is associated with the $k = 0$ site. This joint specifies the overall position and orientation of the chain; it is included in the general formalism described below, but, as will become apparent subsequently, it requires special treatment. A chain of $n_r + 2$ sites has a total of $n_r + 6$ DOFs, of which n_r are internal.

11.3 Kinematic and dynamic relations

If v_k and ω_k are the linear and angular velocities of site k, then the velocities and accelerations of adjacent sites are related by

$$\left.\begin{aligned}
\omega_k &= \omega_{k-1} + \hat{h}_k \dot{\theta}_k \\
v_k &= v_{k-1} + \omega_{k-1} \times b_{k-1} \\
\dot{\omega}_k &= \dot{\omega}_{k-1} + \hat{h}_k \ddot{\theta}_k + \omega_{k-1} \times \hat{h}_k \dot{\theta}_k \\
\dot{v}_k &= \dot{v}_{k-1} + \dot{\omega}_{k-1} \times b_{k-1} + \omega_{k-1} \times (\omega_{k-1} \times b_{k-1})
\end{aligned}\right\} \quad k = 1, \ldots n_r \tag{11.3.1}$$

While the mass elements of the chain are normally (and rather arbitrarily) regarded as residing at the sites, one can equally well associate these masses with the bonds; if

$$r_k^c = r_k + c_k \tag{11.3.2}$$

is the position of the center of mass of the atoms fixed to bond k, then the acceleration of r_k^c is

$$\dot{v}_k^c = \dot{v}_k + \dot{\omega}_k \times c_k + \omega_k \times (\omega_k \times c_k) \tag{11.3.3}$$

Define f_k to be the force and n_k the torque acting on bond k across joint k (an equal but opposite force and torque act on bond $k-1$ across the joint); the rotational and translational equations of motion for bond k are

$$\mathcal{I}_k \dot{\boldsymbol{\omega}}_k + \boldsymbol{\omega}_k \times (\mathcal{I}_k \boldsymbol{\omega}_k) = \boldsymbol{n}_k - \boldsymbol{n}_{k+1} - \boldsymbol{c}_k \times \boldsymbol{f}_k$$
$$- (\boldsymbol{b}_k - \boldsymbol{c}_k) \times \boldsymbol{f}_{k+1} + \boldsymbol{n}_k^e \tag{11.3.4}$$

$$m_k \dot{\boldsymbol{v}}_k^c = \boldsymbol{f}_k - \boldsymbol{f}_{k+1} + \boldsymbol{f}_k^e \tag{11.3.5}$$

where $\boldsymbol{f}_k^e$ and $\boldsymbol{n}_k^e$ are the externally applied force and torque on the bond ($\boldsymbol{n}_k^e$ including, of course, the torque produced by $\boldsymbol{f}_k^e$). The left-hand side of (11.3.4) is just $d(\mathcal{I}_k \boldsymbol{\omega}_k)/dt$; m_k and $\mathcal{I}_k$ are the mass of the bond and its moment of inertia, the latter evaluated relative to the bond center of mass (to be defined later) and expressed in the space-fixed frame[†]. The terms of (11.3.4) and (11.3.5) can be rearranged,

$$\boldsymbol{n}_k = \boldsymbol{n}_{k+1} + \boldsymbol{b}_k \times \boldsymbol{f}_{k+1} + m_k \boldsymbol{c}_k \times \dot{\boldsymbol{v}}_k^c + \mathcal{I}_k \dot{\boldsymbol{\omega}}_k + \boldsymbol{\omega}_k \times (\mathcal{I}_k \boldsymbol{\omega}_k)$$
$$- \boldsymbol{n}_k^e - \boldsymbol{c}_k \times \boldsymbol{f}_k^e \tag{11.3.6}$$

$$\boldsymbol{f}_k = \boldsymbol{f}_{k+1} + m_k \dot{\boldsymbol{v}}_k^c - \boldsymbol{f}_k^e \tag{11.3.7}$$

Finally,

$$t_k - \hat{\boldsymbol{h}}_k \cdot \boldsymbol{n}_k \tag{11.3.8}$$

is the component of the torque along the rotation axis $\hat{\boldsymbol{h}}_k$ at joint k and corresponds to the torsional interaction produced by twisting around bond $k - 1$; the functional dependence of this quantity on the dihedral angle θ_k is known.

11.4 Recursive description of dynamics

Spatial vector formulation

The relations (11.3.1) can be expressed in a more concise form by introducing six-component 'spatial' vectors which combine rotational and translational quantities (such as $\boldsymbol{\omega}$ and $\boldsymbol{v}$). It is also convenient to represent some of the vectors by means of antisymmetric matrices having the form

$$\tilde{u} = \begin{pmatrix} 0 & -u_z & u_y \\ u_z & 0 & -u_x \\ -u_y & u_x & 0 \end{pmatrix} \tag{11.4.1}$$

so that a vector cross product can be expressed as a matrix product

$$\tilde{u}\boldsymbol{v} \equiv \boldsymbol{u} \times \boldsymbol{v} \tag{11.4.2}$$

[†] In dealing with rigid bodies it is often convenient to work in a body-fixed frame; this turns out not to be the case here and vector components will be expressed in the space-fixed coordinate frame.

The result of combining the rotational and translational relations (11.3.1) is

$$\begin{pmatrix} \boldsymbol{\omega}_k \\ \boldsymbol{v}_k \end{pmatrix} = \begin{pmatrix} I & 0 \\ -\tilde{\boldsymbol{b}}_{k-1} & I \end{pmatrix} \begin{pmatrix} \boldsymbol{\omega}_{k-1} \\ \boldsymbol{v}_{k-1} \end{pmatrix} + \begin{pmatrix} \hat{\boldsymbol{h}}_k \\ 0 \end{pmatrix} \dot{\theta}_k \tag{11.4.3}$$

$$\begin{pmatrix} \dot{\boldsymbol{\omega}}_k \\ \dot{\boldsymbol{v}}_k \end{pmatrix} = \begin{pmatrix} I & 0 \\ -\tilde{\boldsymbol{b}}_{k-1} & I \end{pmatrix} \begin{pmatrix} \dot{\boldsymbol{\omega}}_{k-1} \\ \dot{\boldsymbol{v}}_{k-1} \end{pmatrix} + \begin{pmatrix} \hat{\boldsymbol{h}}_k \\ 0 \end{pmatrix} \ddot{\theta}_k + \begin{pmatrix} \boldsymbol{\omega}_{k-1} \times \hat{\boldsymbol{h}}_k \dot{\theta}_k \\ \boldsymbol{\omega}_{k-1} \times (\boldsymbol{\omega}_{k-1} \times \boldsymbol{b}_{k-1}) \end{pmatrix} \tag{11.4.4}$$

or, equivalently,

$$V_k = \phi_{k-1,k}^T V_{k-1} + H_k^T \dot{W}_k \tag{11.4.5}$$

$$A_k = \phi_{k-1,k}^T A_{k-1} + H_k^T \ddot{W}_k + X_k \tag{11.4.6}$$

where V_k and A_k are examples of spatial vectors. The 6×6 matrix

$$\phi_{k-1,k}^T = \begin{pmatrix} I & 0 \\ -\tilde{\boldsymbol{b}}_{k-1} & I \end{pmatrix} \tag{11.4.7}$$

and its untransposed form – appearing in (11.4.10) – are used to propagate kinematic and dynamic information between joints. The spatial vector

$$H_k^T = \begin{pmatrix} \hat{\boldsymbol{h}}_k \\ 0 \end{pmatrix} \tag{11.4.8}$$

is the six-component joint axis vector (in the more general case of a joint with n DOFs, H_k^T becomes a $6 \times n$ matrix); the spatial vector

$$X_k = \begin{pmatrix} \tilde{\boldsymbol{\omega}}_{k-1} & 0 \\ 0 & \tilde{\boldsymbol{\omega}}_{k-1} \end{pmatrix} \begin{pmatrix} \hat{\boldsymbol{h}}_k \dot{\theta}_k \\ \boldsymbol{v}_k - \boldsymbol{v}_{k-1} \end{pmatrix} \tag{11.4.9}$$

contains the remaining terms of the acceleration equation; $W_k \equiv \theta_k$ are the dihedral angles; I and 0 denote appropriately sized unit and zero matrices[†].

In a similar way, we can combine (11.3.6) and (11.3.7),

$$\begin{pmatrix} \boldsymbol{n}_k \\ \boldsymbol{f}_k \end{pmatrix} = \begin{pmatrix} I & \tilde{\boldsymbol{b}}_k \\ 0 & I \end{pmatrix} \begin{pmatrix} \boldsymbol{n}_{k+1} \\ \boldsymbol{f}_{k+1} \end{pmatrix} + \begin{pmatrix} m_k \boldsymbol{c}_k \times \dot{\boldsymbol{v}}_k^c + \mathcal{I}_k \dot{\boldsymbol{\omega}}_k + \boldsymbol{\omega}_k \times (\mathcal{I}_k \boldsymbol{\omega}_k) \\ m_k \dot{\boldsymbol{v}}_k^c \end{pmatrix}$$
$$- \begin{pmatrix} \boldsymbol{n}_k^e + \boldsymbol{c}_k \times \boldsymbol{f}_k^e \\ \boldsymbol{f}_k^e \end{pmatrix} \tag{11.4.10}$$

---

† Italic capitals are used to represent most of the six-component vectors and the associated matrices; the type of quantity will be apparent from the context.

and, in terms of spatial vectors, (11.4.10) and (11.3.8) can be expressed as

$$F_k = \phi_{k,k+1} F_{k+1} + M_k A_k + Y_k \tag{11.4.11}$$

$$T_k = H_k F_k \tag{11.4.12}$$

Here we have used (11.3.3) and defined the symmetric 6×6 mass matrix

$$M_k = \begin{pmatrix} \mathcal{I}_k - m_k \tilde{c}_k \tilde{c}_k & m_k \tilde{c}_k \\ -m_k \tilde{c}_k & m_k I \end{pmatrix} \tag{11.4.13}$$

The spatial vector Y_k in (11.4.11) holds the remaining force contributions; with the aid of the identity

$$c_k \times [\omega_k \times (\omega_k \times c_k)] = -\omega_k \times [c_k \times (c_k \times \omega_k)] \tag{11.4.14}$$

it can be written as

$$Y_k = \begin{pmatrix} \tilde{\omega}_k (\mathcal{I}_k - m_k \tilde{c}_k \tilde{c}_k) \omega_k \\ m_k \tilde{\omega}_k \tilde{\omega}_k c_k \end{pmatrix} - \begin{pmatrix} n_k^e + c_k \times f_k^e \\ f_k^e \end{pmatrix} \tag{11.4.15}$$

For later use, we define a spatial vector corresponding to the external force and torque

$$F_k^e = \begin{pmatrix} n_k^e + c_k \times f_k^e \\ f_k^e \end{pmatrix} \tag{11.4.16}$$

In order to use the recursion relations for V_k, A_k and F_k, the velocity and acceleration of the initial site, V_0 and A_0, must be provided; for the final site, F_{n_r+1} is zero since there are no further bonds. Once the external forces have been evaluated for the current chain state, the recursion relations (11.4.11) and (11.4.6) provide the means for determining the $\ddot{W}_k$ (equivalent to $\ddot{\theta}_k$) in terms of known quantities, which, along with A_0 (which contains $\dot{\omega}_0$ and $\dot{v}_0$), can then be integrated over a single timestep.

Stacked operators

The expressions (11.4.5), (11.4.6), (11.4.11) and (11.4.12) can be rewritten in a concise, 'stacked' form

$$V = \phi^T V + H^T \dot{W} \tag{11.4.17}$$

$$A = \phi^T A + H^T \ddot{W} + X \tag{11.4.18}$$

$$F = \phi F + M A + Y \tag{11.4.19}$$

$$T = H F \tag{11.4.20}$$

that combines all values of k for the chain. A quantity such as V, which contains all the V_k values, is also referred to as a spatial vector, while, for example, the matrix ϕ containing all the $\phi_{k,k+1}$ is known as a spatial operator. The contents of these vectors and matrices are addressed in the next paragraph.

The spatial operator approach was originally developed for the case of a fixed initial bond [rod92] – the base in the example of a robot arm – for which $V_0 = 0$, so that $\dot{W} = (\dot{\theta}_1, \ldots \dot{\theta}_{n_r})^T$ is a vector with just n_r components, and the other vectors and matrices are sized accordingly. In order to remove the restriction of a fixed base [jai95], six extra DOFs are added to the problem. The changes and the resulting vector and matrix sizes are as follows:

- Redefine $\dot{W} = (V_0, \dot{\theta}_1, \ldots \dot{\theta}_{n_r})^T$ as a vector with $n_r + 6$ components; likewise for $\ddot{W}$.
- Increase the size of the block diagonal matrix $H = \mathrm{diag}(H_1, \ldots H_{n_r})$ from $6n_r \times n_r$ to $6(n_r + 1) \times (n_r + 6)$ by including an extra 6×6 block $H_0 = I$, so that now $H = \mathrm{diag}(I, H_1, \ldots H_{n_r})$.
- The block diagonal matrix M is of size $6(n_r + 1) \times 6(n_r + 1)$.
- Matrix ϕ is of similar size and its only nonzero blocks are those to the immediate right of the diagonal, namely $\{\phi_{01}, \ldots \phi_{n_r-1,n_r}\}$.
- Vectors V, A, F, X and Y all have $6(n_r + 1)$ components, for example, $V = (V_0, \ldots V_{n_r})^T$.
- Vector T is organized in the same way as $\dot{W}$, with $n_r + 6$ components, and $T_0 = 0$ because the special $k = 0$ joint exerts no torque.

These enlarged vectors and matrices are used in the subsequent analysis.

The next step is to define

$$\Phi = (I - \phi)^{-1} \tag{11.4.21}$$

so that

$$\Phi = \Phi\phi + I \tag{11.4.22}$$

Since it is easily seen that $\phi^{n_r+1} = 0$, (11.4.21) is equivalent to

$$\Phi = I + \phi + \phi^2 + \cdots + \phi^{n_r} \tag{11.4.23}$$

From (11.4.23), the elements of the upper triangular block matrix Φ, each a 6×6 matrix, are

$$\Phi_{ij} = \begin{cases} I & j = i \\ \phi_{i,i+1} & j = i + 1 \\ \phi_{i,i+1} \cdots \phi_{j-1,j} & j > i + 1 \end{cases} \tag{11.4.24}$$

Then, in terms of Φ, (11.4.17)–(11.4.20) become

$$V = \Phi^T H^T \dot{W} \tag{11.4.25}$$

$$A = \Phi^T (H^T \ddot{W} + X) \tag{11.4.26}$$

$$F = \Phi(MX + Y) \tag{11.4.27}$$

$$T = \mathcal{M}\ddot{W} + H\Phi(M\Phi^T X + Y) \tag{11.4.28}$$

where

$$\mathcal{M} = H\Phi M \Phi^T H^T \tag{11.4.29}$$

While M is a sparse, $6(n_r + 1) \times 6(n_r + 1)$ block diagonal matrix, $\mathcal{M}$ is only of size $(n_r + 6) \times (n_r + 6)$, but, though smaller, it is densely populated. The equation of motion (11.4.28) can, in principle, be integrated numerically, and this is one of the approaches to solving the problem, but the computation required to evaluate $\mathcal{M}^{-1}$ in order to obtain $\ddot{W}$ grows as $(n_r + 6)^3$ and so does not provide a practical approach for any but the shortest of chains. The alternative method, to be described below, requires a computational effort that grows only linearly with n_r, together with what amounts to the inversion of a 6×6 matrix; clearly this will prove to be a much more efficient calculation, even for relatively small n_r.

Mass matrix inversion

The task of obtaining an explicit expression for $\mathcal{M}^{-1}$ begins with the definition of a new 6×6 matrix P_k that is related to M_k in the following way – the motivation for this particular approach is discussed in [rod92] –

$$P_k = \phi_{k,k+1}(I - G_{k+1}H_{k+1})P_{k+1}\phi_{k,k+1}^T + M_k \tag{11.4.30}$$

where

$$G_k = P_k H_k^T D_k^{-1} \tag{11.4.31}$$

$$D_k = H_k P_k H_k^T \tag{11.4.32}$$

For joints with a single DOF, G_k is a six-component vector and D_k a nonzero scalar. Note that P_k is symmetric. Now introduce another new matrix

$$\psi_{k,k+1} = \phi_{k,k+1}(I - G_{k+1}H_{k+1}) \tag{11.4.33}$$

and substitute it in (11.4.30). The stacked forms of (11.4.30)–(11.4.33) are

$$P = \psi P \phi^T + M \tag{11.4.34}$$

$$G = P H^T D^{-1} \tag{11.4.35}$$

$$D = H P H^T \tag{11.4.36}$$

$$\psi = \phi(I - GH) \tag{11.4.37}$$

Here, P and ψ are $6(n_r + 1) \times 6(n_r + 1)$ matrices, and G is a $(n_r + 6) \times 6(n_r + 1)$ block diagonal matrix (so that $G_{k+1}H_{k+1}$ is a square matrix). The matrix D is of size $(n_r + 6) \times (n_r + 6)$; the first 6×6 diagonal block corresponds to D_0, while the remaining n_r diagonal elements are the scalars D_k. From (11.4.34) and (11.4.37),

$$M = P - \phi P \phi^T + \phi G H P \phi^T \tag{11.4.38}$$

and so, by using (11.4.22),

$$\Phi M \Phi^T = P + \Phi \phi P + P \phi^T \Phi^T + \Phi \phi P H^T D^{-1} H P \phi^T \Phi^T \tag{11.4.39}$$

Substituting (11.4.39) in (11.4.29) and then using $GD = PH^T$ from (11.4.35), as well as (11.4.36), leads to

$$
\begin{aligned}
\mathcal{M} &= H P H^T + H \Phi \phi P H^T + H P \phi^T \Phi^T H^T + H \Phi \phi P H^T D^{-1} H P \phi^T \Phi^T H^T \\
&= (I + H \Phi \phi G) D (I + H \Phi \phi G)^T
\end{aligned}
\tag{11.4.40}
$$

Note that the new factorization of $\mathcal{M}$ in (11.4.40) has the form of a product of three $(n_r + 6) \times (n_r + 6)$ square matrices.

Now it is a simple matter to invert $\mathcal{M}$. Use a special case of the Woodbury formula [pre92] for the inverse of a matrix[†]

$$(I + Q_1 Q_2)^{-1} = I - Q_1(I + Q_2 Q_1)^{-1} Q_2 \tag{11.4.41}$$

to obtain

$$(I + H \Phi \phi G)^{-1} = I - H \Phi (I + \phi G H \Phi)^{-1} \phi G \tag{11.4.42}$$

If, by analogy with (11.4.21) for Φ, we define

$$\Psi = (I - \psi)^{-1} \tag{11.4.43}$$

then, from (11.4.37) and (11.4.22),

$$\Psi^{-1} = \Phi^{-1} + \phi G H \tag{11.4.44}$$

so that (11.4.42) becomes

$$(I + H \Phi \phi G)^{-1} = I - H \Psi \phi G \tag{11.4.45}$$

[†] This can be proved by matching the terms of a formal power series expansion.

Thus the inverse of (11.4.40) is

$$\mathcal{M}^{-1} = (I - H\Psi\phi G)^T D^{-1}(I - H\Psi\phi G) \tag{11.4.46}$$

and so, from (11.4.28),

$$\ddot{W} = (I - H\Psi\phi G)^T D^{-1}(I - H\Psi\phi G)[T - H\Phi(M\Phi^T X + Y)]$$
$$= (I - H\Psi\phi G)^T D^{-1}[T - H\Psi(\phi GT + M\Phi^T X + Y)] \tag{11.4.47}$$

where we have used (11.4.44) to replace $H(I - \Psi\phi GH)\Phi$ by $H\Psi$.

To eliminate Ψ rewrite (11.4.47) as

$$(I + H\Phi\phi G)^T \ddot{W} = D^{-1}[T - H\Psi(\phi GT + M\Phi^T X + Y)] \tag{11.4.48}$$

and use (11.4.34) and (11.4.22) to obtain

$$\Psi M\Phi^T = \Psi P(\phi^T \Phi^T + I) - \Psi\psi P\phi^T \Phi^T$$
$$= \Psi P + \Psi(I - \psi)P\phi^T \Phi^T$$
$$= \Psi P + P\phi^T \Phi^T \tag{11.4.49}$$

Then, by defining

$$E = T - HZ \tag{11.4.50}$$

where

$$Z = \Psi(\phi GT + PX + Y) \tag{11.4.51}$$

it follows from the transpose of (11.4.35) that

$$(I + H\Phi\phi G)^T \ddot{W} = D^{-1}E - G^T \phi^T \Phi^T X \tag{11.4.52}$$

Rearranging (11.4.52) and using A from (11.4.26) leads to

$$\ddot{W} = D^{-1}E - G^T \phi^T \Phi^T (H^T \ddot{W} + X)$$
$$= D^{-1}E - G^T \phi^T A \tag{11.4.53}$$

It is also possible to eliminate Ψ from the definition of Z by substituting T from (11.4.50) into (11.4.51),

$$(I - \Psi\phi GH)Z = \Psi(\phi GE + PX + Y) \tag{11.4.54}$$

and then using (11.4.44) to obtain

$$Z = \Phi(\phi GE + PX + Y) \tag{11.4.55}$$

The two equations (11.4.53) and (11.4.55) embody the required new recursion relations. Use (11.4.21) and reintroduce the k indices to obtain

$$Z_k = \phi_{k,k+1}(Z_{k+1} + G_{k+1}E_{k+1}) + P_k X_k + Y_k \tag{11.4.56}$$

$$\ddot{W}_k = D_k^{-1}E_k - G_k^T \phi_{k-1,k}^T A_{k-1} \tag{11.4.57}$$

Note that (11.4.56) and (11.4.57), which are intended to be used in opposite k directions, provide the required results without the need for explicit evaluation of $\mathcal{M}^{-1}$ as implied by (11.4.28).

Recursion relations

The recursion relations for propagating the velocity, force and acceleration values along the chain are as follows. The velocities (translational and rotational) V_k are obtained by starting with V_0 and iterating (11.4.5),

$$V_k = \phi_{k-1,k}^T V_{k-1} + H_k^T \dot{W}_k, \qquad k = 1, \ldots n_r \tag{11.4.58}$$

The forces (including torques), as represented by E_k, as well as the matrices D_k and G_k, are obtained by iterating (11.4.30) and (11.4.56),

$$\left. \begin{aligned}
P_k &= \phi_{k,k+1}(I - G_{k+1}H_{k+1})P_{k+1}\phi_{k,k+1}^T + M_k \\
D_k &= H_k P_k H_k^T \\
G_k &= P_k H_k^T D_k^{-1} \\
Z_k &= \phi_{k,k+1}Z_{k+1}' + P_k X_k + Y_k \\
E_k &= T_k - H_k Z_k \\
Z_k' &= Z_k + G_k E_k
\end{aligned} \right\} \qquad k = n_r, \ldots 0 \tag{11.4.59}$$

where the additional quantity Z_k' has been introduced for computational convenience, and the iteration begins with $P_{n_r+1} = 0$ and $Z_{n_r+1}' = 0$.

Finally, the values of $\ddot{W}_k$ ($\equiv \ddot{\theta}_k$) are determined by starting with A_0, whose evaluation is discussed below, and iterating (11.4.6) and (11.4.57),

$$\left. \begin{aligned}
A_k' &= \phi_{k-1,k}^T A_{k-1} \\
\ddot{W}_k &= D_k^{-1}E_k - G_k^T A_k' \\
A_k &= A_k' + H_k^T \ddot{W}_k + X_k
\end{aligned} \right\} \qquad k = 1, \ldots n_r \tag{11.4.60}$$

where A_k' is also introduced for convenience.

Recall that $k = 0$ is associated with a 'joint' having the full six DOFs, so that $H_0 = I$, $X_0 = 0$ and $\ddot{W}_0 = A_0$. Now because $A_{-1} = 0$, we have $A_0 = D_0^{-1}E_0$ from (11.4.60), and since $T_0 = 0$,

$$D_0 A_0 = -Z_0 \tag{11.4.61}$$

where both D_0 and Z_0 have already been determined by (11.4.59). Thus A_0 can be evaluated numerically by solving the set of six linear equations implicit in (11.4.61).

The overall algorithm for a single timestep entails the following sequence of steps using the two-stage leapfrog integrator:

- integrate the base velocities and coordinates, and the joint angular velocities and angles (first leapfrog stage);
- iterate (11.4.58) to compute the velocities;
- compute the site coordinates;
- compute the external forces acting on the sites and the resulting torques, as well as any other necessary quantities;
- iterate (11.4.59);
- solve the set of linear equations (11.4.61);
- iterate (11.4.60) to compute the accelerations;
- integrate the base velocities and joint angular velocities (second leapfrog stage).

Interactions

Two kinds of interaction are used in this model. The first is the pair interaction used to prevent overlap of the atoms (or atom groups) located at the sites of the chain. Here, a simple soft-sphere repulsion will suffice, and the nearest and next-nearest (and possibly further) neighbor pairs need not be examined because the fixed bond angles preclude their close approach. In the case of a simulation involving multiple chains in solution, similar interactions would be employed between atoms in different chains and also with the atoms of a simple solvent.

The second kind of interaction is the torsion potential associated with each internal DOF. Here we assume the form of this potential to be

$$u(\theta_k) = -u' \cos(\theta_k - \theta') \tag{11.4.62}$$

where u' is the interaction strength, and the value of the dihedral angle θ' can be chosen, together with a suitable value of the fixed bond angle, to ensure that the ground state has the correct amount of twist to produce a helix. The torque appearing in (11.3.8) is then

$$t_k = -u' \sin(\theta_k - \theta') \tag{11.4.63}$$

Note the absence of the intricate vector algebra associated with the torque calculations required (§10.4) when using methods based on cartesian coordinates.

Inertia tensor

The elements of the inertia tensor appearing in (11.4.13) and (11.4.15) are defined as

$$
(\mathcal{I}_k)_{ij} = \begin{cases} \displaystyle\sum_{\kappa \in k} m_\kappa (r_\kappa^2 - r_{\kappa i}^2) & i = j \\[2ex] \displaystyle -\sum_{\kappa \in k} m_\kappa r_{\kappa i} r_{\kappa j} & i \neq j \end{cases} \tag{11.4.64}
$$

where the sum (or, if appropriate, the volume integral) is over all the mass elements κ that are rigidly attached to bond k, and the coordinates are expressed relative to the center of mass in the space-fixed frame. Then, in (11.4.13) and (11.4.15),

$$
(\mathcal{I}_k - m_k \tilde{c}_k \tilde{c}_k)_{ij} = \begin{cases} (\mathcal{I}_k)_{ii} + m_k (c_k^2 - c_{ki}^2) & i = j \\[2ex] (\mathcal{I}_k)_{ij} - m_k c_{ki} c_{kj} & i \neq j \end{cases} \tag{11.4.65}
$$

where

$$
m_k = \sum_{\kappa \in k} m_\kappa \tag{11.4.66}
$$

The (six-component) spatial momentum of the chain is

$$
\sum_{k=0}^{n_r} M_k V_k \tag{11.4.67}
$$

and the kinetic energy is

$$
\frac{1}{2} \sum_{k=0}^{n_r} [m_k v_k^c \cdot v_k^c + \omega_k \cdot (\mathcal{I}_k \omega_k)] = \frac{1}{2} \sum_{k=0}^{n_r} V_k^T M_k V_k \tag{11.4.68}
$$

11.5 Solving the recursion equations

Organizational matters

In the present case study[♠] it is assumed that all bond lengths $|b_k| = b$ and all bond angles $\alpha_k = \alpha$, values referred to as *bondLen* and *bondAng*, respectively. A spherical mass element, with a finite moment of inertia about its own center, is associated with each site. Each bond has a single mass attached to its far $(k + 1)$ site, except for the first bond which has masses attached at both ends (the $k = 0, 1$ sites); thus the external force associated with bond k acts on site $k + 1$, and in the case of the first bond an external force acts on site 0 as well. The mass and inertia matrix (in the space-fixed frame) of each link are denoted by *mass* and *inertiaM*.

♠ pr_11_1

For chains whose monomers are single atoms, the definition of each dihedral angle involves three consecutive bonds. The general formulation described above did not associate a torsion term with the last bond, but did include such a term for the first bond. In order to make this first torsion term physically meaningful, it is necessary to extend the chain with an additional bond and site on the other side of the site with six DOFs. This effectively introduces an extra site with index -1 (although, in the program that follows, the array containing the site information is shifted by unity to avoid the negative index); its relative coordinates, prior to applying the overall rotation $\boldsymbol{R}_0$ in (11.2.3), are obtained from (11.2.2) using a zero dihedral angle, namely $-b\boldsymbol{R}_{0,1}^T\hat{z}$. The chain is now longer by one link, but the initial two bonds and three sites form a rigid unit (with no internal DOFs) having the correct bond angle. This additional site would not be required in the more general case where the monomers are used to represent rigid assemblies of atoms, rather than single atoms, since the first torsion term is then already associated with the relative rotation of extended bodies.

The quantities required to describe the state of each bond and the atoms rigidly attached to it are stored in the structure

```
typedef struct {
  RMat rMatT;
  VecR r, rv, omega, omegah, bV, cV, hV;
  real inertiaM[9], fV[6], gV[6], xV[6], yV[6], mass, s, sv, svh, sa,
      torq;
} Link;
```

The (transposed) rotation matrix $\boldsymbol{R}_k^T$ is denoted by $rMatT$. In the case of the initial link (which is actually a virtual link from the origin to the initial site), $\boldsymbol{R}_k^T$ represents the state of the three rotational DOFs associated with site zero, whereas for subsequent links it is the cumulative product matrix describing the link orientation in the space-fixed frame. Other members of the structure include the site coordinates and velocity, r and rv, the link angular velocity in the space-fixed frame $omega$ ($omegah$ and svh are discussed later), and s, sv and sa corresponding to the dihedral angle and its first two derivatives. The vectors bV, cV and hv represent $\boldsymbol{b}_k$, $\boldsymbol{c}_k$ and $\hat{\boldsymbol{h}}_k$; the six-component quantities fV, gV, xV and yV correspond, respectively, to F_k^e, G_k, X_k and Y_k. Other variables required for a complete description of the chain, including an array of all the Link structures for the bonds, are placed in the structure

```
typedef struct {
  Link *L;
  VecR ra, wa;
```

```
    int nLink;
} Poly;                                                                      5
```

Site coordinates and velocities

Given the current state of the chain, namely, the position and orientation of the initial link, the dihedral angles for all of the subsequent links, and the time derivatives of these quantities, the site positions and velocities can be evaluated using (11.2.4) and (11.4.58). The following function performs this task, including the evaluation of the link rotation matrices; it, and many of the subsequent functions, are just the software renditions of the algebra implicit in the formulation, in which much of the work involves processing the six-component vectors and 6×6 matrices. The site coordinates (and forces) are stored in an array of $Site$ structures; the indices of this array are shifted up by unity (relative to the normal chain link indices) to allow for the additional site at the beginning of the chain, as already mentioned.

```
void ComputeLinkCoordsVels ()
{
  RMat rMat;
  VecR bEx, bVp, hVp;
  real phiT[36], vs[6], vsp[6];                                              5
  int k;

  VSet (bVp, 0., 0., bondLen);
  VSet (hVp, 0., 0., 1.);
  for (k = 0; k < P.nLink; k ++) {                                          10
    if (k > 0) {
      MVMul (P.L[k].hV, P.L[k - 1].rMatT.u, hVp);
      BuildLinkRotmatT (&rMat, P.L[k].s, bondAng);
      MulMat (P.L[k].rMatT.u, P.L[k - 1].rMatT.u, rMat.u, 3);
    }                                                                        15
    MVMul (P.L[k].bV, P.L[k].rMatT.u, bVp);
  }
  for (k = 0; k < P.nLink; k ++) {
    VToLin (vs, 0, P.L[k].omega);
    VToLin (vs, 3, P.L[k].rv);                                              20
    BuildLinkPhimatT (phiT, k);
    MulMatVec (vsp, phiT, vs, 6);
    if (k < P.nLink - 1) {
      VFromLin (P.L[k + 1].omega, vsp, 0);
      VVSAdd (P.L[k + 1].omega, P.L[k + 1].sv, P.L[k + 1].hV);              25
    }
    VFromLin (P.L[k + 1].rv, vsp, 3);
  }
  for (k = 0; k < P.nLink; k ++)
    VAdd (P.L[k + 1].r, P.L[k].r, P.L[k].bV);                              30
  for (k = 0; k < P.nLink + 1; k ++) site[k + 1].r = P.L[k].r;
```

```
    VSet (bEx, 0., - sin (bondAng), - cos (bondAng));
    VScale (bEx, bondLen);
    MVMul (site[0].r, P.L[0].rMatT.u, bEx);
    VVAdd (site[0].r, site[1].r);                                    35
  }
```

Here,

```
#define VToLin(a, n, v)                                              \
    a[(n) + 0] = (v).x,                                              \
    a[(n) + 1] = (v).y,                                              \
    a[(n) + 2] = (v).z
#define VFromLin(v, a, n)                                            \    5
    VSet (v, a[(n) + 0], a[(n) + 1], a[(n) + 2])
```

are used for converting between vectors and array elements. In order to maintain
clarity in an already complex problem, this version of the software is designed to
accommodate just a single chain, but by replacing the single structure *P* by a dynamically allocated array of structures, with one element per chain, this limitation
can be removed.

The following macros are introduced for conciseness:

```
#define MAT(a, n, i, j)   (a)[(i) + n * (j)]
#define M3(a, i, j)       MAT (a, 3, i, j)
#define M6(a, i, j)       MAT (a, 6, i, j)
#define DO(m, n)          for (m = 0; m < n; m ++)
```

The functions referenced by *ComputeLinkCoordsVels* are then

```
void BuildLinkRotmatT (RMat *rMat, real dihedA, real bondA)
{
  real cb, cd, sb, sd;

  cb = cos (bondA);                                                  5
  sb = sin (bondA);
  cd = cos (dihedA);
  sd = sin (dihedA);
  M3 (rMat->u, 0, 0) =   cd;
  M3 (rMat->u, 1, 0) =   sd;                                         10
  M3 (rMat->u, 2, 0) =   0.;
  M3 (rMat->u, 0, 1) = - sd * cb;
  M3 (rMat->u, 1, 1) =   cd * cb;
  M3 (rMat->u, 2, 1) =   sb;
  M3 (rMat->u, 0, 2) =   sd * sb;                                    15
  M3 (rMat->u, 1, 2) = - cd * sb;
  M3 (rMat->u, 2, 2) =   cb;
}
```

for assembling the rotation matrix (11.2.2) used to transform between successive links,

```
void BuildLinkPhimatT (real *phiT, int k)
{
  int i, j;

  DO (i, 6) {                                                      5
    DO (j, 6) M6 (phiT, i, j) = (i == j) ? 1. : 0.;
  }
  M6 (phiT, 3, 1) =   P.L[k].bV.z;
  M6 (phiT, 3, 2) = - P.L[k].bV.y;
  M6 (phiT, 4, 0) = - P.L[k].bV.z;                                 10
  M6 (phiT, 4, 2) =   P.L[k].bV.x;
  M6 (phiT, 5, 0) =   P.L[k].bV.y;
  M6 (phiT, 5, 1) = - P.L[k].bV.x;
}
```

for building the (6×6) matrix ϕ^T (11.4.7), and `MulMatVec` (§18.4) for multiplying a matrix by a vector.

Link inertia and forces

The center of mass position and the inertia matrix associated with each link are evaluated by the following function; it includes special treatment for the initial link that has three sites attached.

```
void BuildLinkInertiaMats ()
{
  VecR d;
  real dd, iBall, inertiaK;
  int k;                                                           5

  inertiaK = 0.1;
  for (k = 0; k < P.nLink; k ++) {
    if (k > 0) {
      P.L[k].mass = 1.;                                            10
      VSub (P.L[k].cV, site[k + 2].r, site[k + 1].r);
    } else {
      P.L[k].mass = 3.;
      VAdd (P.L[k].cV, site[2].r, site[1].r);
      VVAdd (P.L[k].cV, site[0].r);                                15
      VScale (P.L[k].cV, 1./3.);
      VVSub (P.L[k].cV, site[1].r);
    }
    iBall = inertiaK * P.L[k].mass;
    VSub (d, site[k + 2].r, site[k + 1].r);                        20
    dd = VLenSq (d);
    M3 (P.L[k].inertiaM, 0, 0) = dd - Sqr (d.x) + iBall;
    M3 (P.L[k].inertiaM, 1, 1) = dd - Sqr (d.y) + iBall;
```

```
   M3 (P.L[k].inertiaM, 2, 2) = dd - Sqr (d.z) + iBall;
   M3 (P.L[k].inertiaM, 0, 1) = - d.x * d.y;                      25
   M3 (P.L[k].inertiaM, 0, 2) = - d.x * d.z;
   M3 (P.L[k].inertiaM, 1, 2) = - d.y * d.z;
   if (k == 0) {
     VSub (d, site[0].r, site[1].r);
     M3 (P.L[k].inertiaM, 0, 0) += dd - Sqr (d.x);               30
     M3 (P.L[k].inertiaM, 0, 1) -= d.x * d.y;
     ... (similarly for the other matrix elements) ...
   }
   M3 (P.L[k].inertiaM, 1, 0) = M3 (P.L[k].inertiaM, 0, 1);
   M3 (P.L[k].inertiaM, 2, 0) = M3 (P.L[k].inertiaM, 0, 2);      35
   M3 (P.L[k].inertiaM, 2, 1) = M3 (P.L[k].inertiaM, 1, 2);
 }
}
```

Now that the positions of all the chain sites have been determined, the forces
acting on the sites, both due to other sites in the chain that are not close neighbors,
and from other sources – other chains, solvent atoms, walls bounding the region –
can be evaluated; this computation is discussed later. The following function then
uses these site forces to compute the components of (11.4.16); it also evaluates the
torques due to bond torsion and the resulting contribution to the potential energy.

```
void ComputeLinkForces ()
{
  VecR d, fc, tq, tq1;
  real ang;
  int k;                                                          5

  for (k = 1; k < P.nLink; k ++) {
    ang = P.L[k].s - twistAng;
    P.L[k].torq = - uCon * sin (ang);
    uSum -= uCon * cos (ang);                                    10
  }
  for (k = 0; k < P.nLink; k ++) {
    fc = site[k + 2].f;
    VCross (tq, P.L[k].bV, fc);
    if (k == 0) {                                                15
      VVAdd (fc, site[1].f);
      VVAdd (fc, site[0].f);
      VSub (d, site[0].r, site[1].r);
      VCross (tq1, d, site[0].f);
      VVAdd (tq, tq1);                                           20
    }
    VToLin (P.L[k].fV, 0, tq);
    VToLin (P.L[k].fV, 3, fc);
  }
}                                                                 25
```

Here, $uCon$ corresponds to u' in (11.4.62) and $twistAng$ to θ'; the indices of the array $site$ have again been shifted by unity. Note that if f_k^s denotes the force acting on site k, then (11.4.16) becomes

$$F_k^e = \begin{pmatrix} \boldsymbol{b}_k \times f_{k+1}^s \\ f_{k+1}^s \end{pmatrix} \tag{11.5.1}$$

for $k > 0$, with appropriate additional contributions for $k = 0$.

Link accelerations

Computing the link accelerations involves recursion relations that traverse the chain in both directions. Prior to this, the X_k and Y_k vectors defined in (11.4.9) and (11.4.15) must be evaluated,

```
void BuildLinkXYvecs (int k)
{
  VecR dv, w, w1, w2;
  int i;
                                                          5
  if (k > 0) {
    VCross (w, P.L[k - 1].omega, P.L[k].hV);
    VScale (w, P.L[k].sv);
    VToLin (P.L[k].xV, 0, w);
    VSub (dv, P.L[k].rv, P.L[k - 1].rv);                  10
    VCross (w, P.L[k - 1].omega, dv);
    VToLin (P.L[k].xV, 3, w);
  } else {
    DO (i, 6) P.L[k].xV[i] = 0.;
  }                                                       15
  MVMul (w, P.L[k].inertiaM, P.L[k].omega);
  VCross (w1, P.L[k].omega, w);
  VToLin (P.L[k].yV, 0, w1);
  VCross (w, P.L[k].omega, P.L[k].cV);
  VCross (w2, P.L[k].omega, w);                           20
  VScale (w2, P.L[k].mass);
  VToLin (P.L[k].yV, 3, w2);
}
```

Applying the backward (11.4.59) and forward (11.4.60) recursion relations leads to the link accelerations. Once again, the software merely implements the algebra contained in these relations. Vector quantities whose only role is to transfer values between successive iterations, such as Z_k', need not be assigned permanent storage and are represented as arrays, such as zp, that are overwritten during each iteration.

```
void ComputeLinkAccels ()
{
  real as[6], asp[6], h[3], mMat[36], phi[36], phiT[36], pMat[36],
     tMat1[36], tMat2[36], z[6], zp[6], zt[6], dk, e;
  int i, j, k;                                                          5

  DO (i, 6) zp[i] = 0.;
  for (k = P.nLink - 1; k >= 0; k --) {
    BuildLinkPhimatT (phiT, k);
    DO (i, 6) {                                                        10
      DO (j, 6) M6 (phi, i, j) = M6 (phiT, j, i);
    }
    BuildLinkXYvecs (k);
    BuildLinkMmat ((k == P.nLink - 1) ? pMat : mMat, k);
    if (k < P.nLink - 1) {                                             15
      DO (i, 6) {
        DO (j, 6) M6 (tMat1, i, j) = (i == j) ? 1. : 0.;
      }
      VToLin (h, 0, P.L[k + 1].hV);
      DO (i, 6) {                                                      20
        DO (j, 3) M6 (tMat1, i, j) -= P.L[k + 1].gV[i] * h[j];
      }
      MulMat (tMat2, tMat1, pMat, 6);
      MulMat (tMat1, tMat2, phiT, 6);
      MulMat (pMat, phi, tMat1, 6);                                    25
      DO (i, 6) {
        DO (j, 6) M6 (pMat, i, j) += M6 (mMat, i, j);
      }
    }
    if (k > 0) {                                                       30
      VToLin (h, 0, P.L[k].hV);
      dk = 0.;
      DO (i, 3) {
        DO (j, 3) dk += h[i] * M6 (pMat, i, j) * h[j];
      }                                                                35
    }
    MulMatVec (z, phi, zp, 6);
    DO (i, 6) z[i] += P.L[k].yV[i] - P.L[k].fV[i];
    if (k > 0) {
      DO (i, 6) {                                                      40
        P.L[k].gV[i] = 0.;
        DO (j, 3) P.L[k].gV[i] += M6 (pMat, i, j) * h[j];
        P.L[k].gV[i] /= dk;
      }
      MulMatVec (zt, pMat, P.L[k].xV, 6);                              45
      DO (i, 6) z[i] += zt[i];
      e = P.L[k].torq;
      DO (i, 3) e -= h[i] * z[i];
      P.L[k].sa = e / dk;
      DO (i, 6) zp[i] = z[i] + e * P.L[k].gV[i];                       50
```

```
      }
    }
    SolveLineq (pMat, z, 6);
    DO (i, 6) as[i] = - z[i];
    VFromLin (P.wa, as, 0);                                                    55
    VFromLin (P.ra, as, 3);
    for (k = 1; k < P.nLink; k ++) {
      BuildLinkPhimatT (phiT, k - 1);
      MulMatVec (asp, phiT, as, 6);
      DO (i, 6) P.L[k].sa -= P.L[k].gV[i] * asp[i];                            60
      if (k < P.nLink - 1) {
        DO (i, 6) as[i] = asp[i] + P.L[k].xV[i];
        VToLin (h, 0, P.L[k].hV);
        DO (i, 3) as[i] += h[i] * P.L[k].sa;
      }                                                                        65
    }
  }
```

The matrices M_k are constructed by

```
void BuildLinkMmat (real *mMat, int k)
{
  VecR w;
  int i, j;
                                                                               5
  DO (i, 6) {
    DO (j, 6) {
      if (i < 3 && j < 3)
        M6 (mMat, i, j) = M3 (P.L[k].inertiaM, i, j);
      else M6 (mMat, i, j) = (i == j) ? P.L[k].mass : 0.;                      10
    }
  }
  VSCopy (w, P.L[k].mass, P.L[k].cV);
  M6 (mMat, 2, 4) = w.x;
  M6 (mMat, 1, 5) = - w.x;                                                     15
  M6 (mMat, 0, 5) = w.y;
  M6 (mMat, 2, 3) = - w.y;
  M6 (mMat, 1, 3) = w.z;
  M6 (mMat, 0, 4) = - w.z;
  M6 (mMat, 4, 2) = M6 (mMat, 2, 4);                                           20
  ... (fill other entries of symmetric matrix) ...
}
```

The call to $SolveLineq$ (§18.4) solves the set of linear equations in (11.4.61) for D_0 using the standard LU decomposition method.

11.6 Implementation details

If the simulation treats a single chain without solvent, then it is reasonable just to use cells and not a neighbor list (this is a minor issue). Irrespective of approach, there are no direct pair interactions between nearest-, second- or third-nearest neighbor monomers, because these are taken into account by either the rigid structure or the torsional interactions. Thus for the cell method – the function is now called *ComputeSiteForces* – the rule used to select interacting atom pairs in *ComputeForces* (§3.4) is changed to

```
if ((m1 != m2 || j2 < j1) && abs (j1 - j2) > 3)
```

The double loop over the contents of pairs of cells can be skipped if either of the cells is empty, a likely occurrence for the case of just a single chain; this is done by including the test

```
if (cellList[m1] < 0) continue;
```

immediately after the evaluation of the cell index *m1* and likewise for the index *m2*.

The integration routine updates the state of the initial link, as well as the dihedral angles. Part of the following function is borrowed from the one used for the rotation matrix approach to rigid bodies in §8.5, while the rest is just a straightforward treatment of the dihedral angle variables.

```
void LeapfrogStepLinks (int part)
{
  RMat mc, mt;
  VecR t;
  int k;                                                              5

  if (part == 1) {
    VVSAdd (P.L[0].omega, 0.5 * deltaT, P.wa);
    VVSAdd (P.L[0].rv, 0.5 * deltaT, P.ra);
    for (k = 1; k < P.nLink; k ++)                                   10
      P.L[k].sv += 0.5 * deltaT * P.L[k].sa;
    VSCopy (t, 0.5 * deltaT, P.L[0].omega);
    BuildStepRmatT (&mc, &t);
    MulMat (mt.u, mc.u, P.L[0].rMatT.u, 3);
    P.L[0].rMatT = mt;                                               15
    VVSAdd (P.L[0].r, deltaT, P.L[0].rv);
    for (k = 1; k < P.nLink; k ++) P.L[k].s += deltaT * P.L[k].sv;
  } else {
    VVSAdd (P.L[0].omega, 0.5 * deltaT, P.wa);
    VVSAdd (P.L[0].rv, 0.5 * deltaT, P.ra);                          20
    for (k = 1; k < P.nLink; k ++)
```

```
        P.L[k].sv += 0.5 * deltaT * P.L[k].sa;
  }
}
```

In order to ensure that the angular velocity values (*omega* and *sv*) used in the recursion relations correspond to the same instant in time as the coordinates, these values can optionally be updated over an additional half timestep at the end of the first part of the procedure, after storing the current values. Then, at the start of the second part, the original values are restored. This leads to reduced energy fluctuations.

```
  P.L[0].omegah = P.L[0].omega;
  VVSAdd (P.L[0].omega, 0.5 * deltaT, P.wa);
  for (k = 1; k < P.nLink; k ++) {
    P.L[k].svh = P.L[k].sv;
    P.L[k].sv += 0.5 * deltaT * P.L[k].sa;                    5
  }
  ...
  P.L[0].omega = P.L[0].omegah;
  for (k = 1; k < P.nLink; k ++) P.L[k].sv = P.L[k].svh;
```

The dihedral angles are reset to the $[0, 2\pi]$ range by

```
void AdjustLinkAngles ()
{
  int k;

  for (k = 1; k < P.nLink; k ++) {                           5
    if (P.L[k].s >= 2. * M_PI) P.L[k].s -= 2. * M_PI;
    else if (P.L[k].s < 0.) P.L[k].s += 2. * M_PI;
  }
}
```

The following function evaluates the kinetic and total energies; the quantities are normalized per degree of freedom, as is appropriate for an isolated chain[†].

```
void EvalProps ()
{
  VecR w1, w2;
  int k;
                                                             5
  kinEnVal = 0.;
  for (k = 0; k < P.nLink; k ++) {
    MVMul (w2, P.L[k].inertiaM, P.L[k].omega);
```

[†] The values of *rv* and *omega* for links beyond the first could be reevaluated (using *ComputeLinkCoordsVels*) at the end of the timestep before using them here.

```
    VCross (w1, P.L[k].cV, P.L[k].rv);
    kinEnVal += 0.5 * (P.L[k].mass * (VLenSq (P.L[k].rv) +        10
        2. * VDot (P.L[k].omega, w1)) + VDot (P.L[k].omega, w2));
  }
  totEnVal = (kinEnVal + uSum) / nDof;
  kinEnVal /= nDof;
}                                                                  15
```

The function for resetting the temperature to a specified value is

```
void AdjustTemp ()
{
  real vFac;
  int k;
                                                                   5
  vFac = sqrt (temperature / (2. * kinEnVal));
  VScale (P.L[0].rv, vFac);
  VScale (P.L[0].omega, vFac);
  for (k = 1; k < P.nLink; k ++) P.L[k].sv *= vFac;
  ComputeLinkCoordsVels ();                                        10
  EvalProps ();
}
```

The functions introduced above are called by *SingleStep*, which includes the sequence of calls (for a container with hard walls)

```
  LeapfrogStepLinks (1);
  AdjustLinkAngles ();
  ComputeLinkCoordsVels ();
  BuildLinkInertiaMats ();
  ComputeSiteForces ();                                            5
  ComputeWallForces ();
  ComputeLinkForces ();
  ComputeLinkAccels ();
  LeapfrogStepLinks (2);
```

In *SetParams* the following values are set,

```
  nSite = chainLen + 1;
  nDof = chainLen + 4;
  bondAng = 2. * M_PI / helixPeriod;
  twistAng = asin (1.1 * rCut / (helixPeriod * bondLen));
```

where the bond parameters are tailored to produce a helical ground state with periodicity *helixPeriod* and a small amount of space between nearby monomers in adjacent turns of the helix. Memory allocation (*AllocArrays*) includes

```
AllocMem (P.L, chainLen, Link);
AllocMem (site, nSite, Site);
```

The initial state of the chain is generated by the following function; it produces a coiled chain with a relatively large coil radius, and a local zigzag conformation for each successive pair of bonds (as shown later in Figure 11.3). This configuration is far removed from the ground state favored by the torsion potential. Dihedral angle time derivatives are randomly set, the translational velocity of the initial site is adjusted so that the center of mass of the entire chain is at rest and the chain is shifted to the center of the region; velocities are then scaled to correspond to the desired temperature.

```
void InitLinkState ()
{
  VecR rs, vs, w;
  real mSum;
  int j, k;                                                    5

  P.nLink = chainLen - 1;
  VZero (P.L[0].r);
  VZero (P.L[0].rv);
  VZero (P.ra);                                                10
  DO (j, 9) P.L[0].rMatT.u[j] = (j % 4 == 0) ? 1. : 0.;
  VZero (P.L[0].omega);
  VZero (P.wa);
  for (k = 1; k < P.nLink; k ++) {
    P.L[k].s = M_PI + ((k % 2 == 0) ? 0.42 : -0.4);            15
    P.L[k].sv = 0.2 * (1. - 2. * RandR ());
    P.L[k].sa = 0.;
  }
  ComputeLinkCoordsVels ();
  BuildLinkInertiaMats ();                                     20
  VZero (rs);
  VZero (vs);
  mSum = 0.;
  for (k = 0; k < P.nLink; k ++) {
    VAdd (w, P.L[k].r, P.L[k].cV);                             25
    VVSAdd (rs, P.L[k].mass, w);
    VCross (w, P.L[k].omega, P.L[k].cV);
    VVAdd (w, P.L[k].rv);
    VVSAdd (vs, P.L[k].mass, w);
    mSum += P.L[k].mass;                                       30
  }
  VVSAdd (P.L[0].r, -1. / mSum, rs);
  VVSAdd (P.L[0].rv, -1. / mSum, vs);
  ComputeLinkCoordsVels ();
```

```
  BuildLinkInertiaMats ();                                          35
  ComputeSiteForces ();
  ComputeLinkForces ();
  EvalProps ();
  AdjustTemp ();
}                                                                   40
```

The functions *AccumProps* and *PrintSummary* are used to accumulate and output averages, as in previous examples; the former must include

```
  totEnergy.val = totEnVal;
  kinEnergy.val = kinEnVal;
```

Additional variables introduced here are

```
Poly P;
Site *site;
real bondAng, bondLen, kinEnVal, totEnVal, twistAng, uCon;
int chainLen, helixPeriod, nDof, nSite;
```

and the input data must now include (the region size is specified explicitly here instead of via the density)

```
  NameR (bondLen),
  NameI (chainLen),
  NameI (helixPeriod),
  NameR (rcgion),
  NameR (uCon),                                                      5
```

If the bond angles and torsional interactions have been suitably chosen (as is the case here), then at low temperature the chain should collapse into a helical ground state, unless it manages to become entangled in a manner that somehow prevents this occurring. An order parameter quantifying the structure of the ground state can be defined as

$$S = \frac{1}{n_r} \left| \sum_{k=1}^{n_r} \hat{\boldsymbol{d}}_k \right| \qquad (11.6.1)$$

where $\boldsymbol{d}_k = \boldsymbol{b}_{k-1} \times \boldsymbol{b}_k$; for a well-formed helix $S \approx 1$ and any entanglement produces a noticeable reduction in this value. The quantity S is evaluated as follows.

```
void EvalHelixOrder ()
{
  VecR dr1, dr2, rc, rcSum;
  real f;
  int k;                                                            5
```

```
    VZero (rcSum);
    for (k = 0; k < P.nLink; k ++) {
      VSub (dr1, site[k + 1].r, site[k].r);
      VSub (dr2, site[k + 2].r, site[k + 1].r);        10
      VCross (rc, dr1, dr2);
      f = VLen (rc);
      if (f > 0.) VVSAdd (rcSum, 1. / f, rc);
    }
    helixOrder = VLen (rcSum) / P.nLink;               15
  }
```

11.7 Measurements

Equilibrium

In order to test how well energy is conserved, a run was carried out using the following data:

bondLen	1.3
chainLen	80
deltaT	0.001
helixPeriod	8
region	24. 24. 24.
stepAvg	2000
stepEquil	10000
stepLimit	100000
temperature	2.
uCon	1.

The various velocities are adjusted periodically during the equilibration period, and then the system is allowed to proceed with no further intervention. The results are summarized in Table 11.1 and it is apparent that, left unattended, the total energy is subject to a slight amount of drift.

Chain collapse

The goal of this case study♠ is to follow the behavior of the chain as it gradually cools from a high temperature state. The fixed bond angles and the preferred dihedral angles have already been chosen so that the minimum energy ground state of the chain is a neatly coiled helix; the question is whether the chain will be able to collapse into this ordered conformation, or will this process be obstructed by the chain becoming entangled with itself.

♠ pr_11_2

Table 11.1. Energy conservation

timestep	$\langle E \rangle$	$\sigma(E)$	$\langle E_K \rangle$	$\sigma(E_K)$
20 000	0.7466	0.0001	0.9334	0.0283
40 000	0.7472	0.0005	0.9130	0.0307
60 000	0.7456	0.0008	1.0019	0.0109
80 000	0.7561	0.0003	0.9996	0.0131
100 000	0.7547	0.0009	1.0556	0.0165

To achieve progressive temperature reduction, add to `SingleStep`

```
if (stepCount % stepReduceTemp == 0 && temperature > tempFinal) {
    temperature *= tempReduceFac;
    AdjustTemp ();
}
```

and, after incorporating the additional variables used here, run the simulation with data that includes

```
chainLen          80
deltaT            0.004
helixPeriod       6
stepAdjustTemp    2000
stepAvg           2000
stepLimit         1000000
stepReduceTemp    4000
stepSnap          10000
tempFinal         0.001
tempInit          4.
tempReduceFac     0.97
uCon              5.
```

Figure 11.2 shows how the order parameter S and the negative of the potential energy, vary with time; snapshots of initial, intermediate and final states of the run appear in Figure 11.3. The chain behaves as might be expected; other initial states – governed by the choice of random number seed used to set the velocities – produce entirely different folding pathways, most of which still lead to the helical final state (if other parameters are left unchanged); a more extensive treatment of this problem appears in [rap02a].

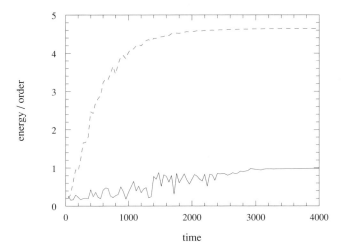

Fig. 11.2.   Time dependence of the order parameter (solid curve) and the negative of the energy (dashed curve) as the chain collapses into an ordered helical state.

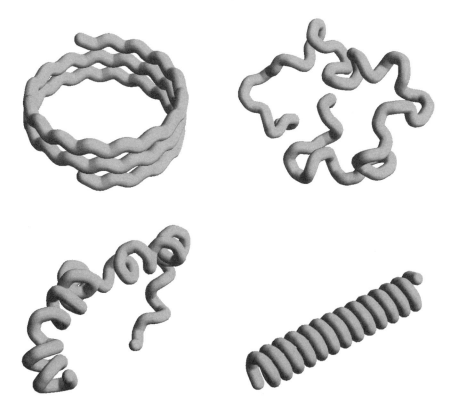

Fig. 11.3.   Helix formation: snapshots of the initial state, a fairly random state early in the run, a subsequent state with helical domains and the eventual well-formed helix.

11.8 Further study

11.1 Extend the recursive approach to handle molecules with more general tree-like topological structures.

11.2 Explore techniques for dealing with molecules that either consist of, or contain, closed ring structures.

12

Many-body interactions

12.1 Introduction

The range of problems amenable to study using MD knows few bounds and as computers become more powerful the range continues to expand. Because of the enormous breadth of the subject, we have chosen to concentrate on the simplest of systems and avoid overly specialized models. Most of the case studies up to this point have been based on short-ranged, two-body interactions; within this framework a considerable variety of problems can be studied, but a few conspicuous gaps remain. Pair potentials have their limitations, and while certain kinds of intermolecular interaction can be imitated by the appropriate combinations of pair potentials, it is sometimes essential to introduce many-body interactions to capture specific features of the 'real' intermolecular force [mai81].

In this chapter we present two different approaches to the introduction of many-body interactions, namely, three-body interactions and the embedded-atom method, each in the form of a case study. We cannot do justice to the range of applications to which these and other enhancements of the MD method, such as those discussed in Chapter 13, have contributed, but in the prevailing culinary atmosphere we hope the reader will gain at least a taste of what is involved.

12.2 Three-body forces

The problem

Even when regarded simply as effective potentials, the capacity of the pair potential to reproduce known behavior has its limitations. We have already encountered situations where the potential extends beyond the basic two-body form: the interaction site method used for rigid molecules, in which forces between molecules (as opposed to the sites in the molecules) involve both distance and orientation (§8.3), and the intramolecular forces associated with the internal degrees of freedom of partially rigid molecules that depend on the relative coordinates of sets of three or four atoms (§10.4). The situation described here is entirely different; the force between two atoms will now depend on the positions of all other atoms in the vicinity. Thus, in addition to the forces between pairs of atoms, a new type of force is introduced that acts on triplets of atoms and whose strength is a function of the three interatomic distances.

The crystalline state of silicon is a four-coordinated diamond lattice whose density increases upon melting; this reflects the fact that the liquid is more closely packed than the solid, exactly the opposite of what normally occurs – with water being another prominent exception. Since LJ-type potentials can only produce closely packed solids, the question is how to augment the potential function so that it incorporates a preferred set of bond directions, such as the tetrahedral arrangement needed for silicon. The simplest way to do this is to introduce three-body interactions, chosen in a manner that stabilizes particular bond angles; this represents yet another attempt to imitate classically what is at heart a quantum effect.

Formulation

The model proposed for liquid silicon [sti85] has been kept as simple as possible – to the extent that any three-body interaction can be considered 'simple' from the computational perspective. The interactions include both two- and three-body contributions, expressed in suitable reduced units.

The two-body part of the potential has the form

$$u_2(r) = \begin{cases} a\left(\dfrac{b}{r^4} - 1\right)\exp\left(\dfrac{1}{r - r_c}\right) & r < r_c \\ 0 & r \geq r_c \end{cases} \tag{12.2.1}$$

and the corresponding force is

$$-\nabla_r u_2 = \frac{a}{r}\left[\frac{4b}{r^5} + \left(\frac{b}{r^4} - 1\right)\frac{1}{(r - r_c)^2}\right]\exp\left(\frac{1}{r - r_c}\right)r \tag{12.2.2}$$

Observe the design of the function: both u_2 and its derivatives are zero at the cutoff range r_c; while this has the advantage of reducing spurious effects associated with cutoffs in general, care is required in the computation to avoid any risk of numerical overflow when $r \approx r_c$.

The three-body part is symmetric under permutations of the atom indices and is, of course, invariant under translation and rotation,

$$u_3(\boldsymbol{r}_1, \boldsymbol{r}_2, \boldsymbol{r}_3) = h(\boldsymbol{r}_{12}, \boldsymbol{r}_{13}) + h(\boldsymbol{r}_{21}, \boldsymbol{r}_{23}) + h(\boldsymbol{r}_{31}, \boldsymbol{r}_{32}) \tag{12.2.3}$$

Each function h has the form

$$h(\mathbf{r}_{12}, \mathbf{r}_{13}) = \lambda \exp\left(\frac{\gamma}{r_{12} - r_c} + \frac{\gamma}{r_{13} - r_c}\right)\left(\cos\theta_{213} + \tfrac{1}{3}\right)^2 \tag{12.2.4}$$

provided both $r_{12} < r_c$ and $r_{13} < r_c$, and is zero if either condition is violated; θ_{213} is the angle between $\mathbf{r}_{12}$ and $\mathbf{r}_{13}$. The cutoff at r_c is smooth, as in the two-body part. An important feature built into the functional form of h is that it has a minimum when θ_{213} equals the tetrahedral angle ($\cos\theta = -1/3$). Not only does this introduce the desired angular correlations, it helps ensure that each atom prefers a considerably smaller number of immediate neighbors (namely four) than close packing allows.

The force contribution of each h function in (12.2.3) is evaluated separately; the forces due to a typical function $h(\mathbf{r}_{12}, \mathbf{r}_{13})$ act on three atoms, and since

$$\nabla_{r_1} h = -\nabla_{r_2} h - \nabla_{r_3} h \tag{12.2.5}$$

only two of the derivatives need be computed. Thus, if

$$c \equiv \cos\theta_{213} = \hat{\mathbf{r}}_{12} \cdot \hat{\mathbf{r}}_{13} \tag{12.2.6}$$

then, for $m = 2$ and 3,

$$-\nabla_{r_m} h(\mathbf{r}_{12}, \mathbf{r}_{13}) = -\lambda\left(c + \tfrac{1}{3}\right)\exp\left(\frac{\gamma}{r_{12} - r_c} + \frac{\gamma}{r_{13} - r_c}\right)$$
$$\times \left(\frac{\gamma\left(c + \tfrac{1}{3}\right)}{(r_{1m} - r_c)^2}\,\hat{\mathbf{r}}_{1m} + 2\nabla_{r_m} c\right) \tag{12.2.7}$$

The derivatives $\nabla_{r_m} c$ are computed in exactly the same way as the bond angle forces discussed in §10.4,

$$\nabla_{r_2} c = (c\hat{\mathbf{r}}_{12} - \hat{\mathbf{r}}_{13})/r_{12} \tag{12.2.8}$$
$$\nabla_{r_3} c = (c\hat{\mathbf{r}}_{13} - \hat{\mathbf{r}}_{12})/r_{13} \tag{12.2.9}$$

Numerical values for the constants appearing in (12.2.1) and (12.2.4) are specified in the function `ComputeForces` below.

Implementation details

The neighbor-list method is once again the preferred choice, although with some modification. In order to identify all interacting atom triplets the list will have to be scanned twice inside a doubly nested loop, and the list must also include each atom pair twice, both as ij and ji. The neighbor list will be stored in an alternative form better suited for this computation: instead of simply listing the possibly interacting pairs, the information will now be stored in two parts, one a table of serial numbers of the neighboring atoms, the other a set of pointers to the first entry in the table corresponding to the neighbors of each atom. Such an approach could have been used for the original treatment in §3.4, although the extra data operations and the low repetition count of the resulting inner loop could reduce the computational efficiency.

Neighbor-list construction♠ is carried out by the following function derived from the original version. Here atoms are scanned first, then the set of cells surrounding the cell in which the particular atom resides and, finally, the contents of each of these cells.

```
#define OFFSET_VALS                                         \
    {{-1,-1,-1}, {0,-1,-1}, {1,-1,-1}, {-1,0,-1},           \
     {0,0,-1}, {1,0,-1}, {-1,1,-1}, {0,1,-1}, {1,1,-1},     \
     {-1,-1,0}, {0,-1,0}, {1,-1,0}, {-1,0,0}, {0,0,0},      \
     {1,0,0}, {-1,1,0}, {0,1,0}, {1,1,0}, {-1,-1,1},        \
     {0,-1,1}, {1,-1,1}, {-1,0,1}, {0,0,1}, {1,0,1},        \       5
     {-1,1,1}, {0,1,1}, {1,1,1}}

void BuildNebrList ()
{                                                                   10
  ...
  nebrTabLen = 0;
  for (j1 = 0; j1 < nMol; j1 ++) {
    VSAdd (rs, mol[j1].r, 0.5, region);
    VMul (m1v, rs, invWid);                                         15
    nebrTabPtr[j1] = nebrTabLen;
    for (offset = 0; offset < 27; offset ++) {
      VAdd (m2v, m1v, vOff[offset]);
      VZero (shift);
      VCellWrapAll ();                                              20
      m2 = VLinear (m2v, cells) + nMol;
      DO_CELL (j2, m2) {
        if (j2 != j1) {
          VSub (dr, mol[j1].r, mol[j2].r);
          VVSub (dr, shift);                                        25
          if (VLenSq (dr) < rrNebr) {
            if (nebrTabLen >= nebrTabMax)
              ErrExit (ERR_TOO_MANY_NEBRS);
            nebrTab[nebrTabLen] = j2;
            ++ nebrTabLen;                                          30
          }
        }
      }
    }
  }                                                                 35
  nebrTabPtr[nMol] = nebrTabLen;
}
```

♠ pr_12_1

The arrays used are

```
int *nebrTab, *nebrTabPtr;
```

and they are allocated (in *AllocArrays* – note that the size of *nebrTab* has been halved) by

```
AllocMem (nebrTab, nebrTabMax, int);
AllocMem (nebrTabPtr, nMol + 1, int);
```

The interaction calculations, including both two- and three-body contributions, are carried out by the following function.

```
void ComputeForces ()
{
  VecR dr, dr12, dr13, w2, w3;
  real aCon = 7.0496, bCon = 0.60222, cr, er, fcVal, gCon = 1.2,
    lCon = 21., p12, p13, ri, ri3, rm, rm12, rm13, rr, rr12,        5
    rr13, rrCut;
  int j1, j2, j3, m2, m3, n;

  rrCut = Sqr (rCut) - 0.001;
  DO_MOL VZero (mol[n].ra);                                        10
  uSum = 0.;
  for (j1 = 0; j1 < nMol; j1 ++) {
    for (m2 = nebrTabPtr[j1]; m2 < nebrTabPtr[j1 + 1]; m2 ++) {
      j2 = nebrTab[m2];
      if (j1 < j2) {                                               15
        VSub (dr, mol[j1].r, mol[j2].r);
        VWrapAll (dr);
        rr = VLenSq (dr);
        if (rr < rrCut) {
          rm = sqrt (rr);                                          20
          er = exp (1. / (rm - rCut));
          ri = 1. / rm;
          ri3 = Cube (ri);
          fcVal = aCon * (4. * bCon * Sqr (ri3) +
            (bCon * ri3 * ri - 1.) * ri / Sqr (rm - rCut)) * er;   25
          VVSAdd (mol[j1].ra, fcVal, dr);
          VVSAdd (mol[j2].ra, - fcVal, dr);
          uSum += aCon * (bCon * ri3 * ri - 1.) * er;
        }
      }                                                            30
    }
  }
  for (j1 = 0; j1 < nMol; j1 ++) {
    for (m2 = nebrTabPtr[j1]; m2 < nebrTabPtr[j1 + 1] - 1; m2 ++) {
      j2 = nebrTab[m2];                                            35
      VSub (dr12, mol[j1].r, mol[j2].r);
```

```
      VWrapAll (dr12);
      rr12 = VLenSq (dr12);
      if (rr12 < rrCut) {
        rm12 = sqrt (rr12);                                             40
        VScale (dr12, 1. / rm12);
        for (m3 = m2 + 1; m3 < nebrTabPtr[j1 + 1]; m3 ++) {
          j3 = nebrTab[m3];
          VSub (dr13, mol[j1].r, mol[j3].r);
          VWrapAll (dr13);                                              45
          rr13 = VLenSq (dr13);
          if (rr13 < rrCut) {
            rm13 = sqrt (rr13);
            VScale (dr13, 1. / rm13);
            cr = VDot (dr12, dr13);                                     50
            er = lCon * (cr + 1./3.) * exp (gCon / (rm12 - rCut) +
                gCon / (rm13 - rCut));
            p12 = gCon * (cr + 1./3.) / Sqr (rm12 - rCut);
            p13 = gCon * (cr + 1./3.) / Sqr (rm13 - rCut);
            VSSAdd (w2, p12 + 2. * cr / rm12, dr12, -2. / rm12, dr13);  55
            VSSAdd (w3, p13 + 2. * cr / rm13, dr13, -2. / rm13, dr12);
            VScale (w2, - er);
            VScale (w3, - er);
            VVSub (mol[j1].ra, w2);
            VVSub (mol[j1].ra, w3);                                     60
            VVAdd (mol[j2].ra, w2);
            VVAdd (mol[j3].ra, w3);
            uSum += (cr + 1./3.) * er;
          }
        }                                                              65
      }
    }
  }
}
```

Measurements

The simulations shown here cursorily examine the RDFs in the crystalline and liquid phases based on input data

deltaT	0.005
density	0.483
initUcell	3 3 3
limitRdf	20
nebrTabFac	50
rangeRdf	2.5
rCut	1.8
rNebrShell	0.4
sizeHistRdf	100
stepAdjustTemp	500

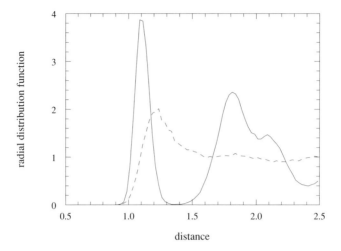

Fig. 12.1. Radial distribution functions for model silicon in the crystalline (solid curve) and liquid (dashed curve) states.

stepAvg	200
stepEquil	0
stepLimit	5000
stepRdf	50
temperature	0.08

The initial state is a diamond lattice, so that $N_m = 216$. Runs are carried out at temperatures 0.08 and 0.12. Constant-temperature MD is used together with PC integration.

The RDF results shown in Figure 12.1 are based on an average over 3000 time-steps, after allowing 2000 timesteps for equilibration. A more detailed discussion of the degree to which this model captures the unusual structural properties of silicon is to be found in [sti85].

12.3 Embedded-atom approach

Interactions

The use of density-independent pair potentials – the Lennard-Jones potential is a well-known example – is justified when the electron clouds responsible for the attractive and repulsive components of the interatomic interactions remain local-ized close to the individual atoms. In metals this is no longer the case and valence electrons may be shared among atoms. This calls for potentials that take the local electron density into account and which, consequently, have a many-body nature.

A convenient way of describing such interactions involves the embedded-atom potential [daw84]. This potential consists of two parts: a pair interaction between metal atoms – actually ions – that does not explicitly depend on density, and a many-body term that depends on the local value of the density at the point where the atom is located. The introduction of many-body interactions can lead to substantial changes in the mechanical properties of the solid, compared with those based on pair interactions alone; examples include the elastic shear moduli and the energy of vacancy formation, properties that cannot be characterized correctly if just pair interactions are involved. A simplified version of the embedded-atom potential, expressed entirely in terms of analytic functions, is described in [hol91]; this form will be considered here.

The two parts of the potential energy are expressed as

$$U = \tfrac{1}{2} \sum_{i=1}^{N_m} \left[\chi \sum_{j \neq i} \phi(r_{ij}) + (1 - \chi) \mathcal{U}(\rho_i) \right] \tag{12.3.1}$$

Here, $\phi(r)$ is a density-independent pair potential and χ is the fractional contribution of this part. The embedding energy $\mathcal{U}$ is a nonlinear function of the local atomic density ρ_i, which is defined in terms of a sum over the neighbors of i of a local weighting function $w(r)$,

$$\rho_i = \sum_{j \neq i} w(r_{ij}) \tag{12.3.2}$$

The function $w(r)$ will be defined in (12.3.7). The nonlinearity of $\mathcal{U}$ is necessary to ensure that it introduces many-body effects, otherwise its contribution would simply be pairwise additive.

The pair potential has the form

$$\phi(r) = \begin{cases} \phi_{lj}(r) & r < r_s \\ \phi_{sp}(r) & r_s \leq r < r_m \\ 0 & r \geq r_m \end{cases} \tag{12.3.3}$$

where $\phi_{lj}(r)$ is the LJ potential (2.2.1) – other potential functions can also be used. The function

$$\phi_{sp}(r) = -a_2(r_m^2 - r^2)^2 + a_3(r_m^2 - r^2)^3 \tag{12.3.4}$$

is a cubic spline (in the variable r^2) that is introduced to ensure the pair interaction drops smoothly to zero over the range r_s to r_m; this is accomplished by requiring that ϕ_{lj} and ϕ_{sp} satisfy

$$\phi_{lj}(r_s) = \phi_{sp}(r_s) \equiv \phi_s$$
$$\phi'_{lj}(r_s) = \phi'_{sp}(r_s) \equiv \phi'_s \qquad (12.3.5)$$
$$\phi''_{lj}(r_s) = \phi''_{sp}(r_s) = 0$$

The cutoff at r_m is introduced, as in other MD studies, to reduce the computational effort. The conditions (12.3.5) lead to relations between the unknowns r_m, a_2 and a_3,

$$r_m^2 = 5r_s^2 \left[1 - \sqrt{1 - (9 - 24\phi_s/r_s\phi'_s)/25} \right]$$

$$a_2 = \frac{5r_s^2 - r_m^2}{8r_s^3(r_m^2 - r_s^2)} \phi'_s \qquad (12.3.6)$$

$$a_3 = \frac{3r_s^2 - r_m^2}{12r_s^3(r_m^2 - r_s^2)^2} \phi'_s$$

The numerical values are determined by the function `EvalEamParams` later on.

The weighting functions $w(r)$ that contribute to the local density (12.3.2) have the spherically symmetric form

$$w(r) = \begin{cases} \dfrac{1}{d(d+1)e} \left(\dfrac{r_m^2 - r^2}{r_m^2 - r_0^2} \right)^2 & r < r_m \\ 0 & r \geq r_m \end{cases} \qquad (12.3.7)$$

where d is the dimensionality of the problem (2 or 3), $r_0 = 2^{1/6}\sigma$ corresponds to the minimum of ϕ_{lj} and $e \equiv \exp(1)$. At $r = r_0$, a value corresponding to the normal (zero pressure) bulk density provided the interaction range encompasses nearest neighbors only,

$$w(r_0) = \frac{1}{d(d+1)e} \qquad (12.3.8)$$

so that at this density $\rho_i = 1/e$; the choice of $w(r)$ is governed by the requirement that the minimum of $\mathcal{U}$ should occur at this density.

The nonlinear embedding energy function $\mathcal{U}$ is chosen to be

$$\mathcal{U}(\rho_i) = d(d+1)\epsilon e\rho_i \log \rho_i \qquad (12.3.9)$$

so that at $r = r_0$

$$U = \frac{N_m}{2} [-\chi d(d+1)\epsilon - (1 - \chi)d(d+1)\epsilon]$$
$$= -N_m n_b \epsilon \qquad (12.3.10)$$

Since there are $n_b = d(d + 1)/2$ bonds per atom in a close packed solid, this amounts to a contribution of $-\epsilon$ per bond; $-U/N_m$ can be regarded as the bulk cohesive energy. The value of χ is chosen so that the energy required to form a vacancy is considerably smaller than the cohesive energy, as occurs in practice; this is very different from the situation at $\chi = 1$, where only the pair potential contributes to the cohesive energy.

The forces acting on the atoms are readily determined, starting from

$$f_k = -\frac{\partial U}{\partial r_k}$$

$$= -\chi \sum_{j \neq k} \phi'(r_{kj})\hat{r}_{kj} - (1 - \chi)n_b e \sum_{i=1}^{N_m}(\log \rho_i + 1)\frac{\partial \rho_i}{\partial r_k} \tag{12.3.11}$$

The second sum in (12.3.11) can be reorganized,

$$\sum_{i=1}^{N_m}(\log \rho_i + 1)\frac{\partial \rho_i}{\partial r_k} = \sum_{i \neq j}(\log \rho_i + 1)w'(r_{ij})(\delta_{ik} - \delta_{jk})\hat{r}_{ij}$$

$$= \sum_{j \neq k} w'(r_{jk})(\log \rho_k + \log \rho_j + 2)\hat{r}_{kj} \tag{12.3.12}$$

since $w'(r_{ij}) = w'(r_{ji})$. Thus

$$f_i = -\sum_{j \neq i}[\chi\phi'(r_{ij}) + (1 - \chi)n_b e w'(r_{ij})](\log \rho_i + \log \rho_j + 2)\hat{r}_{ij} \tag{12.3.13}$$

where

$$n_b e w'(r) = -2r\left(\frac{r_m^2 - r^2}{r_m^2 - r_0^2}\right) \tag{12.3.14}$$

Implementation

The following function[♦] computes the forces due to the embedded-atom potential. It employs a neighbor list, constructed in the usual way for two-body interactions. Standard reduced units are used, in which $\sigma = 1$, $r_0 = 2^{1/6}$ and $\epsilon = 1$. The value of χ is fixed at 1/3. The local densities ρ_i are evaluated by summing over all neighbors j of i for which $r_{ij}^2 < r_m^2$; identification of such pairs is facilitated by the fact that

---

♦ pr_12_2

they all appear in the neighbor list used for computing the pair interactions.

```
void ComputeForces ()
{
  VecR dr;
  real eDim, fcVal, rCutC, rr, rrCdi, rrCut, rrd, rri, rri3,
    rrSwitch, t, uVal;
  int j1, j2, n;

  eDim = NDIM * (NDIM + 1) * exp (1.);
  rCutC = pow (2., 1./6.);
  rrCut = Sqr (rCut);
  rrCdi = 1. / Sqr (rrCut - Sqr (rCutC));
  rrSwitch = Sqr (rSwitch);
  DO_MOL mol[n].logRho = 0.;
  for (n = 0; n < nebrTabLen; n ++) {
    j1 = nebrTab[2 * n];
    j2 = nebrTab[2 * n + 1];
    VSub (dr, mol[j1].r, mol[j2].r);
    VWrapAll (dr);
    rr = VLenSq (dr);
    if (rr < rrCut) {
      t = Sqr (rrCut - rr);
      mol[j1].logRho += t;
      mol[j2].logRho += t;
    }
  }
  DO_MOL {
    if (mol[n].logRho > 0.)
      mol[n].logRho = log ((rrCdi / eDim) * mol[n].logRho);
  }
  DO_MOL VZero (mol[n].ra);
  uSum = 0.;
  for (n = 0; n < nebrTabLen; n ++) {
    j1 = nebrTab[2 * n];
    j2 = nebrTab[2 * n + 1];
    VSub (dr, mol[j1].r, mol[j2].r);
    VWrapAll (dr);
    rr = VLenSq (dr);
    if (rr < rrCut) {
      rrd = rrCut - rr;
      if (rr < rrSwitch) {
        rri = 1. / rr;
        rri3 = Cube (rri);
        fcVal = 48. * rri3 * (rri3 - 0.5) * rri;
        uVal = 4. * rri3 * (rri3 - 1.);
      } else {
        fcVal = (4. * splineA2 + 6. * splineA3 * rrd) * rrd;
        uVal = (splineA2 + splineA3 * rrd) * Sqr (rrd);
      }
      fcVal = embedWt * fcVal + (1. - embedWt) * 2. * rrCdi *
```

```
          (mol[j1].logRho + mol[j2].logRho + 2.) * rrd;              50
      VVSAdd (mol[j1].ra, fcVal, dr);
      VVSAdd (mol[j2].ra, - fcVal, dr);
      uSum += uVal;
    }
  }                                                                   55
  t = 0.;
  DO_MOL t += mol[n].logRho * exp (mol[n].logRho);
  uSum = embedWt * uSum + (1. - embedWt) * 0.5 * eDim * t;
}
```

The structure *Mol* includes an additional real quantity *logRho* standing for $\log \rho_i$. Several new variables are used in defining the potential,

```
real embedWt, rSwitch, splineA2, splineA3;
```

corresponding to χ, r_s, a_2 and a_3, respectively, with *rCut* now denoting r_m; these variables are initialized by a function called from *SetParams*,

```
void EvalEamParams ()
{
  real bb, p, pd, rr, rr3;

  rSwitch = pow (26. / 7., 1. / 6.);                                 5
  rr = Sqr (rSwitch);
  rr3 = Cube (rr);
  p = 4. * (1. / rr3 - 1.) / rr3;
  pd = - 48. * (1. / rr3 - 0.5) / (rSwitch * rr3);
  bb = 4. * (1. - sqrt (1. + 3. * p / (2. * rSwitch * pd)));         10
  splineA2 = (6. * p + bb * rSwitch * pd) / (2. * Sqr (bb * rr));
  splineA3 = - (4. * p + bb * rSwitch * pd) / (2. * Sqr (bb * rr) *
      bb * rr);
  rCut = rSwitch * sqrt (bb + 1.);
  embedWt = 0.3333;                                                  15
}
```

The rest of the program for this version of the embedded-atom method is similar to the two-body case.

Structure measurements

The first of the case studies examines the RDF of a fluid whose atoms interact via the embedded-atom potential, using a system whose initial state is an FCC lattice. The input data are similar to those used in the soft-sphere RDF study (§4.3) at a density of 0.8. The results are shown in Figure 12.2 and compared with the soft-sphere fluid under similar conditions. Apart from a certain amount of softening in

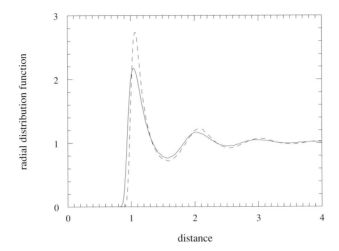

Fig. 12.2.   Radial distribution function for the embedded-atom fluid (solid curve) compared with the soft-sphere fluid (dashed curve).

the peaks, the two curves exhibit similar trends, although this does not prevent other properties from differing substantially, as is indeed the case.

Collision modeling

The second case study[♠] focuses on collisions between extended bodies; this includes the phenomenon of spallation, in which one body (the projectile) impacting at high speed on the surface of another, larger body (the target) causes the ejection of material from the opposite surface [hol91]. The bodies involved are often metallic and it is therefore appropriate to model them using embedded-atom potentials. In order to simplify the task of visualizing the results, as well as reducing the amount of work involved, the simulations are carried out in two dimensions.

The simulated system initially contains two objects: the moving disklike projectile consisting of an array of atoms interacting by means of the embedded-atom potential, and a strip (or wall) of similar atoms forming the stationary target with which the disk is due to collide. A picture of the system appears in Figure 12.3 below. The boundaries of the system are hard walls – to prevent atoms escaping the simulation container – and the atoms at each end of the strip are anchored to prevent the target from moving as a whole.

♠ pr_12_3

The following function initializes the coordinates, using *CoordInRegion* to identify the atoms forming the projectile and target. The atoms belonging to these two bodies are initially arranged on a triangular lattice. Two extra items are included in the *Mol* structure, *inObj* for indicating which atom belongs to which body and *fixed* for identifying anchored atoms.

```
void InitCoords ()
{
  VecR c;
  real wyMax, wyMin;
  int n, nx, ny, p;                                                  5

  n = 0;
  for (ny = 0; ny < initUcell.y; ny ++) {
    for (nx = 0; nx < initUcell.x; nx ++) {
      p = CoordInRegion (nx, ny, &c);                                10
      if (p >= 0) {
        mol[n].r = c;
        mol[n].inObj = p;
        ++ n;
      }                                                              15
    }
  }
  wyMin = 0.5 * region.y;
  wyMax = - wyMin;
  DO_MOL {                                                           20
    if (mol[n].inObj == 1) {
      wyMin = Min (wyMin, mol[n].r.y);
      wyMax = Max (wyMax, mol[n].r.y);
    }
  }                                                                  25
  DO_MOL mol[n].fixed = (mol[n].inObj == 1 &&
    (mol[n].r.y == wyMin || mol[n].r.y == wyMax)) ? 1 : 0;
}


int CoordInRegion (int nx, int ny, VecR *pr)                         30
{
  VecR c, dr;
  int regionCode;

  regionCode = -1;                                                   35
  c.x = nx - initUcell.x / 2 + ((ny % 2 == 0) ? 0.5 : 0.);
  c.y = (ny - initUcell.y / 2) * 0.5 * sqrt (3.);
  VScale (c, initSep);
  if (fabs (c.x) < 0.5 * nMolWall * initSep &&
    fabs (c.y) < 0.5 * region.y) regionCode = 1;                     40
  else {
    VSub (dr, c, diskInitPos);
    if (VLenSq (dr) < Sqr (0.5 * nMolDisk * initSep)) regionCode = 0;
  }
```

```
    if (regionCode >= 0) *pr = c;                                        45
    return (regionCode);
  }
```

Here, *nMolDisk* and *nMolWall* specify the projectile diameter and target thickness in terms of numbers of atoms; the initial spacing of the atoms is determined by *initSep*. The initial *x* coordinate of the projectile is *diskInitPos*, and the target is oriented parallel to the *y* axis and positioned at the origin. The actual number of atoms involved in the simulation is computed by the following function.

```
void EvalMolCount ()
{
  VecR c;
  int nx, ny;
                                                                         5
  nMol = 0;
  for (ny = 0; ny < initUcell.y; ny ++) {
    for (nx = 0; nx < initUcell.x; nx ++) {
      if (CoordInRegion (nx, ny, &c) >= 0) ++ nMol;
    }                                                                   10
  }
}
```

The velocities are initialized as follows, where *diskInitVel* is the initial projectile speed and all mobile atoms are also assigned small random thermal velocities.

```
void InitVels ()
{
  int n;

  velMag = 0.1;                                                          5
  VZero (vSum);
  DO_MOL {
    if (! mol[n].fixed) {
      VRand (&mol[n].rv);
      VScale (mol[n].rv, velMag);                                       10
      VVAdd (vSum, mol[n].rv);
    } else VZero (mol[n].rv);
  }
  DO_MOL {
    if (! mol[n].fixed) VVSAdd (mol[n].rv, - 1. / nMol, vSum);          15
    if (mol[n].inObj == 0) mol[n].rv.x += diskInitVel;
  }
}
```

Other required quantities are determined by *SetParams*, which includes

```
VecR t;

initSep = pow (2., 1./6.);
region.x = 4. * nMolDisk * initSep;
region.y = region.x;                                              5
VSet (t, initSep, 0.5 * sqrt (3.) * initSep);
VDiv (initUcell, region, t);
initUcell.y += 2;
VSet (diskInitPos,
    0.5 * (- 0.5 * region.x - 0.5 * nMolWall * initSep), 0.);     10
```

The function *SingleStep* includes a call to *ApplyWallBoundaryCond* after the first call to the leapfrog integrator and to *ZeroFixedAccels* prior to the second.

```
void ApplyWallBoundaryCond ()
{
  real vSign;
  int n;
                                                                  5
  DO_MOL {
    if (fabs (mol[n].r.x) >= 0.5 * region.x) {
      vSign = (mol[n].r.x > 0.) ? 1. : -1.;
      mol[n].r.x = 0.49999 * vSign * region.x;
      VScale (mol[n].rv, 0.1);                                    10
      if (mol[n].rv.x * vSign > 0.) mol[n].rv.x *= -1.;
    }
    ... (ditto for y component) ...
  }
}                                                                 15

void ZeroFixedAccels ()
{
  int n;
                                                                  20
  DO_MOL {
    if (mol[n].fixed) VZero (mol[n].ra);
  }
}
```

If, after updating the coordinates, an atom is found to be outside the container (in practice by only a very small amount), then its position and velocity are modified as if it had collided with the relevant container wall (similar functions appear in §7.3). The affected coordinate component is altered to bring the atom back inside and the normal component of the velocity reversed. The velocity is also rescaled to

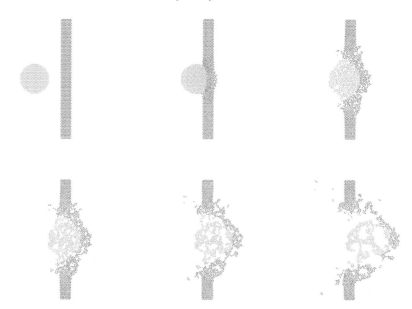

Fig. 12.3. Collision modeling for a small system: successive frames show the initial state, the start of the collision, the ejection of matter from the target surface, and the fragmentation of the projectile.

make collisions of this kind highly inelastic; this helps prevent atoms that bounce off the container walls from interfering with the process under observation.

All atoms interact the same way, although it is straightforward to allow different interactions within the projectile and target, and between the two. New variables and inputs are

```
VecR diskInitPos;
real diskInitVel, initSep;
int nMolDisk, nMolWall;

  NameR (diskInitVel),
  NameI (nMolDisk),
  NameI (nMolWall),
```

and input data

```
diskInitVel      4.
nMolDisk         20
nMolWall         8
```

The results of a single simulation run are shown as a series of images in Figure 12.3. This is another example of the kind of problem where direct visualization of the behavior proves to be highly informative. Different initial projectile speeds will produce different kinds of behavior [hol91]. Since the atoms have small initial thermal velocities, repeated runs with the same parameters, except for the random number seed, will lead to different results.

12.4 Further study

12.1 There exists a method for introducing orientation-dependent forces without resorting to multiple interaction sites; it is based on generalizing the LJ interaction to allow both σ and ϵ to depend on relative molecular alignment (the molecules are treated as rigid bodies) [gay81, sar93]. Ellipsoidal molecules used in the study of liquid crystals can be modeled in this way; examine the spatial and orientational order that occurs in such systems and the expected transitions between the liquid, nematic (orientationally ordered) and crystalline states.

12.2 Examine the collision process as a function of projectile size and velocity, and replace the disk by other shapes; vary the interaction parameters to make the projectile harder and remove the attractive part of the force between projectile and target (to prevent impact welding).

12.3 Study how the collision results are affected if the embedded-atom potential is replaced by, for example, the LJ potential.

12.4 The embedded-atom approach has been used in simulating fracture [hol95]; examine the technical details involved in computations of this kind and, in particular, how suitable boundary conditions can be incorporated.

13

Long-range interactions

13.1 Introduction

The systems examined so far have been based on interactions whose range is comparatively short. Short-range forces exclude a very important group of problems involving electric charges and dipoles; in systems of this type the central role of the long-range forces cannot be neglected. Long-range interactions, which, by definition, are not truncated and therefore extend over the entire system, inherently require $O(N_m^2)$ computational effort. This requirement clearly precludes the treatment of large systems directly and has lead to the development of alternative approaches that are able to deal with the problem efficiently while reducing or avoiding truncation artifacts. In this chapter we introduce some of these techniques.

The first of the methods is based on the Ewald resummation technique familiar from condensed matter physics. By reorganizing the interaction sums over periodic images of the system, the Ewald method is able to incorporate periodic boundaries in a manner that avoids explicit truncation.

The other two methods involve subdividing the simulation region into a hierarchy of cells. At each level in the hierarchy, each of the cells has associated with it a reduced representation of the atoms that it contains, for example, the total charge of the atoms and the position of their center of charge, or, to extend this idea further, several coefficients of the multipole expansion of the charges in the cell relative to the cell center. At ascending levels in the hierarchy – as the cells become larger – this information provides an increasingly coarse representation of the cell contents, but this is compensated by the fact that the information from larger cells will only be used in computing the interactions with more distant atoms.

The simpler of the hierarchical methods considered is an adaptive tree technique, in which subdivision of each cell is repeated until occupancy reaches unity, but only the center of charge is evaluated for each cell. The other method employs a fixed number of cell subdivisions, but several terms of the multipole expansion of each cell's occupants are generated; this can be carried out efficiently because of relations that exist between the expansions for smaller and larger cells. The goal of both approaches is that the interaction computations should require, asymptotically, $O(N_m)$ computational effort. Note that the cell grids used in these methods are for organizational purposes only, unlike other particle–grid methods [hoc88] where the particles experience fields evaluated (from the particle positions) at the sites of a discrete grid.

13.2 Ewald method

Interaction resummation

In the study of ionic crystals, where long-range electrostatic interactions dominate, the Ewald method [zim72] is able to take advantage of the periodic lattice structure to rearrange the expression for the total energy into a form that can be readily evaluated. The same idea can be applied in simulating charged and dipolar fluids with periodic boundaries, where, because the long-range force cannot be truncated without incurring serious error, it continues to act between the periodic replicas of the system as well. The Ewald technique thus eliminates the discontinuity arising from truncated long-range forces, although there are more subtle problems connected with how the properties are affected by the choice of boundary conditions [neu83, del86]. The computational effort depends to some extent on the degree of accuracy required but it grows considerably less rapidly than the all-pairs $O(N_m^2)$ rate.

Consider a system of N_m atoms, each of which now carries a charge (for more on Coulomb systems see [han86a]). A periodic array of replicated systems is created, in the spirit of the periodic boundary conditions used previously, but now, because of the long-range nature of the interactions, the energy of the replicated system includes contributions from all replicas since no truncation is imposed. The total interaction energy is

$$U_{qq} = \frac{1}{2} \sum_{n}' \sum_{i=1}^{N_m} \sum_{j=1}^{N_m} \frac{q_i q_j}{|r_{ij} + Ln|} \tag{13.2.1}$$

where q_i is the charge on atom i and L is the edge length (a cubic region is assumed). The sum is over all integer vectors n, and the prime indicates that terms with $i = j$ are omitted when $|n| = 0$ – self-interaction is prevented but atoms do interact with their replica images.

The Ewald formula is based on reorganizing this replica sum into sums over concentric spherical shells, assuming charge neutrality $\sum_j q_j = 0$,

$$
U_{qq} = \sum_{i \leq i < j \leq N_m} q_i q_j \left[{\sum_n}' \frac{\mathrm{erfc}(\alpha|r_{ij} + Ln|)}{|r_{ij} + Ln|} \right.
$$

$$
\left. + \frac{1}{\pi L} \sum_{n \neq 0} \frac{1}{n^2} \exp\left(-\frac{\pi^2 n^2}{\alpha^2 L^2} + \frac{2\pi i}{L} n \cdot r_{ij} \right) \right]
$$

$$
+ \frac{1}{2} \left[\sum_{n \neq 0} \left(\frac{\mathrm{erfc}(\alpha L n)}{L n} + \frac{1}{\pi L n^2} \exp\left(-\frac{\pi^2 n^2}{\alpha^2 L^2} \right) \right) - \frac{2\alpha}{\sqrt{\pi}} \right] \sum_{j=1}^{N_m} q_j^2
$$

$$
+ \frac{2\pi}{3 L^3} \left| \sum_{j=1}^{N_m} q_j r_j \right|^2
\tag{13.2.2}
$$

where

$$
\mathrm{erfc}(x) = \frac{2}{\sqrt{\pi}} \int_x^\infty e^{-t^2}\, dt
\tag{13.2.3}
$$

is the complementary error function[†]. There are various derivations of (13.2.2); one approach [del80] involves the introduction of a convergence factor into a series that is otherwise only conditionally convergent, followed by a Jacobi theta-function transformation, and then the extraction of the leading order asymptotic terms as the convergence factor tends to zero.

The rearranged sums include a parameter α, whose value must be determined (as shown later) to maximize numerical accuracy; α can be chosen to ensure that terms of order $\exp(-\alpha^2 L^2)$ are negligible, so that (13.2.2) becomes

$$
U_{qq} = \sum_{1 \leq i < j \leq N_m} \frac{q_i q_j\, \mathrm{erfc}(\alpha r_{ij})}{r_{ij}} - \frac{\alpha}{\sqrt{\pi}} \sum_{j=1}^{N_m} q_j^2
$$

$$
+ \frac{1}{2\pi L} \sum_{n \neq 0} \left[\frac{1}{n^2} \exp\left(-\frac{\pi^2 n^2}{L^2 \alpha^2} \right) \left| \sum_{j=1}^{N_m} q_j \exp\left(\frac{2\pi i}{L} n \cdot r_j \right) \right|^2 \right]
\tag{13.2.4}
$$

The real-space terms in (13.2.4) are now short-ranged, so a spherical cutoff (with range $r_c < L/2$) can be used together with periodic boundaries. The sum over Fourier space, $\sum_n$, will also prove amenable to truncation after only a limited number of terms. The squared dipole moment sum in (13.2.2) has been dropped

---

[†] Note that for $x \gg 1$, $\mathrm{erfc}(x) \sim e^{-x^2}$.

from the result; the physical implication of doing this is that the outermost replica shell is effectively surrounded by a conducting medium, whereas including this term would amount to placing the system in a vacuum [del80, del86].

A typical value for the free parameter is $\alpha = 5/L$, though the results turn out to be relatively insensitive to the choice, provided there are sufficient terms in $\sum_n$. A spherical cutoff is imposed on this sum, so that $n \equiv |n| \leq n_c$; typically n_c is about 5. The accuracy of the Ewald result with the chosen parameters is readily checked numerically [kol92]. The invariance under the transformation $n \to -n$ can be used to halve the number of terms in $\sum_n$ by considering $n_z \geq 0$ only; the computational work can be reduced even further by restricting the sum to a single octant and calculating the contributions of the four octants $(\pm n_x, \pm n_y, +n_z)$ at the same time.

Imposing a cutoff on the real-space sum at r_c leads to an error of order $\exp(-\alpha^2 r_c^2)$; truncating the Fourier space sum at n_c produces an error of order $\exp(-\pi^2 n_c^2/\alpha^2 L^2)$. Therefore, to obtain similarly sized errors in both the real and Fourier contributions to U_{qq}, simply set

$$n_c = \alpha^2 r_c L/\pi \qquad (13.2.5)$$

For the case $r_c = L/2$, if $\alpha = 5/L$ we obtain the very modest number

$$n_c = 25/2\pi \approx 4 \qquad (13.2.6)$$

The actual number of terms in the sum (before halving) is roughly $4\pi n_c^3/3$. For large systems it can be shown [per88] that if the optimal α is chosen for a specified numerical accuracy, then the computational effort grows as $N_m^{3/2}$; this represents a considerable saving over the original N_m^2 dependence and, of course, there are no longer any errors due to interaction cutoff.

Dipolar systems can be treated in a similar fashion. Even though the interaction energy of a pair of dipoles falls off with distance as $1/r^3$, this is still sufficiently slow in three dimensions for the sum over replica systems to remain conditionally convergent, so that the use of a cutoff is not possible; thus the same careful consideration given to the charge problem is required for dipoles as well.

The interaction of a pair of dipoles of strength μ is

$$-\mu^2 (s_i \cdot \nabla)(s_j \cdot \nabla)\left(\frac{1}{r_{ij}}\right) \qquad (13.2.7)$$

where s_i is a unit vector along the direction of the dipole. This differential operator can be applied to (13.2.2) to obtain the potential energy of a dipole system [ada76, del80],

$$U_{dd} = \mu^2 \sum_{1 \le i < j \le N_m} \left[\left(\frac{\text{erfc}(\alpha r_{ij})}{r_{ij}^3} + \frac{2\alpha \exp(-\alpha^2 r_{ij}^2)}{\sqrt{\pi} r_{ij}^2} \right) (s_i \cdot s_j) \right.$$

$$\left. - \left(\frac{3 \, \text{erfc}(\alpha r_{ij})}{r_{ij}^5} + \left(2\alpha^2 + \frac{3}{r_{ij}^2}\right) \frac{2\alpha \exp(-\alpha^2 r_{ij}^2)}{\sqrt{\pi} r_{ij}^2} \right) (s_i \cdot r_{ij})(s_j \cdot r_{ij}) \right]$$

$$+ \frac{2\pi \mu^2}{L^3} \sum_{n \ne 0} \left[\frac{1}{n^2} \exp\left(-\frac{\pi^2 n^2}{L^2 \alpha^2}\right) \left| \sum_{j=1}^{N_m} (n \cdot s_j) \exp\left(\frac{2\pi i}{L} n \cdot r_j\right) \right|^2 \right]$$

$$- \frac{2\alpha^3 \mu^2 N_m}{3\sqrt{\pi}} + \frac{2\pi \mu^2}{3 L^3} \left| \sum_{j=1}^{N_m} s_j \right|^2 \qquad (13.2.8)$$

The derivation of (13.2.8) makes use of the results

$$\nabla(s \cdot r) = s \qquad (13.2.9)$$

$$\frac{d}{dx} \text{erfc}(\alpha x) = -\frac{2\alpha}{\sqrt{\pi}} e^{-\alpha^2 x^2} \qquad (13.2.10)$$

Terms of order $\exp(-\alpha^2 L^2)$ will be dropped because α can be chosen to ensure their extreme smallness, and since the system is again assumed to be surrounded by a conducting medium the squared sum over s_j is also dropped. The energy (13.2.8) can then be written concisely as

$$U_{dd} = \mu^2 \sum_{1 \le i < j \le N_m} \left[a_1(r_{ij})(s_i \cdot s_j) - a_2(r_{ij})(s_i \cdot r_{ij})(s_j \cdot r_{ij}) \right]$$

$$+ \frac{2\pi \mu^2}{L^3} \sum_{n \ne 0} e(n) \left[C(n)^2 + S(n)^2 \right] - \frac{2\alpha^3 \mu^2 N_m}{3\sqrt{\pi}} \qquad (13.2.11)$$

where several new functions have been introduced,

$$a_n(r) = \begin{cases} \dfrac{\text{erfc}(\alpha r)}{r^3} + \dfrac{2\alpha \exp(-\alpha^2 r^2)}{\sqrt{\pi} r^2} & n = 1 \\[2ex] -\dfrac{1}{r} \dfrac{d}{dr} a_{n-1}(r) & n = 2, 3 \end{cases} \qquad (13.2.12)$$

$$e(n) = \frac{1}{n^2} \exp\left(\frac{-\pi^2 n^2}{L^2 \alpha^2}\right) \qquad (13.2.13)$$

$$\begin{Bmatrix} C \\ S \end{Bmatrix}(n) = \sum_{j=1}^{N_m} (n \cdot s_j) \begin{Bmatrix} \cos \\ \sin \end{Bmatrix}\left(\frac{2\pi}{L} n \cdot r_j\right) \qquad (13.2.14)$$

In the limit $\alpha \rightarrow 0$, (13.2.11) reduces to the simple dipole result

$$U_{dd} = \mu^2 \sum_{i<j} \frac{1}{r_{ij}^3}\left[s_i \cdot s_j - \frac{3}{r_{ij}^2}(s_i \cdot r_{ij})(s_j \cdot r_{ij})\right] \qquad (13.2.15)$$

Dynamics

The MD model used in this case study deals with dipoles that are attached to soft-sphere atoms; if the LJ potential is used for this problem instead [mai81], the interaction is known as the Stockmayer potential. The Lagrange equations of motion for translation involve the usual soft-sphere interaction together with a contribution from the force produced by the dipolar interactions. The latter is obtained by evaluating $-\nabla_{r_i} U_{dd}$, and consists of a sum over atoms truncated at the cutoff range r_c, together with a Fourier space sum truncated at n_c. The dipole contribution to the force on a single atom is thus

$$f_i = \mu^2 \sum_{j(\neq i)}\left[\left(a_2(r_{ij})(s_i \cdot s_j) - a_3(r_{ij})(s_i \cdot r_{ij})(s_j \cdot r_{ij})\right)r_{ij}\right.$$

$$\left. + a_2(r_{ij})\left((s_j \cdot r_{ij})s_i + (s_i \cdot r_{ij})s_j\right)\right] \qquad (13.2.16)$$

$$+ \frac{8\pi^2\mu^2}{L^4} \sum_{n\neq 0} e(n)(n \cdot s_i)n\left[C(n)\sin\left(\frac{2\pi}{L}n \cdot r_i\right) - S(n)\cos\left(\frac{2\pi}{L}n \cdot r_i\right)\right]$$

The equation of motion for the dipole vector s_i is just the rotation equation for a linear rigid molecule given in §8.2,

$$\ddot{s}_i = I^{-1}g_i - \left(I^{-1}(s_i \cdot g_i) + \dot{s}_i^2\right)s_i \qquad (13.2.17)$$

where $g_i = -\nabla_{s_i} U_{dd}$ also can be expressed as two sums,

$$g_i = \mu^2 \sum_{j(\neq i)}[-a_1(r_{ij})s_j + a_2(r_{ij})(s_j \cdot r_{ij})r_{ij}] \qquad (13.2.18)$$

$$- \frac{4\pi\mu^2}{L^3} \sum_{n\neq 0} e(n)n\left[C(n)\cos\left(\frac{2\pi}{L}n \cdot r_i\right) + S(n)\sin\left(\frac{2\pi}{L}n \cdot r_i\right)\right]$$

In (13.2.17) we demonstrate the use of the second-order form of the rotational equation of motion; there is also an alternative method based on a pair of first-order equations for each dipole (see §8.2).

The real-space contributions to each of the force terms f_i and g_i, and to the potential energy, are evaluated♠ by the following function; a cubic simulation region is assumed throughout, with the cutoff at $r_c = L/2$.

♠ pr_13_1

```
void ComputeForcesDipoleR ()
{
  VecR dr, w;
  real a1, a2, a3, alpha2, d, irPi, rr, rrCut, rri, sr1, sr2, ss, t;
  int j1, j2, n;

  rrCut = Sqr (0.5 * region.x);
  irPi = 1. / sqrt (M_PI);
  alpha2 = Sqr (alpha);
  DO_MOL VZero (mol[n].sa);
  for (j1 = 0; j1 < nMol - 1; j1 ++) {
    for (j2 = j1 + 1; j2 < nMol; j2 ++) {
      VSub (dr, mol[j1].r, mol[j2].r);
      VWrapAll (dr);
      rr = VLenSq (dr);
      if (rr < rrCut) {
        d = sqrt (rr);
        rri = 1. / rr;
        t = 2. * dipoleInt * alpha * exp (- alpha2 * rr) * rri * irPi;
        a1 = dipoleInt * erfc (alpha * d) * rri / d + t;
        a2 = 3. * a1 * rri + 2. * alpha2 * t;
        a3 = 5. * a2 * rri + 4. * Sqr (alpha2) * t;
        ss = VDot (mol[j1].s, mol[j2].s);
        sr1 = VDot (mol[j1].s, dr);
        sr2 = VDot (mol[j2].s, dr);
        VSSAdd (w, sr2, mol[j1].s, sr1, mol[j2].s);
        t = (a2 * ss - a3 * sr1 * sr2);
        VSSAdd (w, t, dr, a2, w);
        VVAdd (mol[j1].ra, w);
        VVSub (mol[j2].ra, w);
        VVSAdd (mol[j1].sa, - a1, mol[j2].s);
        VVSAdd (mol[j1].sa, a2 * sr2, dr);
        VVSAdd (mol[j2].sa, - a1, mol[j1].s);
        VVSAdd (mol[j2].sa, a2 * sr1, dr);
        uSum += a1 * ss - a2 * sr1 * sr2;
      }
    }
  }
  uSum -= 2. * dipoleInt * Cube (alpha) * nMol * irPi / 3.;
}
```

The version of the structure *Mol* used here has the additional elements

```
VecR s, sv, sa, sa1, sa2, so, svo;
```

The next function handles the Fourier space part of the interactions. Here there is a choice between versions that either do or do not take advantage of the full symmetries in the sum over **n**. The latter is more concise, the former more efficient but even more tedious to read. We will leave the faster version as an exercise for the concerned reader and discuss the shorter form that only makes use of the shortcut provided by the $\pm\boldsymbol{n}$ symmetry. In the computation, the sines and cosines of the dot products are each expanded as triple-product sums

$$\sin(x + y + z) = \sin x \cos y \cos z + \ldots \qquad (13.2.19)$$

and the trigonometric functions are tabulated by *EvalSinCos* prior to the calculation using multiple-angle recursion relations, as in §5.4.

```
void ComputeForcesDipoleF ()
{
  VecR vc, vn, vs;
  real fMult, gr, gs, gu, pc, ps, sumC, sumS, t, w;
  int n, nvv, nx, ny, nz;                                          5

  gu = 2. * M_PI * dipoleInt / Cube (region.x);
  gr = 4. * M_PI * gu / region.x;
  gs = 2. * gu;
  EvalSinCos ();                                                   10
  w = Sqr (M_PI / (region.x * alpha));
  for (nz = 0; nz <= fSpaceLimit; nz ++) {
    for (ny = - fSpaceLimit; ny <= fSpaceLimit; ny ++) {
      for (nx = - fSpaceLimit; nx <= fSpaceLimit; nx ++) {
        VSet (vn, nx, ny, nz);                                     15
        nvv = VLenSq (vn);
        if (nvv == 0 || nvv > Sqr (fSpaceLimit)) continue;
        fMult = 2. * exp (- w * nvv) / nvv;
        if (nz == 0) fMult *= 0.5;
        sumC = 0.;                                                 20
        sumS = 0.;
        DO_MOL {
          VSet (vc, tCos[abs (nx)][n].x, tCos[abs (ny)][n].y,
            tCos[nz][n].z);
          VSet (vs, tSin[abs (nx)][n].x, tSin[abs (ny)][n].y,      25
            tSin[nz][n].z);
          if (nx < 0) vs.x = - vs.x;
          if (ny < 0) vs.y = - vs.y;
          pc = vc.x * vc.y * vc.z - vc.x * vs.y * vs.z -
            vs.x * vc.y * vs.z - vs.x * vs.y * vc.z;               30
          ps = vs.x * vc.y * vc.z + vc.x * vs.y * vc.z +
            vc.x * vc.y * vs.z - vs.x * vs.y * vs.z;
          sumC += VDot (vn, mol[n].s) * pc;
          sumS += VDot (vn, mol[n].s) * ps;
        }                                                          35
        DO_MOL {
          VSet (vc, tCos[abs (nx)][n].x, tCos[abs (ny)][n].y,
            tCos[nz][n].z);
          VSet (vs, tSin[abs (nx)][n].x, tSin[abs (ny)][n].y,
```

```
                      tSin[nz][n].z);                                        40
            if (nx < 0) vs.x = - vs.x;
            if (ny < 0) vs.y = - vs.y;
            pc = vc.x * vc.y * vc.z - vc.x * vs.y * vs.z -
                vs.x * vc.y * vs.z - vs.x * vs.y * vc.z;
            ps = vs.x * vc.y * vc.z + vc.x * vs.y * vc.z +      45
                vc.x * vc.y * vs.z - vs.x * vs.y * vs.z;
            t = gr * fMult * VDot (vn, mol[n].s) *
                (sumC * ps - sumS * pc);
            VVSAdd (mol[n].ra, t, vn);
            t = gs * fMult * (sumC * pc + sumS * ps);          50
            VVSAdd (mol[n].sa, - t, vn);
        }
        uSum += gu * fMult * (Sqr (sumC) + Sqr (sumS));
      }
    }                                                          55
  }
}
```

Here, `fSpaceLimit` corresponds to n_c. The function that fills the sine and cosine arrays is

```
void EvalSinCos ()
{
  VecR t, tt, u, w;
  int j, n;
                                                               5
  VSetAll (t, 2. * M_PI);
  VDiv (t, t, region);
  DO_MOL {
    VMul (tt, t, mol[n].r);
    VSetAll (tCos[0][n], 1.);                                  10
    VSetAll (tSin[0][n], 0.);
    VSet (tCos[1][n], cos (tt.x), cos (tt.y), cos (tt.z));
    VSet (tSin[1][n], sin (tt.x), sin (tt.y), sin (tt.z));
    VSCopy (u, 2., tCos[1][n]);
    VMul (tCos[2][n], u, tCos[1][n]);                          15
    VMul (tSin[2][n], u, tSin[1][n]);
    VSetAll (tt, 1.);
    VVSub (tCos[2][n], tt);
    for (j = 3; j <= fSpaceLimit; j ++) {
      VMul (w, u, tCos[j - 1][n]);                             20
      VSub (tCos[j][n], w, tCos[j - 2][n]);
      VMul (w, u, tSin[j - 1][n]);
      VSub (tSin[j][n], w, tSin[j - 2][n]);
    }
  }                                                            25
}
```

The right-hand sides of the equations of motion (13.2.17) are computed by the following function; prior to calling the function, the variables $mol[].sa$ contain the values of g_i.

```
void ComputeDipoleAccel ()
{
  real t;
  int n;

  DO_MOL {
    t = VDot (mol[n].sa, mol[n].s) + mInert * VLenSq (mol[n].sv);
    VVSAdd (mol[n].sa, - t, mol[n].s);
    VScale (mol[n].sa, 1. / mInert);
  }
}
```

New variables appearing in this program are

```
VecR **tCos, **tSin;
real alpha, dipoleInt, mInert, vvsSum;
int fSpaceLimit, stepAdjustTemp;
```

where $dipoleInt$ corresponds to μ^2; input data include

```
NameR (alpha),
NameR (dipoleInt),
NameI (fSpaceLimit),
NameR (mInert),
NameI (stepAdjustTemp),
```

and the extra array allocations ($AllocArrays$) are

```
AllocMem2 (tCos, fSpaceLimit + 1, nMol, VecR);
AllocMem2 (tSin, fSpaceLimit + 1, nMol, VecR);
```

Initialization ($SetupJob$) requires

```
InitAngCoords ();
InitAngVels ();
InitAngAccels ();
```

where, assuming the dipoles are initially aligned, these functions are

```
void InitAngCoords ()
{
  int n;
```

```
    DO_MOL VSet (mol[n].s, 0., 0., 1.);                              5
  }

  void InitAngVels ()
  {
    real ang, angvFac;                                             10
    int n;

    angvFac = velMag / sqrt (1.5 * mInert);
    DO_MOL {
      ang = 2. * M_PI * RandR ();                                  15
      VSet (mol[n].sv, cos (ang), sin (ang), 0.);
      VScale (mol[n].sv, angvFac);
    }
  }
```

and *InitAngAccels* is based on the PC version of *InitAccels* (§3.6). The initial values ensure that $\dot{s}_i = \omega_i \times s_i$, and the factor of 1.5 in *InitAngVels* arises from the fact that each molecule has only two rotational degrees of freedom.

The sequence of function calls in *SingleStep* is the following:

```
    PredictorStep ();
    PredictorStepS ();
    ComputeForces ();
    ComputeForcesDipoleR ();
    ComputeForcesDipoleF ();                                        5
    ComputeDipoleAccel ();
    ApplyThermostat ();
    CorrectorStep ();
    CorrectorStepS ();
    AdjustDipole ();                                               10
    ApplyBoundaryCond ();
    EvalProps ();
    if (stepCount % stepAdjustTemp == 0) AdjustTemp ();
```

The integration functions *PredictorStepS* and *CorrectorStepS* for the rotational motion are the same as the corresponding (three-dimensional) translational functions – simply change the variable names. Calculations of kinetic energy, the Lagrange multiplier used for the thermostat, and the temperature adjustment, must all allow for the rotational motion, as described in §8.2. The function *AdjustDipole* is needed to restore the *s* vectors to unit length because of the small numerical drift,

```
  void AdjustDipole ()
  {
    real sFac;
    int n;
```

```
DO_MOL {
  sFac = VLen (mol[n].s);
  VScale (mol[n].s, 1. / sFac);
}
}
```

Properties

In addition to the spatial correlations present in the fluid, the fact that each molecule has a dipole moment means that orientational order can be studied as well. The magnitude of the dipole directional order parameter

$$M = \frac{1}{N_m} \left| \sum_{i=1}^{N_m} s_i \right| \qquad (13.2.20)$$

is evaluated by an addition to `EvalProps`,

```
VecR w;
...
VZero (w);
DO_MOL VVAdd (w, mol[n].s);
dipoleOrder.val = VLen (w) / nMol;
```

The variable used in computing $\langle M \rangle$ and its fluctuations – the latter are related to the dielectric constant [del86, han86b] – is

```
Prop dipoleOrder;
```

and the additions to `AccumProps` for this calculation are

```
if (icode == 0) {
  ...
  PropZero (dipoleOrder);
} else if (icode == 1) {
  ...
  PropAccum (dipoleOrder);
} else if (icode == 2) {
  ...
  PropAvg (dipoleOrder, stepAvg);
}
```

The order parameter M is a measure of long-range orientational order, but the local dipole alignment is a separate issue. The liquid state is characterized by short-range structural order, so that for a dipolar fluid it becomes possible to examine certain combinations of positional and orientational order to determine, for example, how the average relative orientation of the dipoles in the neighborhood of a given dipole depends on range. To be more specific, given the definition of some orientational quantity, compute its average values over the atoms in a series of concentric spherical shells centered on the atom of interest; this amounts to an extension of the RDF computation in §4.3 where only shell occupancy was considered.

The two quantities of interest here [del86, han86b] are the relative orientation of the dipole vectors, $s_i \cdot s_j$, and the angular part of the dipole energy (13.2.15), namely, $3(s_i \cdot \hat{r}_{ij})(s_j \cdot \hat{r}_{ij}) - s_i \cdot s_j$. The principal change to $EvalRdf$, apart from having to initialize and later normalize three sets of data rather than one, just as in the case of rigid molecules (§8.4), is the addition

```
real sr1, sr2, ss;
...
if (rr < Sqr (rangeRdf)) {
  ss = VDot (mol[j1].s, mol[j2].s);
  sr1 = VDot (mol[j1].s, dr);                              5
  sr2 = VDot (mol[j2].s, dr);
  n = sqrt (rr) / deltaR;
  ++ histRdf[0][n];
  histRdf[1][n] += ss;
  histRdf[2][n] += 3. * sr1 * sr2 / rr - ss;              10
}
```

Measurements

We begin with a test[♠] that provides some idea of the accuracy of the Ewald method for dipole systems. The computations use a set of 216 dipoles that are randomly positioned and oriented in a cube of edge $L = 10$. In order to prevent the results from being dominated by a few very close pairs, instead of using completely random coordinates, the dipoles are placed at the sites of a simple cubic lattice and then each coordinate component is randomly shifted in either direction by up to a quarter of the lattice spacing. The dipole energy and the sum of the absolute values of the force components (per dipole and with $\mu = 1$) are then computed for various values of α and n_c. The results of the force calculations are shown in Figure 13.1; even with $n_c = 5$ there is no problem in obtaining forces with relative error of order 10^{-4} and energies (not shown) with error 10^{-3}.

The MD run described here – compare [pol80, kus90] which use slightly different potentials – is carried out for $N_m = 108$, with input data

```
alpha            5.5
deltaT           0.0025
density          0.8
```

♠ pr_ewaldtest

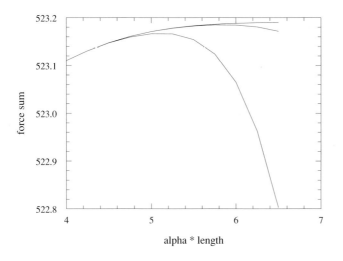

Fig. 13.1. Sum of force components for a random dipole system as a function of the parameter α, for $n_c = 5, 6$ and 7.

```
dipoleInt        4.
fSpaceLimit      5
initUcell        3 3 3
limitRdf         250
mInert           0.025
rangeRdf         2.5
sizeHistRdf      125
stepAdjustTemp   1000
stepAvg          200
stepEquil        1000
stepLimit        86000
stepRdf          20
temperature      1.35
```

The initial state is an FCC lattice with parallel dipoles. We use constant-temperature dynamics, PC integration and a timestep whose value is influenced by the moment of inertia; the temperature drift over 1000 timesteps is typically 0.5% (there is also a very slight drift in the center of mass momentum due to the way the forces are evaluated, which must also be corrected).

The results for $g(r)$ and the two functions used to measure short-range orientational order are shown in Figure 13.2. The function $h_{110}(r)$ measures the distance dependence of $\langle s_i \cdot s_j \rangle$ and $h_{112}(r)$ does the same for $\langle 3(s_i \cdot \hat{r}_{ij})(s_j \cdot \hat{r}_{ij}) - s_i \cdot s_j \rangle$. Equilibration tends to be relatively slow for this system, so measurements made during the first 50 000 timesteps are ignored and the rest are averaged.

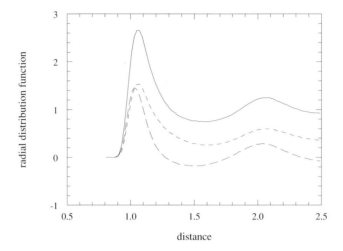

Fig. 13.2. Radial distribution function (solid curve) for the dipole fluid, together with the correlation functions $h_{112}(r)$ (short dashes) and $h_{110}(r)$ (long dashes).

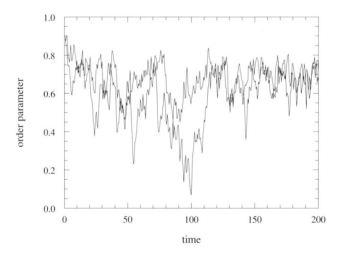

Fig. 13.3. Time dependence of the order parameter for two different runs.

Properties associated with the presence of long-range order, in particular the dielectric constant, are more difficult to evaluate reliably [kus90]. One reason for this is the very long fluctuation timescale. Figure 13.3 shows how M varies over the course of two separate runs, with different values of `randSeed` used for the initial state as in §2.5 (see also [pay93]). Attempting to estimate a mean value from data that is subject to relatively large fluctuations over long time intervals is risky at best; while the results indicate $\langle M \rangle = 0.62$, a block variance analysis (§4.2) designed to obtain $\sigma(\langle M \rangle)$ shows no hint of convergence. In order to arrive at a reliable estimate – not the goal of this pedagogical demonstration – substantially longer runs are required. The question of the nature of the surrounding medium alluded to earlier, as well as the possible size dependence of the results, must also be addressed.

13.3 Tree-code approach

Design considerations

The adaptive approach to handling long-range interactions [mak89, pfa94] can be described as follows. Subdivide the entire region into eight cells (for the case of three dimensions, four cells in two dimensions); repeat the subdivision for each of the cells that contains more than a single atom, and continue this recursive process until no cells are multiply occupied. The interaction computation then considers each atom in turn and pairs it with all cells, beginning with the cells at the top level in the subdivision hierarchy. If the center of the cell is sufficiently distant from the atom – the distance criterion is chosen to be proportional to the size of the cell under examination, so the relevant quantity is the angle subtended by the cell at the atom position – then its total charge, positioned at the center of charge, is used to represent the entire contents of the cell. If this criterion is not met, each of the occupied descendant cells must be examined in a similar manner. This process may need to continue until a singly-occupied cell is eventually reached, at which stage the pair of atoms can be treated without approximation.

The accuracy of the interaction calculation depends on the minimal range at which a cell is no longer required to undergo further subdivision; the greater the range, the higher the accuracy. Since the computation speed drops as the range is increased, a compromise between accuracy and performance must be established empirically. A similar issue is encountered in the fast-multipole method described in the next section. Clearly, a certain minimal number of atoms is required for this technique to be more efficient computationally than the much simpler all-pairs approach.

Tree construction

The most natural way of structuring the data used to organize the atoms into a hierarchical set of cells is the 'oct-tree', a tree structure [knu68, knu73] in which each node represents a cell[†], and – assuming the cell contains more than one atom – includes pointers to the nodes at the next level in the tree representing the cells into which it is divided. The C structure associated with each cell node is[♠]

```
typedef struct {
  VecR cm, midPt;
  int atomPtr, nOcc, nSub, subPtr;
} TCell;
```

It contains the coordinates of the cell's center of charge `cm` and midpoint `midPt`. A singly-occupied cell uses `atomPtr` to identify the atom involved; during the subdivision phase `atomPtr` is employed as the start of a temporary linked list of atoms belonging to the cell that have yet to be assigned to smaller cells at the next level in the subdivision. Other elements are the number of atoms in the cell `nOcc`, the number of occupied descendant cells `nSub` and a pointer `subPtr` showing where the first of the occupied descendant cells at the next lower level is stored; additional descendants will be stored consecutively – indeed all occupied cells at a given level are stored consecutively – and cells found to be empty are ignored.

The tree is constructed (at each timestep) by successively subdividing occupied cells, one level of the hierarchy at a time, until no multiple occupancies remain; the process is detailed below. Additional information, required to complete the description of each level in the hierarchy and assist in the processing, is placed in the structure

```
typedef struct {
  VecR coordSum;
  int cCur, cLast, fCell, lCell;
} TLevel;
```

The roles of these quantities will become apparent in due course. Additionally, the `Mol` structure associated with each atom includes a pointer `next`, used for building the linked lists of atoms associated with the cells during tree construction (the list whose header is stored in `atomPtr`).

The interaction computations involve a backtracking algorithm that traverses the oct-tree, level by level, starting from the top level; each cell is accessed in turn, and a decision is made whether it is sufficiently far from the atom under consideration that its center of charge coordinates can be used to evaluate the forces, or whether further subdivision is required[‡]. Computing the centers of charge of the cell contents uses a similar scheme. While there are, of course, a variety of ways

---

[†] These cells are distinct from the cells used elsewhere for short-range interactions.

[♠] pr_13_2

[‡] Recursion is not used explicitly in the program – in the sense that functions repeatedly call themselves – to avoid the computational overhead involved; instead, an equivalent iterative approach is used.

of organizing both the computation and the supporting data structures, the method
used here has been chosen for its simplicity; efficiency is unlikely to be impacted
significantly on account of this choice.

The individual functions required for the force computation are called from
ComputeForces,

```
void ComputeForces ()
{
  BuildIntTree ();
  LocateIntTreeCellCm ();
  ScanIntTree ();                                                5
}
```

Since this approach does not attempt to address the interaction truncation issue, periodic boundaries are not used and the system will have hard-wall boundaries (note
that due to wall collisions total momentum is not conserved in this case). Clearly,
a large region size is required if finite-size effects are to be kept to a minimum; if
the system is energetically bound, placing the walls at a sufficiently large distance
could produce an effectively open system, provided no atoms manage to travel far
enough to encounter the walls.

The function *BuildIntTree* constructs the tree structure. The subdivision process continues until cells contain only a single atom; limits (assumed to be adequately large) have been set on the numbers of levels and cells allowed, but these
restrictions could be relaxed.

```
void BuildIntTree ()
{
  VecR cEdge;
  int c, cFree, m, mm, n, nOccMax, nv, p, pp;
                                                                 5
  nv = 0;
  cFree = 1;
  tLevel[nv].fCell = 0;
  tLevel[nv].lCell = 0;
  tCell[0].nOcc = nMol;                                          10
  VZero (tCell[0].midPt);
  tCell[0].atomPtr = 0;
  DO_MOL mol[n].next = n + 1;
  mol[nMol - 1].next = -1;
  nOccMax = nMol;                                                15
  cEdge = region;
  while (nOccMax > 1) {
    ++ nv;
    if (nv > maxLevel) ErrExit (ERR_TOO_MANY_LEVELS);
    VScale (cEdge, 0.5);                                         20
    tLevel[nv].fCell = cFree;
```

```
      nOccMax = 1;
      for (c = tLevel[nv - 1].fCell; c <= tLevel[nv - 1].lCell; c ++) {
        if (tCell[c].nOcc > 1) {
          tCell[c].subPtr = cFree;                                          25
          if (cFree + 8 >= maxCells) ErrExit (ERR_TOO_MANY_CELLS);
          for (m = 0; m < 8; m ++) {
            tCell[cFree + m].nOcc = 0;
            tCell[cFree + m].atomPtr = -1;
          }                                                                 30
          for (p = tCell[c].atomPtr; p >= 0; p = pp) {
            m = ((mol[p].r.x >= tCell[c].midPt.x) ? 1 : 0) +
                ((mol[p].r.y >= tCell[c].midPt.y) ? 2 : 0) +
                ((mol[p].r.z >= tCell[c].midPt.z) ? 4 : 0);
            pp = mol[p].next;                                               35
            mol[p].next = tCell[cFree + m].atomPtr;
            tCell[cFree + m].atomPtr = p;
            ++ tCell[cFree + m].nOcc;
          }
          mm = 0;                                                          40
          for (m = 0; m < 8; m ++) {
            if (tCell[cFree + m].nOcc > 0) {
              tCell[cFree + mm].atomPtr = tCell[cFree + m].atomPtr;
              tCell[cFree + mm].nOcc = tCell[cFree + m].nOcc;
              nOccMax = Max (nOccMax, tCell[cFree + mm].nOcc);             45
              VSAdd (tCell[cFree + mm].midPt, tCell[c].midPt,
                  -0.5, cEdge);
              if (m & 1) tCell[cFree + mm].midPt.x += cEdge.x;
              if (m & 2) tCell[cFree + mm].midPt.y += cEdge.y;
              if (m & 4) tCell[cFree + mm].midPt.z += cEdge.z;             50
              ++ mm;
            }
          }
          tCell[c].nSub = mm;
          cFree += mm;                                                     55
        }
      }
      tLevel[nv].lCell = cFree - 1;
    }
    tLevel[nv + 1].lCell = -1;                                             60
  }
```

After the cell assignment, the centers of charge of the cell occupants are computed using a backtracking algorithm to traverse the tree. In this example all atoms are assumed to have identical (unit) charge, so the centers of charge and mass coincide; this is easily changed to handle the more general case.

```
void LocateIntTreeCellCm ()
{
  int nv, p;
```

```
    nv = 0;
    tLevel[nv].cCur = 0;
    tLevel[nv].cLast = tCell[0].nSub;
    VZero (tLevel[nv].coordSum);
    do {
      if (tLevel[nv].cCur < tLevel[nv].cLast) {
        if (tCell[tLevel[nv].cCur].nOcc == 1) {
          p = tCell[tLevel[nv].cCur].atomPtr;
          tCell[tLevel[nv].cCur].cm = mol[p].r;
          VVAdd (tLevel[nv].coordSum, mol[p].r);
          ++ tLevel[nv].cCur;
        } else {
          ++ nv;
          tLevel[nv].cCur = tCell[tLevel[nv - 1].cCur].subPtr;
          tLevel[nv].cLast = tLevel[nv].cCur +
              tCell[tLevel[nv - 1].cCur].nSub;
          VZero (tLevel[nv].coordSum);
        }
      } else {
        VVAdd (tLevel[nv - 1].coordSum, tLevel[nv].coordSum);
        VSCopy (tCell[tLevel[nv - 1].cCur].cm,
            1. / tCell[tLevel[nv - 1].cCur].nOcc, tLevel[nv].coordSum);
        -- nv;
        ++ tLevel[nv].cCur;
      }
    } while (nv > 0);
}
```

New variables introduced for this computation are

```
TLevel *tLevel;
TCell *tCell;
real distFac;
int maxCells, maxLevel, nPair;
```

and the required memory allocation (in `AllocArrays`)

```
    AllocMem (tLevel, maxLevel, TLevel);
    AllocMem (tCell, maxCells, TCell);
```

with `maxLevel` supplied as input data, and the rather arbitrary limit (in `SetParams`)

```
    maxCells = 2 * nMol;
```

Interaction computations

The interactions are evaluated by *ScanIntTree*, again using a backtracking approach. This function considers each atom in turn and pairs it with the averaged contents of all the cells at suitable levels in hierarchy. The level at which each cell is utilized depends on its distance from the atom, the further away it is, the larger the cell can be (in other words, it can be at a higher level in the tree) while maintaining a specified degree of accuracy. Note that Newton's third law is not used here; indeed, because of the way the atoms contribute to the combined cell sums, the interaction of atom i on j and that of j on i at a given instant can differ, so that the total momentum is not strictly conserved (even without the effects of the hard walls).

```
void ScanIntTree ()
{
  VecR dr;
  real b, edgeLen, rr, rri;
  int n, nv, p;                                              5

  DO_MOL VZero (mol[n].ra);
  uSum = 0.;
  nPair = 0;
  DO_MOL {                                                   10
    nv = 0;
    edgeLen = region.x;
    tLevel[nv].cCur = 0;
    tLevel[nv].cLast = tCell[0].nSub;
    do {                                                     15
      if (tLevel[nv].cCur < tLevel[nv].cLast) {
        VSub (dr, mol[n].r, tCell[tLevel[nv].cCur].cm);
        rr = VLenSq (dr);
        if (rr > Sqr (distFac * edgeLen)) {
          rri = 1. / rr;                                     20
          b = tCell[tLevel[nv].cCur].nOcc * sqrt (rri);
          VVSAdd (mol[n].ra, b * rri, dr);
          uSum += b;
          ++ nPair;
          ++ tLevel[nv].cCur;                                25
        } else {
          if (tCell[tLevel[nv].cCur].nOcc == 1) {
            p = tCell[tLevel[nv].cCur].atomPtr;
            if (p != n) {
              VSub (dr, mol[n].r, mol[p].r);                 30
              rri = 1. / VLenSq (dr);
              b = sqrt (rri);
              VVSAdd (mol[n].ra, b * rri, dr);
              uSum += b;
              ++ nPair;                                      35
            }
```

```
            ++ tLevel[nv].cCur;
        } else {
            ++ nv;
            tLevel[nv].cCur = tCell[tLevel[nv - 1].cCur].subPtr;        40
            tLevel[nv].cLast = tLevel[nv].cCur +
                tCell[tLevel[nv - 1].cCur].nSub;
            edgeLen *= 0.5;
        }
    }                                                                    45
    } else {
        -- nv;
        ++ tLevel[nv].cCur;
        edgeLen *= 2.;
    }                                                                    50
    } while (nv > 0);
  }
}
```

The overall accuracy of the calculation is determined by the parameter *distFac*.
When multiplied by the cell size *edgeLen* at the level of subdivision being tested,
it specifies the minimum distance from the atom at which a (multiply occupied)
cell of that particular size can be used without further subdivision. The variable
nPair counts how many interaction terms are computed. Filling in the remaining
details of the calculation is left as an exercise for the reader.

13.4 Fast-multipole method

Background

The tree-code approach regards all the charged particles in each cell as being positioned at a single point; in terms of multipole expansions, this amounts to ignoring
all but the lowest-order monopole term. Accuracy can be improved by extending
the expansions to higher order; furthermore, once it is realized that there are ways
of transforming the coefficients of the multipole expansions that allow evaluation
of cell contributions at the different levels of the hierarchy without having to recompute the expansions in their entirety, the computations involved can be carried
out very efficiently [gre88, gre89a, whi94]. The discussion begins with a review of
multipole expansions from a computational perspective.

Multipole expansions

The familiar multipole expansion of electrostatics – involving one large distance
variable and one small – can be written as

$$\frac{1}{|\boldsymbol{r} - \boldsymbol{r}'|} = \sum_{l \geq 0} \sum_{|m| \leq l} \frac{r'^l}{r^{l+1}} \frac{(l - m)!}{(l + m)!} P_l^m(\cos\theta) \, P_l^m(\cos\theta') \, e^{im(\phi - \phi')} \qquad (13.4.1)$$

for $r > r'$, where $\boldsymbol{r} = (r, \theta, \phi)$ and $\boldsymbol{r}' = (r', \theta', \phi')$ in spherical coordinates; the functions $P_l^m(u)$, where $m \le l$, are associated Legendre functions[†] [abr68]

$$P_l^m(u) = \frac{(-1)^m}{2^l \, l!} (1 - u^2)^{m/2} \frac{d^{l+m}}{dx^{l+m}} (u^2 - 1)^l \tag{13.4.2}$$

The expansion (13.4.1) can be expressed more compactly as

$$\frac{1}{|\boldsymbol{r} - \boldsymbol{r}'|} = \sum_{l \ge 0} \sum_{|m| \le l} M_{lm}(\boldsymbol{r}) \, L_{lm}(\boldsymbol{r}') \tag{13.4.3}$$

where

$$M_{lm}(\boldsymbol{r}) = \frac{(l - m)!}{r^{l+1}} P_l^m(\cos \theta) \, e^{im\phi} \tag{13.4.4}$$

$$L_{lm}(\boldsymbol{r}) = \frac{r^l}{(l + m)!} P_l^m(\cos \theta) \, e^{-im\phi} \tag{13.4.5}$$

Note that both M_{lm} and L_{lm} satisfy

$$M_{l,-m}(\boldsymbol{r}) = (-1)^m M_{lm}^*(\boldsymbol{r}) \tag{13.4.6}$$

$$M_{lm}(-\boldsymbol{r}) = (-1)^l M_{lm}(\boldsymbol{r}) \tag{13.4.7}$$

For computational work it is convenient to treat the real and imaginary parts of these quantities separately [per96]. Let

$$M_{lm}(\boldsymbol{r}) = M_{lm}^c(\boldsymbol{r}) + i M_{lm}^s(\boldsymbol{r}) \tag{13.4.8}$$

$$L_{lm}(\boldsymbol{r}) = L_{lm}^c(\boldsymbol{r}) - i L_{lm}^s(\boldsymbol{r}) \tag{13.4.9}$$

then

$$\frac{1}{|\boldsymbol{r} - \boldsymbol{r}'|} = \sum_{l \ge 0} \sum_{m=0}^{l} (2 - \delta_{m0}) \left[M_{lm}^c(\boldsymbol{r}) \, L_{lm}^c(\boldsymbol{r}') + M_{lm}^s(\boldsymbol{r}) \, L_{lm}^s(\boldsymbol{r}') \right] \tag{13.4.10}$$

where all the coefficients $M_{lm}^{c,s}$ and $L_{lm}^{c,s}$ can be evaluated by recursion relations, as described below; from (13.4.6), $M_{lm}^{c,s}$ and $L_{lm}^{c,s}$ both satisfy

$$M_{l,-m}^c(\boldsymbol{r}) = (-1)^m M_{lm}^c(\boldsymbol{r}) \tag{13.4.11}$$

$$M_{l,-m}^s(\boldsymbol{r}) = (-1)^{m+1} M_{lm}^s(\boldsymbol{r}) \tag{13.4.12}$$

A two-center multipole expansion exists that generalizes (13.4.1) to the case of one large and two small variables. It appears in various forms in the literature, originally in [car50], and can be expressed compactly [whi94] (omitting obvious summation limits) as

$$\frac{1}{|\boldsymbol{c} - (\boldsymbol{a} + \boldsymbol{b})|} = \sum_{lm} \sum_{l'm'} L_{lm}(\boldsymbol{a}) \, M_{l+l',m+m'}(\boldsymbol{c}) \, L_{l'm'}(\boldsymbol{b}) \tag{13.4.13}$$

[†] There is more than one definition of $P_l^m(u)$ to be found in the literature.

This expansion is convergent for $|\boldsymbol{a}| + |\boldsymbol{b}| < (\sqrt{2} - 1)|\boldsymbol{c}|$.

Recursion relations and derivatives

The associated Legendre functions can be evaluated with the aid of recursion relations [pre92], starting from $P_0^0 = 1$,

$$P_m^m = -(2m - 1)(1 - u^2)^{1/2} P_{m-1}^{m-1} \tag{13.4.14}$$

$$P_l^m = \frac{1}{l - m}[(2l - 1)u P_{l-1}^m - (l + m - 1) P_{l-2}^m] \tag{13.4.15}$$

The corresponding relations for $M_{lm}^{c,s}(\boldsymbol{r})$ and $L_{lm}^{c,s}(\boldsymbol{r})$, where $\boldsymbol{r} = (x, y, z)$, are to be found in [per96] (allowing for notation changes); starting with

$$M_{00}^c = \frac{1}{r}, \quad M_{00}^s = 0, \quad L_{00}^c = 0, \quad L_{00}^s = 0 \tag{13.4.16}$$

the coefficients with identical indices ($m = l$) are generated by

$$M_{mm}^c = -\frac{2m - 1}{r^2}[x M_{m-1,m-1}^c - y M_{m-1,m-1}^s] \tag{13.4.17}$$

$$M_{mm}^s = -\frac{2m - 1}{r^2}[y M_{m-1,m-1}^c + x M_{m-1,m-1}^s] \tag{13.4.18}$$

$$L_{mm}^c = -\frac{1}{2m}[x L_{m-1,m-1}^c - y L_{m-1,m-1}^s] \tag{13.4.19}$$

$$L_{mm}^s = -\frac{1}{2m}[y L_{m-1,m-1}^c + x L_{m-1,m-1}^s] \tag{13.4.20}$$

while the remaining ($m < l$) coefficients satisfy (the same expressions apply for both the c and s superscripts)

$$M_{lm}^{c,s} = \frac{1}{r^2}[(2l - 1)z M_{l-1,m}^{c,s} - (l - 1 + m)(l - 1 - m) M_{l-2,m}^{c,s}] \tag{13.4.21}$$

$$L_{lm}^{c,s} = \frac{1}{(l + m)(l - m)}[(2l - 1)z L_{l-1,m}^{c,s} - r^2 L_{l-2,m}^{c,s}] \tag{13.4.22}$$

The force calculations will require evaluation of the x, y and z derivatives of L_{lm}; these can be derived with the aid of the formula [hob31]

$$\partial_\pm^m \partial_z^{l-m}\left(\frac{1}{r}\right) = (-1)^{l-m}\frac{(l - m)!}{r^{l+1}} P_l^m(\cos\theta)\, e^{\pm i m\phi} \tag{13.4.23}$$

where the partial derivatives are

$$\partial_\pm \equiv \frac{\partial}{\partial x} \pm i\frac{\partial}{\partial y}, \quad \partial_z \equiv \frac{\partial}{\partial z} \tag{13.4.24}$$

From (13.4.5) and (13.4.23) we obtain

$$L_{lm} = \frac{(-1)^{l-m} r^{2l+1}}{(l+m)! \, (l-m)!} \, \partial_-^m \, \partial_z^{l-m} \left(\frac{1}{r}\right) \tag{13.4.25}$$

The derivatives $\partial_{\pm} L_{lm}$ and $\partial_z L_{lm}$ can be evaluated by using

$$\partial_+ \partial_- = -\partial_z^2 \tag{13.4.26}$$

together with

$$\partial_{\pm} r^{2l+1} = (2l+1) r^{2l} \sin\theta \, e^{\pm i\phi} \tag{13.4.27}$$

$$\partial_z r^{2l+1} = (2l+1) r^{2l} \cos\theta \tag{13.4.28}$$

Then apply two of the standard recursion relations for $P_l^m(\cos\theta)$, namely,

$$(2l+1) \sin\theta \, P_l^m = P_{l-1}^{m+1} - P_{l+1}^{m+1} \tag{13.4.29}$$

$$= (l-m+1)(l-m+2) P_{l+1}^{m-1}$$
$$- (l+m)(l+m-1) P_{l-1}^{m-1} \tag{13.4.30}$$

to obtain

$$\partial_{\pm} L_{lm} = \mp L_{l-1,m\mp 1} \tag{13.4.31}$$

$$\partial_z L_{lm} = L_{l-1,m} \tag{13.4.32}$$

The derivatives of the real and imaginary components follow immediately,

$$\frac{\partial}{\partial x} L_{lm}^{c,s} = \frac{1}{2} [L_{l-1,m+1}^{c,s} - L_{l-1,m-1}^{c,s}] \tag{13.4.33}$$

$$\frac{\partial}{\partial y} L_{lm}^{c,s} = \pm \frac{1}{2} [L_{l-1,m+1}^{s,c} + L_{l-1,m-1}^{s,c}] \tag{13.4.34}$$

$$\frac{\partial}{\partial z} L_{lm}^{c,s} = L_{l-1,m}^{c,s} \tag{13.4.35}$$

where, in (13.4.34), $\partial L^c / \partial y$ depends on L^s and vice versa, and the minus sign applies to L^s.

Hierarchical subdivision

The goal of the method is to use multipole expansions to reduce the computational effort from $O(N_m^2)$ to $O(N_m)$. As with the tree-code approach, this is accomplished by means of a hierarchical subdivision of the region into cells, with the cell size at each level being half that of the level above. Interactions between atoms in neighboring cells at the lowest level are computed directly, but for atoms in more distant cells, evaluating the interactions involves the use of multipole expansions to represent the cell occupants. The coefficients of the multipole expansions for the atoms in each of the cells at the lowest level are computed; this information is then transferred, in ways to be specified below, both up and down the hierarchy, and between cells at the same level of the hierarchy.

Depending on the order at which the multipole expansions are truncated, and the maximum distance between directly treated neighboring cells, any desired degree of accuracy can be achieved, although the computational cost increases with accuracy. The amount of work grows linearly with N_m, but the system size at which the method becomes more efficient than the all-pairs approach depends on the required accuracy [whi94, ess95].

The present treatment assumes a fixed level of subdivision; the method can be extended to allow the level of subdivision to adapt locally according to the density (in a manner reminiscent of the tree-code method), a useful approach for problems where significant spatial inhomogeneity occurs. Since each level in the hierarchy contains eight times as many cells as the level above (in three dimensions), not too many levels are required; the total number of levels, starting with a single cell for the entire region at the top level, need not exceed $\lfloor \log N_m / \log 8 \rfloor$. Periodic boundaries are not taken into account here, but the method can be extended to incorporate them [cha97].

Operations on multipole expansions

The fast-multipole method [gre88, whi94] relies on the fact that when the origin of the multipole expansion (13.4.1) is changed, the coefficients of the original and shifted expansions can be related using (13.4.13). Furthermore, (13.4.1) can be converted into a local power series around a distant origin, and the origin of this new expansion can itself be changed; in both instances the expansion coefficients can be related, again with the aid of (13.4.13). In each of the three cases considered below, it is assumed that the distances involved are chosen to ensure the expansion converges – conditions that are indeed satisfied when the expansions are applied.

In order to shift the origin of a multipole expansion, rewrite (13.4.3) as

$$\frac{1}{|\boldsymbol{r} - \boldsymbol{d} - (\boldsymbol{r}' - \boldsymbol{d})|} = \sum_l \sum_m L_{lm}(\boldsymbol{r}' - \boldsymbol{d}) \, M_{lm}(\boldsymbol{r} - \boldsymbol{d}) \tag{13.4.36}$$

By making the replacements

$$\boldsymbol{a} \to \boldsymbol{r}', \quad \boldsymbol{b} \to -\boldsymbol{d}, \quad \boldsymbol{c} \to \boldsymbol{r} - \boldsymbol{d} \tag{13.4.37}$$

in (13.4.13) and then matching the terms with those of (13.4.36) we obtain new expansion coefficients

$$L_{lm}(r' - d) = \sum_{l'm'} L_{l'm'}(r') L_{l-l',m-m'}(-d) \tag{13.4.38}$$

Insofar as the terms of (13.4.38) are concerned, the set of coefficients $L_{l'm'}(r')$ are already known and the $L_{l-l',m-m'}(-d)$ need only be computed once for a given shift vector d. The l' sum in (13.4.38) is over the range $[0, l]$, whereas the limits of the m' sum must satisfy both $|m'| \leq l'$ and $|m - m'| \leq l - l'$, or equivalently,

$$\max(-l', m - (l - l')) \leq m' \leq \min(l', m + (l - l')) \tag{13.4.39}$$

For computational purposes, the coefficients (13.4.38) of the shifted expansion are readily expressed in terms of their real and imaginary components; each product can be written schematically as

$$(L_1^c - iL_1^s)(L_2^c - iL_2^s) = L_1^c L_2^c - L_1^s L_2^s - i(L_1^c L_2^s + L_1^s L_2^c)$$
$$\equiv L_3^c - iL_3^s \tag{13.4.40}$$

and the components then summed separately.

The multipole expansion (13.4.3) can also be converted into a local expansion about a distant origin. Replacing

$$a \to r - d, \quad b \to -r', \quad c \to -d \tag{13.4.41}$$

in (13.4.13), and then applying (13.4.7) to $L_{l'm'}(-r')$, we obtain

$$M_{lm}(r' - d) = \sum_{l'm'} (-1)^{l'} L_{l'm'}(r') M_{l+l',m+m'}(-d) \tag{13.4.42}$$

in which the coefficients $L_{l'm'}(r')$ are known and the $M_{l+l',m+m'}(-d)$ can be computed. The l' sum in (13.4.42) is over the range $[0, l_{max} - l]$, where l_{max} is the maximum expansion order retained; the limits of the m' sum must satisfy both $|m'| \leq l'$ and $|m + m'| \leq l + l'$, or

$$\max(-l', -m - (l + l')) \leq m' \leq \min(l', -m + (l + l')) \tag{13.4.43}$$

The products appearing in (13.4.42) can, once again, be expressed in terms of real and imaginary components,

$$(L_1^c - iL_1^s)(M_2^c + iM_2^s) = L_1^c M_2^c + L_1^s M_2^s + i(L_1^c M_2^s - L_1^s M_2^c)$$
$$\equiv M_3^c + iM_3^s \tag{13.4.44}$$

The origin of a local expansion can be shifted. Substitute

$$a \to r - d, \quad b \to d, \quad c \to r' \tag{13.4.45}$$

in (13.4.13) to obtain

$$M_{lm}(\mathbf{r}' - \mathbf{d}) = \sum_{l'm'} M_{l+l',m+m'}(\mathbf{r}') L_{l'm'}(\mathbf{d}) \tag{13.4.46}$$

in which the $M_{l+l',m+m'}(\mathbf{r}')$ are known and the $L_{l'm'}(\mathbf{d})$ can be computed. Here, too, the m' sum limits are given by (13.4.43).

The relations (13.4.38), (13.4.42) and (13.4.46) are used to convert the expansions when transferring the information they represent among different cells in the hierarchy. Each cell i at the lowest (nth) level has associated with it a set $\mathcal{L}_{ni}$ of L coefficients describing a multipole expansion about its midpoint for all the atoms in the cell. If cell i contains a set of atoms denoted by $\{q_s, \mathbf{r}_s\}$, where, for each atom s, q_s is the charge and $\mathbf{r}_s$ the position, and if $\boldsymbol{\rho}_{ni}$ is the position of the cell midpoint, then the multipole expansion for the entire cell is obtained by replacing $L_{lm}(\mathbf{r}')$ in the original expansion (13.4.3) by a sum over cell occupants

$$\sum_s q_s L_{lm}(\mathbf{r}_s) \tag{13.4.47}$$

where $\mathbf{r}_s$ in (13.4.47) and $\mathbf{r}$ in (13.4.3) are expressed relative to $\boldsymbol{\rho}_{ni}$. Thus, $\mathcal{L}_{ni}$ actually denotes the set of sums (13.4.47) for all l and m; the various transformations described above are then applied to these sums in their entirety.

Traversing the cell hierarchy

The first part of the algorithm [gre88, gre89a] (after evaluating $\mathcal{L}_{ni}$ for the lowest level) entails ascending through the cell hierarchy, and for each cell at level l, summing the L coefficients of cells at the next lower level $l+1$, after shifting each of the expansions to the midpoint of the current cell with the aid of (13.4.38); these sets of cumulative L coefficients, denoted by $\mathcal{L}_{li}$, can be used in evaluating multipole expansions associated with cells at the lth level in the hierarchy. The second part of the algorithm is a descent through the cell hierarchy, during which two computations are made. The first is to consider all non-neighbor cells i' (soon to be defined) whose parent cells are neighbors of the parent of the current cell i, convert their $\mathcal{L}_{li}$ coefficient sets to M coefficients of a local expansion about the midpoint of cell i' using (13.4.42), and denote their sum, together with the contribution received from the parent cell in level $l-1$, by $\mathcal{M}_{li}$. The second computation propagates the $\mathcal{M}_{li}$ to the descendants of the cell at level $l+1$ (except when $l=n$), by shifting the local expansions to the midpoints of each of these cells using (13.4.46). Later, when computing the interactions, it will be the $\mathcal{M}_{ni}$ coefficient sets that concisely describe the interactions with all other atoms, apart from those in neighboring cells.

Cells that are not neighbors are referred to as well-separated – see Figure 13.4. The minimum number of intervening cells between well-separated cells n_w is a

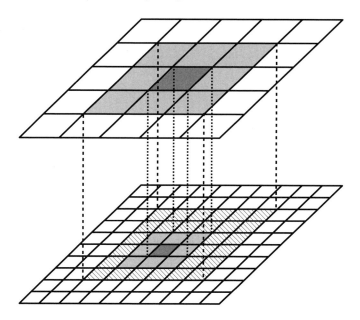

Fig. 13.4. Two levels in the cell hierarchy (two-dimensional version) showing the different cell categories (neighbors, neighbor parents, well-separated) for the case $n_w = 1$.

parameter of the calculation, with a larger value of n_w requiring a smaller value of l_{max} for similar accuracy. The smallest value of n_w is unity, meaning that well-separated cells do not touch, even at their corners; $n_w = 2$ implies that well-separated cells are separated by at least two intermediaries; larger values are unlikely to be needed. The computational tradeoff is related to the fact that while the operations on multipole and local expansions scale as $O(l_{max}^4)$, the number of pairs of atoms whose interactions must be evaluated directly depends on n_w^3 and the number of cells whose L coefficients must be converted to M coefficients depends on $(4n_w + 2)^3 - (2n_w + 1)^3$; for $n_w = 1,2$ the values of the latter are 189 and 875, respectively.

The full algorithm can be summarized as follows (note that the top two levels in the subdivision hierarchy need not be considered, since there are no well-separated cells):

- Expansion coefficients: Evaluate $\mathcal{L}_{ni}$ for all cells i at the lowest (nth) level in the hierarchy ($i = 1, \ldots 8^n$).
- Upward pass: For levels $l = n - 1, \ldots 2$, and for all cells i ($i = 1, \ldots 8^l$) at level l, zero the set of coefficients $\mathcal{L}_{li}$ and then use $\mathcal{L}_{li}$ to accumulate the shifted $\mathcal{L}_{l+1,i'}$ from the descendants i' of i at level $l + 1$.
- Downward pass: Zero $\mathcal{M}_{2,i}$ ($i = 1, \ldots 64$). Then, for levels $l = 2, \ldots n$, for all cells i at level l, and for each cell i' at level l that is well-separated from i but whose parent and that of i are not well-separated, convert $\mathcal{L}_{li'}$ to M coefficients

for use in an expansion about the midpoint of cell i and accumulate in $\mathcal{M}_{li}$. Next, if $l < n$, for all descendants i'' of i at level $l + 1$, shift $\mathcal{M}_{li}$ and use to initialize $\mathcal{M}_{l+1,i''}$.

- Far interactions: For all cells i at the nth level and for each atom in cell i, use $\mathcal{M}_{ni}$ to compute the interaction contribution.
- Near interactions: For all pairs of cells at the nth level that are not well-separated, compute the interactions between all pairs of atoms in the cells.

The software in the next section implements this algorithm.

13.5 Implementing the fast-multipole method

Data organization

Define two C structures

```
typedef struct {
  real c[I(MAX_MPEX_ORD, MAX_MPEX_ORD) + 1],
       s[I(MAX_MPEX_ORD, MAX_MPEX_ORD) + 1];
} MpTerms;

typedef struct {
  MpTerms le, me;
  int occ;
} MpCell;
```

that will contain, respectively, a set of multipole or local expansion coefficients, and the accumulated multipole and local expansions for a cell together with an indication of whether the cell is occupied. It is assumed, for programming⬥ convenience, that the multipole and local expansions are to be carried out to a certain order not exceeding a limit *MAX_MPEX_ORD* that is compiled into the program; the value actually used is *maxOrd* which, after a certain amount of experimentation to determine the preferred value, is likely to be set equal to *MAX_MPEX_ORD* (or vice versa). The required storage increases as the square of this value. Although the expansion coefficients have two indices, since $|m| \le l$ they only occupy a triangular array; storage is saved by introducing

```
#define I(i, j)  ((i) * ((i) + 1) / 2 + (j))
```

as a convenient way of combining the array indices, and for readability

```
#define c(i, j)  c[I(i, j)]
#define s(i, j)  s[I(i, j)]
```

Generating expansion coefficients

The following two functions implement (13.4.16)–(13.4.22) in order to generate full sets of coefficients $M_{lm}^{c,s}(r)$ and $L_{lm}^{c,s}(r)$ up to the specified order, where v is a pointer to $r^{\dagger}$.

```
void EvalMpM (MpTerms *me, VecR *v, int maxOrd)
{
  real a, a1, a2, rri;
  int j, k;

  rri = 1./ VLenSq (*v);
  me->c(0, 0) = sqrt (rri);
  me->s(0, 0) = 0.;
  for (j = 1; j <= maxOrd; j ++) {
    k = j;
    a = - (2 * k - 1) * rri;
    me->c(j, k) = a * (v->x * me->c(j - 1, k - 1) -
      v->y * me->s(j - 1, k - 1));
    me->s(j, k) = a * (v->y * me->c(j - 1, k - 1) +
      v->x * me->s(j - 1, k - 1));
    for (k = j - 1; k >= 0; k --) {
      a1 = (2 * j - 1) * v->z * rri;
      a2 = (j - 1 + k) * (j - 1 - k) * rri;
      me->c(j, k) = a1 * me->c(j - 1, k);
      me->s(j, k) = a1 * me->s(j - 1, k);
      if (k < j - 1) {
        me->c(j, k) -= a2 * me->c(j - 2, k);
        me->s(j, k) -= a2 * me->s(j - 2, k);
      }
    }
  }
}

void EvalMpL (MpTerms *le, VecR *v, int maxOrd)
{
  real a, a1, a2, rr;
  int j, k;

  rr = VLenSq (*v);
  le->c(0, 0) = 1.;
  le->s(0, 0) = 0.;
  for (j = 1; j <= maxOrd; j ++) {
    k = j;
    a = - 1. / (2 * k);
    le->c(j, k) = a * (v->x * le->c(j - 1, k - 1) -
      v->y * le->s(j - 1, k - 1));
    le->s(j, k) = a * (v->y * le->c(j - 1, k - 1) +
      v->x * le->s(j - 1, k - 1));
```

$\dagger$ Indices l and m are represented here by variables j and k for readability; similarly in later functions.

```
    for (k = j - 1; k >= 0; k --) {
      a = 1. / ((j + k) * (j - k));                          45
      a1 = (2 * j - 1) * v->z * a;
      a2 = rr * a;
      le->c(j, k) = a1 * le->c(j - 1, k);
      le->s(j, k) = a1 * le->s(j - 1, k);
      if (k < j - 1) {                                       50
        le->c(j, k) -= a2 * le->c(j - 2, k);
        le->s(j, k) -= a2 * le->s(j - 2, k);
      }
    }
  }
}                                                            55
}
```

Products of expansion coefficients

At various points in the computation – when shifting or converting the expansions –
sums of products of expansion coefficients must be evaluated. The two functions
listed below are used for this task; both contain quadruply nested loops that give
the overall calculation its $O(l_{max}^4)$ behavior.

The product sum in (13.4.38), organized into separate real and imaginary parts
as indicated by (13.4.40), is computed as follows. The limits of what corresponds
to the sum over m' satisfy (13.4.39).

```
void EvalMpProdLL (MpTerms *le1, MpTerms *le2, MpTerms *le3,
    int maxOrd)
{
  real s2, s3, vlc2, vlc3, vls2, vls3;
  int j1, j2, j3, k1, k2, k3;                                5

  for (j1 = 0; j1 <= maxOrd; j1 ++) {
    for (k1 = 0; k1 <= j1; k1 ++) {
      le1->c(j1, k1) = 0.;
      le1->s(j1, k1) = 0.;                                   10
      for (j2 = 0; j2 <= j1; j2 ++) {
        j3 = j1 - j2;
        for (k2 = Max (- j2, k1 - j3);
          k2 <= Min (j2, k1 + j3); k2 ++) {
          k3 = k1 - k2;                                      15
          vlc2 = le2->c(j2, abs (k2));
          vls2 = le2->s(j2, abs (k2));
          if (k2 < 0) vls2 = - vls2;
          vlc3 = le3->c(j3, abs (k3));
          vls3 = le3->s(j3, abs (k3));                       20
          if (k3 < 0) vls3 = - vls3;
          s2 = (k2 < 0 && IsOdd (k2)) ? -1. : 1.;
          s3 = (k3 < 0 && IsOdd (k3)) ? -1. : 1.;
          le1->c(j1, k1) += s2 * s3 * (vlc2 * vlc3 - vls2 * vls3);
```

```
            le1->s(j1, k1) += s2 * s3 * (vls2 * vlc3 + vlc2 * vls3);      25
          }
        }
      }
    }
  }                                                                       30
```

Here, `IsOdd` (§18.2) is unity if its argument is odd and zero otherwise.

The other form of product sum appears in both (13.4.42) and (13.4.46); its computation follows. The limits of the sum over m' satisfy (13.4.43).

```
void EvalMpProdLM (MpTerms *me1, MpTerms *le2, MpTerms *me3,
    int maxOrd)
{
  real s2, s3, vlc2, vls2, vmc3, vms3;
  int j1, j2, j3, k1, k2, k3;                                            5

  for (j1 = 0; j1 <= maxOrd; j1 ++) {
    for (k1 = 0; k1 <= j1; k1 ++) {
      me1->c(j1, k1) = 0.;
      me1->s(j1, k1) = 0.;                                               10
      for (j2 = 0; j2 <= maxOrd - j1; j2 ++) {
        j3 = j1 + j2;
        for (k2 = Max (- j2, - k1 - j3);
            k2 <= Min (j2, - k1 + j3); k2 ++) {
          k3 = k1 + k2;                                                  15
          vlc2 = le2->c(j2, abs (k2));
          vls2 = le2->s(j2, abs (k2));
          if (k2 < 0) vls2 = - vls2;
          vmc3 = me3->c(j3, abs (k3));
          vms3 = me3->s(j3, abs (k3));                                   20
          if (k3 < 0) vms3 = - vms3;
          s2 = (k2 < 0 && IsOdd (k2)) ? -1. : 1.;
          s3 = (k3 < 0 && IsOdd (k3)) ? -1. : 1.;
          me1->c(j1, k1) += s2 * s3 * (vlc2 * vmc3 + vls2 * vms3);
          me1->s(j1, k1) += s2 * s3 * (vlc2 * vms3 - vls2 * vmc3);       25
        }
      }
    }
  }
}                                                                       30
```

Multipole force evaluation

The function `MultipoleCalc` is responsible for the long-range force evaluation. After computing the multipole coefficients at the lowest level in the cell hierarchy, it proceeds upward through the levels, combining the shifted expansions to produce expansions for progressively larger cells. This is followed by the downward pass in which local expansions are evaluated by merging the contributions of well-separated cells. The interaction contributions from well-separated cells are evaluated using these expansions, while contributions of nearby pairs are evaluated by treating the atoms directly. The functions called here are listed subsequently.

```
void MultipoleCalc ()
{
  int j, k, m1;

  VSetAll (mpCells, maxCellsEdge);                                      5
  AssignMpCells ();
  VDiv (cellWid, region, mpCells);
  EvalMpCell ();
  curCellsEdge = maxCellsEdge;
  for (curLevel = maxLevel - 1; curLevel >= 2; curLevel --) {          10
    curCellsEdge /= 2;
    VSetAll (mpCells, curCellsEdge);
    VDiv (cellWid, region, mpCells);
    CombineMpCell ();
  }                                                                    15
  for (m1 = 0; m1 < 64; m1 ++) {
    for (j = 0; j <= maxOrd; j ++) {
      for (k = 0; k <= j; k ++) {
        mpCell[2][m1].me.c(j, k) = 0.;
        mpCell[2][m1].me.s(j, k) = 0.;                                 20
      }
    }
  }
  curCellsEdge = 2;
  for (curLevel = 2; curLevel <= maxLevel; curLevel ++) {             25
    curCellsEdge *= 2;
    VSetAll (mpCells, curCellsEdge);
    VDiv (cellWid, region, mpCells);
    GatherWellSepLo ();
    if (curLevel < maxLevel) PropagateCellLo ();                      30
  }
  ComputeFarCellInt ();
  ComputeNearCellInt ();
}
```

The arrays referenced in the multipole calculations are declared as

```
MpCell **mpCell;
int *mpCellList;
```

and are allocated (*AllocArrays*) by

```
AllocMem (mpCell, maxLevel + 1, MpCell *);
maxCellsEdge = 2;
```

```
for (n = 2; n <= maxLevel; n ++) {
  maxCellsEdge *= 2;
  VSetAll (mpCells, maxCellsEdge);                                    5
  AllocMem (mpCell[n], VProd (mpCells), MpCell);
}
AllocMem (mpCellList, nMol + VProd (mpCells), int);
```

The function *AssignMpCells* uses the same cell assignment as for short-range
interactions (§3.4), except that *mpCells* and *mpCellList* are now the relevant
quantities. Once this has been done, the sets of multipole coefficients $\mathcal{L}_{ni}$ representing the combined contributions of the atoms in the cells at the lowest level in
the hierarchy are evaluated as follows. The charge of each atom is represented by
an additional element *chg* in the *Mol* structure.

```
void EvalMpCell ()
{
  MpTerms le;
  VecR cMid, dr;
  VecI m1v;                                                          5
  int j, j1, k, m1, m1x, m1y, m1z;

  for (m1z = 0; m1z < mpCells.z; m1z ++) {
    for (m1y = 0; m1y < mpCells.y; m1y ++) {
      for (m1x = 0; m1x < mpCells.x; m1x ++) {                       10
        VSet (m1v, m1x, m1y, m1z);
        m1 = VLinear (m1v, mpCells);
        mpCell[maxLevel][m1].occ = 0;
        for (j = 0; j <= maxOrd; j ++) {
          for (k = 0; k <= j; k ++) {                                15
            mpCell[maxLevel][m1].le.c(j, k) = 0.;
            mpCell[maxLevel][m1].le.s(j, k) = 0.;
          }
        }
        if (mpCellList[m1 + nMol] >= 0) {                            20
          VAddCon (cMid, m1v, 0.5);
          VMul (cMid, cMid, cellWid);
          VVSAdd (cMid, - 0.5, region);
          DO_MP_CELL (j1, m1) {
            ++ mpCell[maxLevel][m1].occ;                             25
            VSub (dr, mol[j1].r, cMid);
            EvalMpL (&le, &dr, maxOrd);
            for (j = 0; j <= maxOrd; j ++) {
              for (k = 0; k <= j; k ++) {
                mpCell[maxLevel][m1].le.c(j, k) +=                   30
                    mol[j1].chg * le.c(j, k);
                mpCell[maxLevel][m1].le.s(j, k) +=
                    mol[j1].chg * le.s(j, k);
              }
            }
          }                                                         35
```

```
              }
            }
          }
        }
      }                                                                    40
    }
```

For brevity,

```
#define DO_MP_CELL(j, m)                                              \
    for (j = mpCellList[m + nMol]; j >= 0; j = mpCellList[j])
```

Shifting and combining the multipole expansions at each level are carried out by
the following function. The shift vector is the offset of the midpoints of the cells
when moving between two adjacent levels, and its direction depends on whether
it is the upward pass (in this function) or the downward pass (treated later) that is
being carried out. This function makes use of (13.4.38) and the shifted expansion
coefficients are evaluated by *EvalMpProdLL* which appeared earlier.

```
void CombineMpCell ()
{
  MpTerms le, le2;
  VecR rShift;
  VecI m1v, m2v, mpCellsN;                                               5
  int iDir, j, k, m1, m1x, m1y, m1z, m2;

  VSCopy (mpCellsN, 2, mpCells);
  for (m1z = 0; m1z < mpCells.z; m1z ++) {
    for (m1y = 0; m1y < mpCells.y; m1y ++) {                             10
      for (m1x = 0; m1x < mpCells.x; m1x ++) {
        VSet (m1v, m1x, m1y, m1z);
        m1 = VLinear (m1v, mpCells);
        for (j = 0; j <= maxOrd; j ++) {
          for (k = 0; k <= j; k ++) {                                   15
            mpCell[curLevel][m1].le.c(j, k) = 0.;
            mpCell[curLevel][m1].le.s(j, k) = 0.;
          }
        }
        mpCell[curLevel][m1].occ = 0;                                   20
        for (iDir = 0; iDir < 8; iDir ++) {
          VSCopy (m2v, 2, m1v);
          VSCopy (rShift, -0.25, cellWid);
          if (IsOdd (iDir)) {
            ++ m2v.x;                                                   25
            rShift.x *= -1.;
          }
          if (IsOdd (iDir / 2)) {
            ++ m2v.y;
            rShift.y *= -1.;                                            30
```

```
      }
      if (IsOdd (iDir / 4)) {
        ++ m2v.z;
        rShift.z *= -1.;
      }                                                                    35
      m2 = VLinear (m2v, mpCellsN);
      if (mpCell[curLevel + 1][m2].occ == 0) continue;
      mpCell[curLevel][m1].occ += mpCell[curLevel + 1][m2].occ;
      EvalMpL (&le2, &rShift, maxOrd);
      EvalMpProdLL (&le, &mpCell[curLevel + 1][m2].le, &le2,             40
        maxOrd);
      for (j = 0; j <= maxOrd; j ++) {
        for (k = 0; k <= j; k ++) {
          mpCell[curLevel][m1].le.c(j, k) += le.c(j, k);
          mpCell[curLevel][m1].le.s(j, k) += le.s(j, k);              45
        }
      }
    }
  }
                                                                          50
  }
}
```

At each level, the contributions from well-separated cells are gathered by the
following function, and these form part of the local expansion for each cell. This
function (and functions it calls) is where most of the computational effort is usually expended; `wellSep` corresponds to n_w. Conversion from multipole to local
expansions using (13.4.42) is carried out by *EvalMpProdLM*.

```
void GatherWellSepLo ()
{
  MpTerms le, me, me2;
  VecR rShift;
  VecI m1v, m2v;                                                          5
  real s;
  int j, k, m1, m1x, m1y, m1z, m2, m2x, m2y, m2z;

  for (m1z = 0; m1z < mpCells.z; m1z ++) {
    for (m1y = 0; m1y < mpCells.y; m1y ++) {                           10
      for (m1x = 0; m1x < mpCells.x; m1x ++) {
        VSet (m1v, m1x, m1y, m1z);
        m1 = VLinear (m1v, mpCells);
        if (mpCell[curLevel][m1].occ == 0) continue;
        for (m2z = LoLim (z); m2z <= HiLim (z); m2z ++) {             15
          for (m2y = LoLim (y); m2y <= HiLim (y); m2y ++) {
            for (m2x = LoLim (x); m2x <= HiLim (x); m2x ++) {
              VSet (m2v, m2x, m2y, m2z);
              if (m2v.x < 0 || m2v.x >= mpCells.x ||
                  m2v.y < 0 || m2v.y >= mpCells.y ||                  20
```

```
                  m2v.z < 0 || m2v.z >= mpCells.z) continue;
            if (abs (m2v.x - m1v.x) <= wellSep &&
                abs (m2v.y - m1v.y) <= wellSep &&
                abs (m2v.z - m1v.z) <= wellSep) continue;
            m2 = VLinear (m2v, mpCells);                                    25
            if (mpCell[curLevel][m2].occ == 0) continue;
            for (j = 0; j <= maxOrd; j ++) {
              s = (IsOdd (j)) ? -1. : 1.;
              for (k = 0; k <= j; k ++) {
                le.c(j, k) = s * mpCell[curLevel][m2].le.c(j, k);          30
                le.s(j, k) = s * mpCell[curLevel][m2].le.s(j, k);
              }
            }
            VSub (rShift, m2v, m1v);
            VMul (rShift, rShift, cellWid);                                 35
            EvalMpM (&me2, &rShift, maxOrd);
            EvalMpProdLM (&me, &le, &me2, maxOrd);
            for (j = 0; j <= maxOrd; j ++) {
              for (k = 0; k <= j; k ++) {
                mpCell[curLevel][m1].me.c(j, k) += me.c(j, k);             40
                mpCell[curLevel][m1].me.s(j, k) += me.s(j, k);
              }
            }
          }
        }                                                                  45
      }
    }
  }
}
}                                                                          50
```

Here,

```
#define HiLim(t)   IsEven (m1v.t) + 2 * wellSep + 1
#define LoLim(t)   IsEven (m1v.t) - 2 * wellSep
```

and *IsEven* (§18.2) is the converse of *IsOdd*.

The local expansions are then shifted and propagated down to the cells at the next level in the hierarchy; use is made of (13.4.46) and the shifted expansion coefficients are evaluated by *EvalMpProdLM*.

```
void PropagateCellLo ()
{
  MpTerms le;
  VecR rShift;
  VecI m1v, m2v, mpCellsN;                                                 5
  int iDir, m1, m1x, m1y, m1z, m2;

  VSCopy (mpCellsN, 2, mpCells);
```

```
      for (m1z = 0; m1z < mpCells.z; m1z ++) {
        for (m1y = 0; m1y < mpCells.y; m1y ++) {                          10
          for (m1x = 0; m1x < mpCells.x; m1x ++) {
            VSet (m1v, m1x, m1y, m1z);
            m1 = VLinear (m1v, mpCells);
            if (mpCell[curLevel][m1].occ == 0) continue;
            for (iDir = 0; iDir < 8; iDir ++) {                          15
              VSCopy (m2v, 2, m1v);
              VSCopy (rShift, -0.25, cellWid);
              if (IsOdd (iDir)) {
                ++ m2v.x;
                rShift.x *= -1.;                                         20
              }
              if (IsOdd (iDir / 2)) {
                ++ m2v.y;
                rShift.y *= -1.;
              }                                                          25
              if (IsOdd (iDir / 4)) {
                ++ m2v.z;
                rShift.z *= -1.;
              }
              m2 = VLinear (m2v, mpCellsN);                              30
              EvalMpL (&le, &rShift, maxOrd);
              EvalMpProdLM (&mpCell[curLevel + 1][m2].me, &le,
                &mpCell[curLevel][m1].me, maxOrd);
            }
          }                                                              35
        }
      }
    }
```

Interaction contributions from well-separated cells are computed as follows[†].

```
void ComputeFarCellInt ()
{
  MpTerms le;
  VecR cMid, dr, f;
  VecI m1v;                                                              5
  real u;
  int j1, m1, m1x, m1y, m1z;

  for (m1z = 0; m1z < mpCells.z; m1z ++) {
    for (m1y = 0; m1y < mpCells.y; m1y ++) {                            10
      for (m1x = 0; m1x < mpCells.x; m1x ++) {
        VSet (m1v, m1x, m1y, m1z);
        m1 = VLinear (m1v, mpCells);
        if (mpCell[maxLevel][m1].occ == 0) continue;
        VAddCon (cMid, m1v, 0.5);                                        15
```

[†] As with the tree-code method (§13.3), Newton's third law does not appear in this part of the calculation; it is used in the direct pair evaluations later.

```
        VMul (cMid, cMid, cellWid);
        VVSAdd (cMid, -0.5, region);
        DO_MP_CELL (j1, m1) {
          VSub (dr, mol[j1].r, cMid);
          EvalMpL (&le, &dr, maxOrd);
          EvalMpForce (&f, &u, &mpCell[maxLevel][m1].me, &le, maxOrd);
          VVSAdd (mol[j1].ra, - mol[j1].chg, f);
          uSum += 0.5 * mol[j1].chg * u;
        }
      }
    }
  }
}
```

The function *EvalMpForce* evaluates the multipole contribution to the force using (13.4.33)–(13.4.35) and the interaction energy using (13.4.10).

```
void EvalMpForce (VecR *f, real *u, MpTerms *me, MpTerms *le,
   int maxOrd)
{
  VecR fc, fs;
  real a;
  int j, k;

  VZero (*f);
  for (j = 1; j <= maxOrd; j ++) {
    for (k = 0; k <= j; k ++) {
      if (k < j - 1) {
        fc.x = le->c(j - 1, k + 1);
        fc.y = le->s(j - 1, k + 1);
        fs.x = le->s(j - 1, k + 1);
        fs.y = - le->c(j - 1, k + 1);
      } else {
        fc.x = 0.;
        fc.y = 0.;
        fs.x = 0.;
        fs.y = 0.;
      }
      if (k < j) {
        fc.z = le->c(j - 1, k);
        fs.z = le->s(j - 1, k);
      } else {
        fc.z = 0.;
        fs.z = 0.;
      }
      if (k > 0) {
        fc.x -= le->c(j - 1, k - 1);
        fc.y += le->s(j - 1, k - 1);
        fc.z *= 2.;
        fs.x -= le->s(j - 1, k - 1);
```

```
        fs.y -= le->c(j - 1, k - 1);
        fs.z *= 2.;                                              35
      }
      VVSAdd (*f, me->c(j, k), fc);
      VVSAdd (*f, me->s(j, k), fs);
    }
  }                                                              40
  *u = 0.;
  for (j = 0; j <= maxOrd; j ++) {
    for (k = 0; k <= j; k ++) {
      a = me->c(j, k) * le->c(j, k) + me->s(j, k) * le->s(j, k);
      if (k > 0) a *= 2.;                                        45
      *u += a;
    }
  }
}
```

Direct evaluation of the interactions is required for pairs of atoms in the same and nearby (more precisely, not well-separated) cells; this represents the other heavy part of the calculation since, even for $n_w = 1$, each cell has 189 neighbor cells of this kind.

```
void ComputeNearCellInt ()
{
  VecR dr, ft;
  VecI m1v, m2v;
  real qq, ri;                                                  5
  int j1, j2, m1, m1x, m1y, m1z, m2, m2x, m2xLo, m2y, m2yLo, m2z;

  for (m1z = 0; m1z < mpCells.z; m1z ++) {
    for (m1y = 0; m1y < mpCells.y; m1y ++) {
      for (m1x = 0; m1x < mpCells.x; m1x ++) {                  10
        VSet (m1v, m1x, m1y, m1z);
        m1 = VLinear (m1v, mpCells);
        if (mpCell[maxLevel][m1].occ == 0) continue;
        for (m2z = m1z; m2z <= HiLimI (z); m2z ++) {
          m2yLo = (m2z == m1z) ? m1y : Max (m1y - wellSep, 0);  15
          for (m2y = m2yLo; m2y <= HiLimI (y); m2y ++) {
            m2xLo = (m2z == m1z && m2y == m1y) ?
              m1x : Max (m1x - wellSep, 0);
            for (m2x = m2xLo; m2x <= HiLimI (x); m2x ++) {
              VSet (m2v, m2x, m2y, m2z);                        20
              m2 = VLinear (m2v, mpCells);
              if (mpCell[maxLevel][m2].occ == 0) continue;
              DO_MP_CELL (j1, m1) {
                DO_MP_CELL (j2, m2) {
                  if (m1 != m2 || j2 < j1) {                    25
                    VSub (dr, mol[j1].r, mol[j2].r);
                    ri = 1. / VLen (dr);
```

```
        qq = mol[j1].chg * mol[j2].chg;
        VSCopy (ft, qq * Cube (ri), dr);
        VVAdd (mol[j1].ra, ft);                    30
        VVSub (mol[j2].ra, ft);
        uSum += qq * ri;
      }
    }
  }                                                35
  }
  }
  }
  }                                                40
  }
}
```

Here,

```
#define HiLimI(t)  Min (m1v.t + wellSep, mpCells.t - 1)
```

Since both positive and negative charges are present, soft-sphere interactions are also included in the simulation in order to prevent oppositely charged atoms approaching too closely. The function *SingleStep* calls *ComputeForces* (§3.4) to evaluate these forces, using the neighbor-list method; the acceleration values are initialized there. The charges of the atoms are randomly assigned by

```
void InitCharges ()
{
  int n;

  DO_MOL mol[n].chg = (RandR () > 0.5) ? chargeMag : - chargeMag;    5
}
```

Other variables and input data are

```
VecR cellWid;
VecI mpCells;
real chargeMag;
int curCellsEdge, curLevel, maxCellsEdge, maxLevel, maxOrd, wellSep;
                                                                   5
  NameR (chargeMag),
  NameI (maxLevel),
  NameI (wellSep),
```

Table 13.1. Estimates of relative error in the energy and force calculations for random charge configurations, with $N_m = 8000$ and a tree depth of three.

l_{max}	energy	force
2	1.25e-04	1.00e-03
4	3.92e-06	5.64e-04
10	2.61e-07	4.13e-06

13.6 Results

Accuracy and performance

The accuracy of the fast-multipole method is readily demonstrated: simply compute♠ the interaction energy and forces for random charge distributions and then compare the estimates with the exact values obtained by considering all pairs of charges. Table 13.1 shows the average error over five different distributions for $n_w = 1$ and three values of the expansion order l_{max}. Theoretical estimates are also available from an analysis of the terms omitted when the multipole expansions are truncated. (The worst-case error is not considered here, but it can be estimated numerically by positioning the charges near the cell boundaries.)

Timing measurements demonstrate how the various parameters impact performance[†]. Example values for a single set of interaction calculations (with $n_w = 1$), together with comparisons with the all-pairs approach for the smaller systems, appear in Table 13.2. It is apparent that the optimal tree depth depends on N_m, that the choice of l_{max} has a strong effect on the performance and that the all-pairs approach is much slower even for moderately sized systems. The parameter n_w also influences the performance (not shown), since it determines how many pairs of atoms must be treated directly rather than by means of the cell expansions.

Radial distribution function

The case study considered here deals with a fluid of charged, soft-sphere atoms; it begins with an initial state arranged on a simple cubic lattice and uses hard walls in order to avoid those issues that the Ewald method addresses. The neighbor-list method is used for the soft-sphere interactions, together with leapfrog integration and constant-temperature dynamics. The RDF is evaluated, but contributions from atom pairs with charges of the same and opposite signs are accumulated separately in order to examine the tendency of the charged atoms to surround themselves with

---

♠ `pr_mpoletest`
† As always, times depend on processor, compiler and optimization level.

Table 13.2. Times (s) for a single set of interaction calculations.

N_m	depth	l_{max}	time	all pairs
8 000	3	2	0.2	2.7
8 000	3	4	0.4	
8 000	3	8	1.8	
32 000	3	2	2.6	44.0
32 000	4	2	1.2	
32 000	5	2	4.1	
128 000	3	2	39.3	
128 000	4	2	6.4	
128 000	5	2	8.9	
512 000	4	2	97.5	
512 000	5	2	20.0	
512 000	6	2	59.4	

neighbors of opposite sign. The cumulative numbers of neighbors as a function of separation *cumRdf* are also evaluated.

The function for evaluating the RDF is a modified form of *EvalRdf* in §4.3.

```
void EvalRdf ()
{
  ...
  if (countRdf == 0) {
    for (n = 0; n < sizeHistRdf; n ++) {
      histRdf[0][n] = 0.;
      histRdf[1][n] = 0.;
    }
  }
  deltaR = rangeRdf / sizeHistRdf;
  for (j1 = 0; j1 < nMol - 1; j1 ++) {
    for (j2 = j1 + 1; j2 < nMol; j2 ++) {
      VSub (dr, mol[j1].r, mol[j2].r);
      rr = VLenSq (dr);
      if (rr < Sqr (rangeRdf)) {
        n = sqrt (rr) / deltaR;
        k = (mol[j1].chg * mol[j2].chg > 0.);
        ++ histRdf[k][n];
      }
    }
  }
  ++ countRdf;
  if (countRdf == limitRdf) {
```

```
        normFac = VProd (region) / (2. * M_PI * Cube (deltaR) *
           Sqr (nMol) * countRdf);                                    25
        for (k = 0; k < 2; k ++) {
          cumRdf [k] [0] = 0.;
          for (n = 1; n < sizeHistRdf; n ++)
             cumRdf [k] [n] = cumRdf [k] [n - 1] + histRdf [k] [n];
          for (n = 0; n < sizeHistRdf; n ++) {                        30
             histRdf [k] [n] *= normFac / Sqr (n - 0.5);
             cumRdf [k] [n] /= nMol;
          }
        }
        ...                                                           35
      }
    }
```

The new arrays require

```
real **histRdf, **cumRdf;

  AllocMem2 (histRdf, 2, sizeHistRdf, real);
  AllocMem2 (cumRdf, 2, sizeHistRdf, real);
```

The run includes the following data:

```
chargeMag        4.
density          0.8
initUcell        20 20 20
limitRdf         50
maxLevel         3
rangeRdf         6.
sizeHistRdf      200
stepEquil        2000
stepInitlzTemp   200
stepLimit        8000
stepRdf          20
temperature      1.
wellSep          1
```

The resulting distributions, averaged over the last 1000 timesteps, appear in Figure 13.5; cumulative values are shown in Table 13.3. The most conspicuous feature of these results is – not surprisingly – the strong preference for oppositely charged neighbors at close range; because of overall charge neutrality this effect is limited to relatively short distances.

Table 13.3. Cumulative distributions of unlike and like charges as functions of distance d.

d	unlike	like
0.82	0.00	0.00
1.00	1.11	0.08
1.18	4.43	1.18
1.36	5.75	2.89
1.54	6.66	4.91
1.72	7.77	7.28
1.90	9.77	10.34
2.08	13.38	14.29
2.26	18.13	18.04
2.44	23.13	21.57
2.62	28.20	25.73

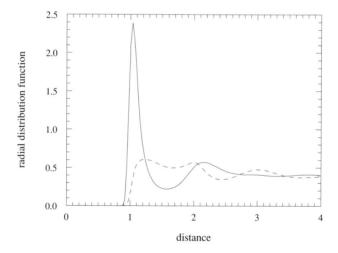

Fig. 13.5. Radial distribution functions for atoms with unlike (solid curve) and like (dashed curve) charges; the functions drop below the expected 0.5 at large distances due to hard-wall boundaries.

13.7 Further study

13.1 Compute the dielectric constant for a fluid of dipoles [kus90]; consider how to account for the surrounding medium.

13.2 Investigate how the adaptive subdivision and fast-multipole approaches can be combined.

13.3   Introduce Ewald summation into the fast-multipole method to handle periodic boundaries.

13.4   An example of an entirely different approach is the reaction-field method, in which interactions are considered in full up to a certain range, and beyond this a mean-field approximation is used [fri75]; explore the benefits and disadvantages of methods of this kind.

14

Step potentials

14.1 Introduction

Practically all the case studies in this book involve systems whose interactions are expressed in terms of continuous potentials. As a consequence, the dynamical equations can be solved numerically with constant-timestep integration methods. If one is prepared to dispense with this continuity another route is available that offers several advantages, although it has its weak points as well. The alternative method is based on step potentials; hard spheres are the simplest example, but the method can be extended to include potentials that have the shape of square wells or barriers, and even flexible 'molecules' can be built. Quantitative comparisons with specific real substances are obviously not the goal here, although comparisons with simple analytical models are possible. In fact, the earliest MD simulations [ald58, ald59, ald62] were of this kind, motivated by a desire to test basic theory.

A limitation of the methods used for continuous potentials, all of which involve a constant timestep Δt, is that they require the changes in interactions over each timestep to be small, otherwise uncontrolled numerical errors can suddenly appear. While this does not usually affect equilibrium studies, because Δt can be made sufficiently (but not too) small that for a particular simulation (namely, a given potential function, temperature and density) the results are predictably stable, systems that are inhomogeneous because of, for example, a large imposed temperature gradient, may prove problematic unless Δt is made unacceptably small. As will become apparent in due course, the step-potential method is unaware of this problem because it advances the system by a series of discrete events. The need for explicit numerical integration is avoided by employing impulsive collisions whenever atoms interact, and between such collisions each atom follows its own linear trajectory. There is no numerical integration error, because trajectories are evaluated to the full precision of the computer hardware.

391

The use of step potentials introduces its own problems. First, there is the increased complexity of the algorithm, since dealing with large numbers of collision events in an efficient manner requires careful attention to the question of data organization. The second problem is memory; storing the information describing events in a manner that is readily accessible and alterable tends to demand a good deal of extra memory.

The case study described here deals with the most basic kind of event-driven MD simulation, namely a hard-sphere fluid subject to periodic boundary conditions. Extensions of the basic method are discussed in §14.5 and appear in a case study in §15.3.

14.2 Computational approach

Dynamics

The physically interesting dynamics of the hard-sphere system is embodied in the collision rules; between collisions nothing of note happens and the atoms (as the spheres will subsequently be called) move in straight lines. Consider two identical atoms i and j currently separated by a distance $r = r_i - r_j$ and having a relative velocity $v = v_i - v_j$. These atoms will collide if and when their separation becomes equal to the atomic diameter σ; if this happens, it will occur at some time τ in the future, where τ is the smaller positive solution of

$$|r + v\tau| = \sigma \tag{14.2.1}$$

The solution, if it exists, is

$$\tau = \frac{-b - \sqrt{b^2 - v^2(r^2 - \sigma^2)}}{v^2} \tag{14.2.2}$$

where

$$b = r \cdot v \tag{14.2.3}$$

In order for a solution to exist, b must be negative and the argument of the square root positive. Solutions come in pairs; the larger positive solution reflects the fact that the separation σ occurs twice if the trajectories are extended beyond the collision point, although this is irrelevant for hard spheres. A negative solution corresponds to trajectories that (apparently) intersected in the past. The outcome of a collision between atoms is a simple change of velocities that preserves energy and momentum, namely,

$$\Delta v_i = -\Delta v_j = -\frac{b}{\sigma^2} r \tag{14.2.4}$$

If the analysis is extended to atoms that consist of a hard repulsive core surrounded by an attractive square well [ald59], then the generalized 'collision' corresponds to one of several kinds of event:

- a collision between the hard cores, as above;
- two atoms either entering or leaving their mutual potential well;
- two atoms bound together in a well bouncing as the well boundary is reached.

We will discuss this extension briefly, although it is not part of the case study.

If σ_c and σ_w are the core and well diameters, then the most general expression for when a collision might occur is

$$\tau = \frac{-b + s\sqrt{b^2 - v^2(r^2 - \sigma^2)}}{v^2} \qquad (14.2.5)$$

For a core collision, only possible for atoms already inside their mutual well, $s = -1$ and $\sigma = \sigma_c$, as in (14.2.2). If the future event is to be either a bounce within a well or an exit from it – this depends on whether there is sufficient kinetic energy – $s = 1$ and $\sigma = \sigma_w$. Finally, when a well is about to be entered, $s = -1$ and $\sigma = \sigma_w$.

The velocity changes when the generalized collision occurs are

$$\Delta v_i = -\Delta v_j = \phi r \qquad (14.2.6)$$

where

$$\phi = \frac{-b + s\sqrt{b^2 - 4r^2\Delta u/m}}{2r^2} \qquad (14.2.7)$$

In (14.2.7), Δu denotes the potential energy change and m is the mass. For a core collision $\Delta u = 0$, so that $\phi = -b/\sigma_c^2$ as in (14.2.4). For well entry $\Delta u = -w$, where w is the well depth, and $s = -1$. For well bounce and exit events, if $b^2 < 4\sigma_w^2 w/m$ the event is a bounce, so $\phi = -b/\sigma_w^2$, otherwise it is an exit, with $\Delta u = +w$ and $s = 1$. In order to obtain a true bound state, a third atom is required to remove the excess kinetic energy from a pair of atoms whose separation has dropped below σ_w, otherwise, because their combined energy is positive, they are not bound together; similarly, a bound pair can only escape from the well if the missing energy is provided by a third atom.

Cell subdivision

The simulation progresses by means of a time ordered sequence of collision events [erp77]. Assuming that all possible future collisions have been examined, it is a simple matter to determine which collision will occur first and advance all the atoms to that point in time. Such a scheme is correct in principle; in practice, the fact that determining the next collision of a given atom requires $O(N_m)$ work, because all atoms must be considered potential collision partners, rules out this simple approach. As in the case of soft-sphere MD, the use of cells (as well as a selective record of possible future collisions that have already been examined – see further) provides the means for reducing the number of atoms examined following a collision to a small value independent of N_m.

Assuming that the simulation region has been divided into cells whose edge length exceeds the sphere diameter σ, it is clear that collisions can only occur between atoms in the same and adjacent cells. By using relatively small cells, the average occupancy can be reduced to just one atom, or even less, per cell – the lower bound depends on density – so the gain in computational efficiency is apparent. All that is needed is to keep track of which atoms belong to which cells, and, since it is hardly appropriate to recompute this information after each collision, the way this is done is to introduce a new kind of event that occurs whenever an atom moves from one cell to another. Determining which cell face an atom will cross next, and when this is due to occur, are simple computational problems that are solved in the program listing later on. It is true the cell crossing events introduce additional work, but the overall reduction in effort more than compensates for this. Periodic boundaries are readily incorporated into the cell-based computations.

Another labor saving device is the use of a local time variable associated with each atom (or, alternatively, with each cell). When a collision occurs, only atoms in the immediate neighborhood are of concern and there is no point in updating the coordinates of atoms much further away. The use of a 'personal' time for each atom provides a record of when its coordinates were last updated, so that one of the few occasions when an update of the entire system is really necessary is prior to recording a snapshot of the system configuration.

Event calendar

We have tacitly assumed that the system always knows when the next event is due, whether it is a collision or a cell crossing and the atom(s) involved. This implies the existence of an event calendar. Such a calendar must not only produce the next event but must also be easily modifiable: the calendar will include many future events and, as collisions occur, changes must be made to its contents, both to incorporate newly predicted collisions and to remove previously predicted collisions that are no longer relevant because a participant has in the meantime undergone a different collision. Once the effort has been made to find a possible future collision, this information should be retained for as long as it is potentially useful, but it must be recognized that if the calendar includes a few collision events involving each atom, it is likely that most of this information will become obsolete before it has a chance to be used. Thus the calendar organization is central to the viability of this method; we will discuss its implementation in §14.3.

When two atoms collide their velocities are changed, so that any information stored in the calendar regarding future events involving these atoms ceases to be valid. Such events will have to be erased from the calendar and replaced by whatever new events are predicted. The potential collisions that must be examined to determine these new events are between each of the colliding atoms and all other atoms in the neighboring cells, including the cell that the atom presently occupies. Similarly, when an atom crosses a cell boundary, the newly adjacent cells also contain potential collision partners that must be considered; in this case, however, existing calendar entries are still relevant and are retained. Following both kinds of event it is necessary to determine the next cell crossing event for the atom(s) involved. If these details are taken care of correctly there is no way in which a collision can be overlooked.

Program details

At this point we describe those parts of the program♠ that deal with the collisions and cell crossings. The handling of the event calendar will be discussed separately. Unlike the previous case studies that were built upon one another's programs, the hard-sphere simulation is organized completely differently and so, with the exception of certain common utility functions, a separate program will be constructed. The reduced MD units used in this simulation are defined so that the atoms have unit mass and diameter. Much of the program is equally suitable for spheres in three dimensions (as in this chapter) and disks in two (see §15.3); the differences between the three- and two-dimensional versions are relatively minor.

The description begins with the main program.

```
int main (int argc, char **argv)
{
  GetNameList (argc, argv);
  PrintNameList (stdout);
  SetParams ();                                          5
  SetupJob ();
  moreCycles = 1;
  eventCount = 0;
  while (moreCycles) {
    SingleEvent ();                                      10
    ++ eventCount;
    if (eventCount >= limitEventCount) moreCycles = 0;
  }
}
```

♠ pr_14_1

Job initialization requires

```
void SetupJob ()
{
  AllocArrays ();
  InitCoords ();
  InitVels ();                                                    5
  timeNow = 0.;
  nextSumTime = 0.;
  collCount = 0.;
  crossCount = 0.;
  StartRun ();                                                    10
  ScheduleEvent (0, MOL_LIMIT + 6, nextSumTime);
}
```

where *MOL_LIMIT* is a constant that exceeds the maximum possible number of atoms in the system.

The function responsible for processing a single event is

```
void SingleEvent ()
{
  real vvSum;
  int n;
                                                                 5
  NextEvent ();
  if (evIdB < MOL_LIMIT) {
    ProcessCollision ();
    ++ collCount;
  } else if (evIdB >= MOL_LIMIT + 100) {                         10
    ProcessCellCrossing ();
    ++ crossCount;
  } else if (evIdB == MOL_LIMIT + 6) {
    UpdateSystem ();
    nextSumTime += intervalSum;                                  15
    ScheduleEvent (0, MOL_LIMIT + 6, nextSumTime);
    VZero (vSum);
    vvSum = 0.;
    DO_MOL {
      VVAdd (vSum, mol[n].rv);                                    20
      vvSum += VLenSq (mol[n].rv);
    }
    kinEnVal = vvSum * 0.5 / nMol;
    PrintSummary (stdout);
  }                                                              25
}
```

The call to *NextEvent* obtains the details of the next event; these include the two values *evIdA* and *evIdB*, that are examined to determine the event type and atom(s) involved. If the event is a collision, recognizable because *evIdB* is less than *MOL_LIMIT*, the two values identify the colliding atoms. For a cell crossing, signaled by a value of *evIdB* not less than *MOL_LIMIT+100*, *evIdA* is the atom and *evIdB* describes the cell face crossed. The only other kind of event expected here, corresponding to the value *MOL_LIMIT+6*, is one which outputs a summary of the properties of the system. Other event classes are readily accommodated (see §15.3).

The structure *Mol* used in this program has the form

```
typedef struct {
  VecR r, rv;
  real time;
  VecI inCell;
} Mol;                                                              5
```

and the following variables (those requiring explanation will receive it in due course) are used,

```
Mol *mol;
EvTree *evTree;
VecR region, vSum;
VecI cellRange[2], cells, initUcell;
real collCount, crossCount, density, intervalSum, kinEnVal,         5
    nextSumTime, temperature, timeNow, velMag;
int *cellList, eventCount, eventMult, evIdA, evIdB, limitEventCount,
    moreCycles, nMol, poolSize;
```

The list of input data items for the program is

```
NameList nameList[] = {
  NameR (density),
  NameI (eventMult),
  NameI (initUcell),
  NameR (intervalSum),                                              5
  NameI (limitEventCount),
  NameR (temperature),
};
```

The function *ProcessCollision* is used to process a single collision event. The hard-sphere collision dynamics are as described in (14.2.4). A new array *cellRange* appears here; its values are used to determine which of the neighboring cells, out of the total of 27 in three dimensions, should be examined for future collision events. Collisions between atoms on opposite sides of the periodic boundaries are treated correctly by pretending that the collision is with one of the periodic replica atoms.

```
void ProcessCollision ()
{
  VecR dr, dv;
  real fac;
                                                                      5
  UpdateMol (evIdA);
  UpdateMol (evIdB);
  VSetAll (cellRange[0], -1);
  VSetAll (cellRange[1], 1);
  VSub (dr, mol[evIdA].r, mol[evIdB].r);                              10
  VWrapAll (dr);
  VSub (dv, mol[evIdA].rv, mol[evIdB].rv);
  fac = - VDot (dr, dv) / VLenSq (dr);
  VVSAdd (mol[evIdA].rv, fac, dr);
  VVSAdd (mol[evIdB].rv, - fac, dr);                                 15
  PredictEvent (evIdA, -1);
  PredictEvent (evIdB, evIdA);
}
```

The function `ProcessCellCrossing` deals with a single cell boundary crossing event. Linked lists are used to record the atoms belonging to each cell; the atom concerned is removed from the list of the cell just exited and added to that of the newly entered cell. Periodic wraparound is applied where necessary and the values in `cellRange` are used here to limit the cells examined for possible future collisions to the newly adjacent cells only.

```
#define VWrapEv(t)                                          \
  if (mol[evIdA].rv.t > 0.) {                               \
    cellRange[0].t = 1;                                      \
    ++ mol[evIdA].inCell.t;                                 \
    if (mol[evIdA].inCell.t == cells.t) {                   \    5
      mol[evIdA].inCell.t = 0;                              \
      mol[evIdA].r.t = -0.5 * region.t;                     \
    }                                                        \
  } else {                                                   \
    cellRange[1].t = -1;                                    \   10
    -- mol[evIdA].inCell.t;                                 \
    if (mol[evIdA].inCell.t == -1) {                        \
      mol[evIdA].inCell.t = cells.t - 1;                    \
      mol[evIdA].r.t = 0.5 * region.t;                      \
    }                                                        \   15
  }

void ProcessCellCrossing ()
{
  int n;                                                          20

  UpdateMol (evIdA);
  n = VLinear (mol[evIdA].inCell, cells) + nMol;
```

```
    while (cellList[n] != evIdA) n = cellList[n];
    cellList[n] = cellList[evIdA];                              25
    VSetAll (cellRange[0], -1);
    VSetAll (cellRange[1], 1);
    switch (evIdB - MOL_LIMIT - 100) {
      case 0:
        VWrapEv (x);                                            30
        break;
      case 1:
        VWrapEv (y);
        break;
      case 2:                                                   35
        VWrapEv (z);
        break;
    }
    PredictEvent (evIdA, evIdB);
    n = VLinear (mol[evIdA].inCell, cells) + nMol;              40
    cellList[evIdA] = cellList[n];
    cellList[n] = evIdA;
  }
```

Predicting future events after a collision or cell crossing is carried out by the function *PredictEvent*. The first part of this function looks at possible cell boundary crossings in all directions and schedules the earliest one. The second part examines every atom in the cells that must be scanned for potential collisions and determines whether a collision is possible using (14.2.2). Much of the code handles the special requirements of periodic boundaries. The reason why two arguments are needed by this function (the first is an atom number, the second either an atom number or a code with values −1 or −2) should be apparent from the listing.

```
#define WhenCross(t)                                              \
    tm.t = (mol[na].rv.t != 0.) ? w.t / mol[na].rv.t : 1e12;
#define VWrapEvC(t)                                               \
    m2v.t = mol[na].inCell.t + m1v.t;                            \
    shift.t = 0.;                                                \    5
    if (m2v.t == -1) {                                          \
      m2v.t = cells.t - 1;                                      \
      shift.t = - region.t;                                     \
    } else if (m2v.t == cells.t) {                              \
      m2v.t = 0;                                                \
      shift.t = region.t;                                       \    10
    }

void PredictEvent (int na, int nb)
{                                                                    15
  VecR dr, dv, rs, shift, tm, w;
  VecI m1v, m2v;
```

```
  real b, d, t, tInt, vv;
  int dir, evCode, m1x, m1y, m1z, n;

  VCopy (w, mol[na].inCell);
  if (mol[na].rv.x > 0.) ++ w.x;
  if (mol[na].rv.y > 0.) ++ w.y;
  if (mol[na].rv.z > 0.) ++ w.z;
  VMul (w, w, region);
  VDiv (w, w, cells);
  VSAdd (rs, mol[na].r, 0.5, region);
  VVSub (w, rs);
  WhenCross (x);
  WhenCross (y);
  WhenCross (z);
  if (tm.y < tm.z) dir = (tm.x < tm.y) ? 0 : 1;
  else dir = (tm.x < tm.z) ? 0 : 2;
  evCode = 100 + dir;
  ScheduleEvent (na, MOL_LIMIT + evCode, timeNow + VComp (tm, dir));
  for (m1z = cellRange[0].z; m1z <= cellRange[1].z; m1z ++) {
    m1v.z = m1z;
    VWrapEvC (z);
    for (m1y = cellRange[0].y; m1y <= cellRange[1].y; m1y ++) {
      m1v.y = m1y;
      VWrapEvC (y);
      for (m1x = cellRange[0].x; m1x <= cellRange[1].x; m1x ++) {
        m1v.x = m1x;
        VWrapEvC (x);
        n = VLinear (m2v, cells) + nMol;
        for (n = cellList[n]; n >= 0; n = cellList[n]) {
          if (n != na && n != nb && (nb >= -1 || n < na)) {
            tInt = timeNow - mol[n].time;
            VSub (dr, mol[na].r, mol[n].r);
            VVSAdd (dr, - tInt, mol[n].rv);
            VVSub (dr, shift);
            VSub (dv, mol[na].rv, mol[n].rv);
            b = VDot (dr, dv);
            if (b < 0.) {
              vv = VLenSq (dv);
              d = Sqr (b) - vv * (VLenSq (dr) - 1.);
              if (d >= 0.) {
                t = - (sqrt (d) + b) / vv;
                ScheduleEvent (na, n, timeNow + t);
              }
            }
          }
        }
      }
    }
  }
}
```

The macro *VCopy* (§18.2) copies vector components while allowing for a change of component type.

The following function is called at the start of the computation to create the cell lists and produce the initial event calendar.

```
void StartRun ()
{
  VecR rs;
  int j, n;

  for (j = 0; j < VProd (cells) + nMol; j ++) cellList[j] = -1;
  DO_MOL {
    mol[n].time = timeNow;
    VSAdd (rs, mol[n].r, 0.5, region);
    VMul (rs, rs, cells);
    VDiv (mol[n].inCell, rs, region);
    j = VLinear (mol[n].inCell, cells) + nMol;
    cellList[n] = cellList[j];
    cellList[j] = n;
  }
  InitEventList ();
  VSetAll (cellRange[0], -1);
  VSetAll (cellRange[1], 1);
  DO_MOL PredictEvent (n, -2);
}
```

The function that updates an atom's coordinates and time variable is

```
void UpdateMol (int id)
{
  real tInt;

  tInt = timeNow - mol[id].time;
  VVSAdd (mol[id].r, tInt, mol[id].rv);
  mol[id].time = timeNow;
}
```

and the entire system can be updated by

```
void UpdateSystem ()
{
  int n;

  DO_MOL UpdateMol (n);
}
```

The only other kind of event included in this version of the program simply outputs

the measured energy and momentum of the system, as well as a report on the
number of events that have occurred so far. The output function is

```
void PrintSummary (FILE *fp)
{
  fprintf (fp, "%.2f %.10g %.10g %.3f %.3f\n",
     timeNow, collCount, crossCount, VCSum (vSum) / nMol),
     kinEnVal);
}
```

Additional quantities that are set in *SetParams* are

```
poolSize = eventMult * nMol;
VCopy (cells, region);
```

The variable *poolSize* determines how much space will be allocated for the event
calendar. The size of the cell array assumes that the smallest possible cell is wanted
(recall that the atoms have unit diameter). Memory allocation in *AllocArrays*
includes, in addition to the usual *mol* and *cellList* arrays,

```
AllocMem (evTree, poolSize, EvTree);
```

an array that appears only in the event processing functions described in §14.3. The
initial state is defined in the same way as for soft spheres (§3.6).

Properties

Equilibrium and transport properties for models based on impulsive interactions
can be defined in an analogous way to the continuous case. The only difference
occurs in those quantities that depend directly on the interactions, such as the pressure. The virial expression (2.3.8) must be replaced [erp77] by its impulsive limit,
namely, a sum over the collisions occurring during the measurement period t_m,

$$PV = \tfrac{1}{3}\left[\left\langle \sum_i v_i^2 \right\rangle + \frac{1}{t_m} \sum_c r_{i_c j_c} \cdot \Delta v_{i_c}\right] \tag{14.2.8}$$

where i_c and j_c are the atoms involved in a particular collision c; the separation
at collision $r_{i_c j_c}$ – where $|r_{i_c j_c}| = \sigma$ – allows for periodic wraparound. Transport
properties follow a similar approach [ald70a].

Computation of the RDF is the same as for soft spheres (§4.3); the only difference is that, instead of the measurement being performed at fixed multiples of the
timestep, a new class of measurement event is required. New variables, in addition
to those needed for the RDF computation, and input data are

```
real intervalRdf, nextRdfTime;

   NameR (intervalRdf),
```

and the initialization (in *SetupJob*) requires

```
nextRdfTime = 0.;
ScheduleEvent (0, MOL_LIMIT + 7, nextRdfTime);
countRdf = 0;
```

The event processing (in *SingleEvent*) now includes a test for the new event type,

```
} else if (evIdB == MOL_LIMIT + 7) {
   UpdateSystem ();
   EvalRdf ();
   nextRdfTime += intervalRdf;
   ScheduleEvent (0, MOL_LIMIT + 7, nextRdfTime);          5
} else ...
```

14.3 Event management

Calendar design

We have already alluded to the central role played by the event calendar. The calendar contains a list of future collisions and cell crossings, as well as events corresponding to measurements of various kinds conducted at fixed time intervals. For a large system, the calendar will hold a great deal of information and it is imperative that it be managed in an efficient way. Efficiency focuses principally on execution time, but space requirements are not neglected.

The scheme we describe here [rap80] is based on a binary tree data structure [knu68, knu73]. The binary tree is a generalization of the linked list[†] to the case where each node (or list member) has pointers to two successors rather than just one; the analogy with an inverted tree is obvious, hence the name. Other data structures could serve the purpose, but the binary tree is relatively straightforward to implement. More significantly, its performance in situations relevant to MD can be analyzed theoretically and, to within a constant factor, can be shown to be optimal.

Each scheduled event is represented by a node in the tree. The information contained within the node identifies the time at which the event is scheduled to occur and the event details: if the event is a collision, then the atoms involved are specified; if it is a cell crossing, then the atom is specified, together with an indication

---

† The oct-tree in §13.3 was another of these generalizations, although it used an entirely different approach to storage and was not designed to be modified.

of which cell boundary is crossed; for other event types, typically measurement events, the details are given by a suitable numerical code. The operations that are performed on the tree data are the following:

- retrieve the earliest event;
- add a new event;
- delete an existing event;
- initialize the tree contents.

After retrieving an event the node containing its description must be deleted from the tree and, whenever a collision occurs, all other event nodes involving either of the participants must also be deleted. A 'pool' of spare nodes exists from which withdrawals are made when events are added to the calendar and to which the nodes are returned once no longer needed; this pool must never be allowed to become empty.

Three pointers are used to link event nodes into the tree; these point to the left and right descendant nodes and to the parent node. The time ordering is such that all the left-hand descendants are events scheduled to occur before the event at the current node, while those on the right are due to occur after it. The pointer to the parent is not essential, but its presence simplifies algorithms for navigating the tree. The actual tree representation of a given set of events is far from unique and depends on the order in which the event nodes are added.

To support rapid deletion of related event nodes, the storage for each event provides additional pointers needed for linking the node into two circular lists [knu68], one list for each of the atoms involved if the event is a collision, two distinct lists for the same atom if the event is a cell crossing (the explanation follows); these pointers are unused in other cases. The reason for two linked lists per atom is again one of convenience: for a given atom j there is one list joining all collision nodes in which j appears as the first partner in the pair and another for those in which j is the second partner. The pointers belonging to the cell crossing node associated with atom j (there is always exactly one such node) are used to access these two lists. To improve performance even further, the circular lists are doubly linked, each having pointers that traverse the list in both directions.

Theoretical performance

The amount of work required to perform certain elementary operations on the data in the binary tree can be estimated theoretically [knu73]; these operations lie at the heart of the calendar management functions to be described shortly. Here we summarize the relevant results.

Consider a binary tree with N nodes. If the tree is balanced, in the sense that all paths from the root (the node from which all others descend) to the nodes at the ends of all the branches are essentially the same length, then it can be shown that an average of 1.39 log N nodes must be tested to find the correct insertion point for a new node (assuming that the value determining the node position – the scheduled time in the case of MD – is randomly chosen). This represents the optimal value. In circumstances more relevant to MD, namely that the entire tree is constructed from a series of events whose scheduled times are (from the tree's point of view) randomly distributed, the average number of tests increases to 2 log N, a value still not too far from optimal. Measurements using actual MD simulations confirm this result [rap80] and lay to rest any concern that the tree might degenerate into a near-linear list (for which average insertion time is proportional to N) over the duration of the run.

Another theoretical result deals with the average number of cycles in a search loop required to relink the neighbors of a particular node after that node is deleted. While this could also have shown a certain amount of N dependence, in actual fact the value is a constant less than 0.5, and is thus completely independent of tree size.

Program details

We have already introduced the pointers associated with the event nodes: there are three pointers for linking each node into the tree, and every node (except for measurement events) also belongs to two circular lists, each of which is doubly linked. Thus the total number of pointers per node is seven, and these, together with the two values specifying what the event actually is, and the scheduled time of the event, are stored in the structure

```
typedef struct {
  real time;
  int left, right, up, circAL, circAR, circBL, circBR, idA, idB;
} EvTree;
```

The first three integer values are used as pointers for traversing the tree. The next four are associated with the two circular lists to which each collision and cell crossing event node belongs; the two lists are denoted *A* and *B*, and the two pointers for each list *L* and *R*. The final two values describe the event, as explained previously. An array *evTree* of structures of this type constitutes the event tree.

The tree fluctuates in size over the course of the simulation as nodes are added and removed. A pool of spare nodes is provided to accommodate these size variations. The first node of *evTree* is not used to hold events, but serves as a fixed root from which the rest of the tree grows; one of its pointers is used to access the pool, the nodes of which are joined into a linked list (using the pointer *circAR*).

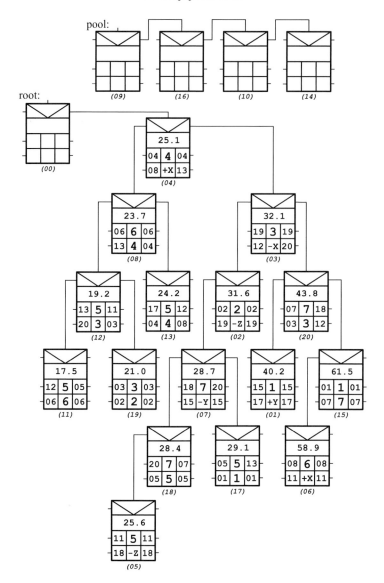

Fig. 14.1. A hand-crafted miniature event tree (the real one is much larger) – the tree links are shown, but for clarity the circular lists are omitted and the four pointer values are shown instead (on either side of the node). Each node includes the event time, the atom(s) involved, and the cell face crossed (if relevant); the value beneath the node is its 'address' in the tree.

The nodes corresponding to cell crossings also occupy reserved locations, the N_m nodes immediately following the root, since there is always one such event scheduled per atom; these nodes also serve as anchors for the circular lists associated with each atom. The remaining nodes are dynamically assigned to collisions and other events as necessary. Figure 14.1 shows an example of a very small event tree.

The functions for scheduling an event, determining the next event and deleting events follow. The list and tree manipulations are entirely standard [knu73] and, with just a little effort, it should be possible to follow the logic of the algorithms.

The first of these functions inserts an event node at the correct location in the tree and then links it into the two circular lists; the node is taken from the pool, with a check being made to ensure that the pool is not empty. The tests (here and subsequently) to determine event type make provision for event types that are not included in the present case study, in particular, collisions with impenetrable boundary walls – see §15.3 – which, from the point of view of event management, combine features of both collisions and cell crossings.

```
void ScheduleEvent (int idA, int idB, real tEvent)
{
  int id, idNew, more;

  id = 0;                                                              5
  if (idB < MOL_LIMIT || idB >= MOL_LIMIT + 2 * NDIM &&
     idB < MOL_LIMIT + 100) {
    if (evTree[0].idA < 0) ErrExit (ERR_EMPTY_EVPOOL);
    idNew = evTree[0].idA;
    evTree[0].idA = evTree[evTree[0].idA].circAR;                     10
  } else idNew = idA + 1;
  if (evTree[id].right < 0) evTree[id].right = idNew;
  else {
    more = 1;
    id = evTree[id].right;                                            15
    while (more) {
      if (tEvent < evTree[id].time) {
        if (evTree[id].left >= 0) id = evTree[id].left;
        else {
          more = 0;                                                   20
          evTree[id].left = idNew;
        }
      } else {
        if (evTree[id].right >= 0) id = evTree[id].right;
        else {                                                        25
          more = 0;
          evTree[id].right = idNew;
        }
      }
    }                                                                 30
  }
  if (idB < MOL_LIMIT) {
    evTree[idNew].circAR = evTree[idA + 1].circAR;
    evTree[idNew].circAL = idA + 1;
    evTree[evTree[idA + 1].circAR].circAL = idNew;                    35
    evTree[idA + 1].circAR = idNew;
    evTree[idNew].circBR = evTree[idB + 1].circBR;
```

```
    evTree[idNew].circBL = idB + 1;
    evTree[evTree[idB + 1].circBR].circBL = idNew;
    evTree[idB + 1].circBR = idNew;                              40
  }
  evTree[idNew].time = tEvent;
  evTree[idNew].idA = idA;
  evTree[idNew].idB = idB;
  evTree[idNew].left = evTree[idNew].right = -1;                 45
  evTree[idNew].up = id;
}
```

We mentioned earlier that the average number of times the loop in this function is executed grows logarithmically with tree size.

The second function determines the next event about to occur, removes it from the tree and ensures that other, now invalid, events are also removed; nodes removed are returned to the pool.

```
void NextEvent ()
{
  int idNow;

  idNow = evTree[0].right;                                        5
  while (evTree[idNow].left >= 0) idNow = evTree[idNow].left;
  timeNow = evTree[idNow].time;
  evIdA = evTree[idNow].idA;
  evIdB = evTree[idNow].idB;
  if (evIdB < MOL_LIMIT + 2 * NDIM) {                            10
    DeleteAllMolEvents (evIdA);
    if (evIdB < MOL_LIMIT) DeleteAllMolEvents (evIdB);
  } else {
    DeleteEvent (idNow);
    if (evIdB < MOL_LIMIT + 100) {                               15
      evTree[idNow].circAR = evTree[0].idA;
      evTree[0].idA = idNow;
    }
  }
}                                                                20
```

The third function does the deletions required following a collision event; it removes all other scheduled events involving the affected atoms from the tree (and returns them to the pool) by traversing all four circular lists to which the atoms belong.

```
void DeleteAllMolEvents (int id)
{
  int idd;

  ++ id;                                                          5
```

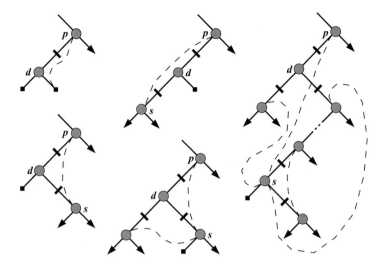

Fig. 14.2.   The various kinds of pointer rearrangements that can occur following node removal from the tree; d is the deleted node, a pointer is added from p to s, and other pointers may need alteration.

```
DeleteEvent (id);
for (idd = evTree[id].circAL; idd != id; idd = evTree[idd].circAL) {
  evTree[evTree[idd].circBL].circBR = evTree[idd].circBR;
  evTree[evTree[idd].circBR].circBL = evTree[idd].circBL;
  DeleteEvent (idd);                                              10
}
evTree[evTree[id].circAL].circAR = evTree[0].idA;
evTree[0].idA = evTree[id].circAR;
evTree[id].circAL = evTree[id].circAR = id;
for (idd = evTree[id].circBL; idd != id; idd = evTree[idd].circBL) { 15
  evTree[evTree[idd].circAL].circAR = evTree[idd].circAR;
  evTree[evTree[idd].circAR].circAL = evTree[idd].circAL;
  DeleteEvent (idd);
  evTree[idd].circAR = evTree[0].idA;
  evTree[0].idA = idd;                                           20
}
evTree[id].circBL = evTree[id].circBR = id;
}
```

The last of these functions rearranges the tree pointers following the removal of a node. All possible eventualities are handled and, as indicated previously, the average number of times the short loop in this function is executed is less than unity. The operations involved are illustrated in Figure 14.2.

```
void DeleteEvent (int id)
{
  int idp, idq, idr;
```

```
  idr = evTree[id].right;                                           5
  if (idr < 0) idq = evTree[id].left;
  else {
    if (evTree[id].left < 0) idq = idr;
    else {
      if (evTree[idr].left < 0) idq = idr;                         10
      else {
        idq = evTree[idr].left;
        while (evTree[idq].left >= 0) {
          idr = idq;
          idq = evTree[idr].left;                                  15
        }
        evTree[idr].left = evTree[idq].right;
        evTree[evTree[idq].right].up = idr;
        evTree[idq].right = evTree[id].right;
        evTree[evTree[id].right].up = idq;                         20
      }
      evTree[evTree[id].left].up = idq;
      evTree[idq].left = evTree[id].left;
    }
  }                                                                25
  idp = evTree[id].up;
  evTree[idq].up = idp;
  if (evTree[idp].right != id) evTree[idp].left = idq;
  else evTree[idp].right = idq;
}                                                                  30
```

One further function is required for initializing the data structures. All nodes are placed in the pool, and all circular lists are constructed so that each contains just the one node associated with the cell crossing for the particular atom.

```
void InitEventList ()
{
  int id;

  evTree[0].left = evTree[0].right = -1;                            5
  evTree[0].idA = nMol + 1;
  for (id = evTree[0].idA; id < poolSize - 1; id ++)
      evTree[id].circAR = id + 1;
  evTree[poolSize].circAR = -1;
  for (id = 1; id < nMol + 1; id ++) {                             10
    evTree[id].circAL = evTree[id].circBL = id;
    evTree[id].circAR = evTree[id].circBR = id;
  }
}
```

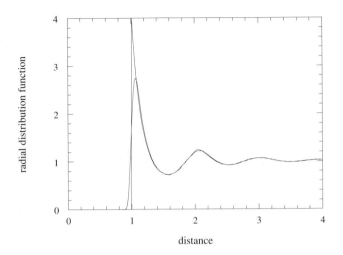

Fig. 14.3.   Radial distribution functions for hard- and soft-sphere fluids.

14.4 Properties

Radial distribution function

The hard-sphere RDF is obtained from a run using input data

density	0.8
eventMult	4
initUcell	8 8 8
intervalRdf	0.25
intervalSum	5.
limitEventCount	1500000
limitRdf	100
rangeRdf	4.
sizeHistRdf	200
temperature	1.

The initial state is an FCC lattice, so that $N_m = 2048$; the conditions are the same as those used in §4.3 for measuring the soft-sphere RDF.

With these input values, a set of RDF results is produced roughly every 6×10^5 collisions. The first set is discarded, the second is shown in Figure 14.3. The soft-sphere RDF is included for comparison; apart from the sharp first peak in the case of hard spheres the two curves are practically identical.

Free-path distribution

The distribution of path lengths between collisions is a characteristic that can only be studied within the hard-sphere framework, since, for continuous potentials, the notion of a collision event is not precisely defined. The average of this distribution is just the familiar mean free path. The new variables required for this analysis[♦] are

```
real *histFreePath, rangeFreePath;
int countFreePath, limitFreePath, sizeHistFreePath;
```

an addition to structure `Mol`,

```
VecR rCol;
```

together with input data items, an additional array allocation (in `AllocArrays`) and initialization (in `SetupJob`),

```
NameI (limitFreePath),
NameR (rangeFreePath),
NameI (sizeHistFreePath),

AllocMem (histFreePath, sizeHistFreePath, real);          5

InitFreePath ();
```

After processing each collision, a call is made to the function that calculates the path lengths while taking into account the periodic boundaries.

```
void EvalFreePath ()
{
  VecR dr;
  int j, n;
                                                          5
  for (n = evIdA; n <= evIdB; n += evIdB - evIdA) {
    if (countFreePath == 0) {
      for (j = 0; j < sizeHistFreePath; j ++) histFreePath[j] = 0.;
    }
    VSub (dr, mol[n].r, mol[n].rCol);                     10
    VWrapAll (dr);
    mol[n].rCol = mol[n].r;
    j = VLen (dr) * sizeHistFreePath / rangeFreePath;
    ++ histFreePath[Min (j, sizeHistFreePath - 1)];
    ++ countFreePath;                                     15
    if (countFreePath == limitFreePath) {
      for (j = 0; j < sizeHistFreePath; j ++)
```

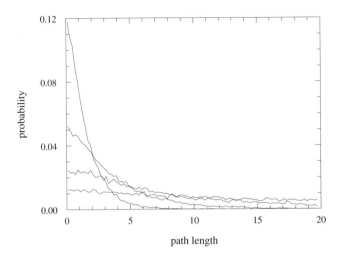

Fig. 14.4.   Free-path distributions for hard spheres at densities 0.025, 0.05, 0.1 and 0.2.

```
      histFreePath[j] /= countFreePath;
    PrintFreePath (stdout);
    countFreePath = 0;                                              20
  }
 }
}
```

The output function *PrintFreePath* (not shown) just prints the path distribution histogram; the initialization function is

```
void InitFreePath ()
{
  int n;

  countFreePath = 0;                                                 5
  DO_MOL mol[n].rCol = mol[n].r;
}
```

The runs are for $N_m = 2048$ and include input data

```
limitFreePath      20000
rangeFreePath      20.
sizeHistFreePath   100
```

with density values ranging from 0.025 to 0.2. The results from the first 50 000 (approximately) collisions are discarded and those for the next 20 000 are shown in Figure 14.4. The distributions all decrease with path length (ignoring noise);

if each is scaled by the mean free path at the corresponding density, all should collapse onto a single curve [ein68].

Efficiency

Any comparison between the computational efficiency of hard- and soft-sphere systems requires some way of quantifying equivalent amounts of computation. One could, for example, measure the computational effort needed to determine pressure to a given degree of accuracy. The results will certainly depend on the density, with soft spheres having the advantage at high density (where the hard-sphere collision rate is high) and the hard spheres at low density (where the mean free path is relatively long). The whole question is often irrelevant, however, because the nature of the problem may dictate which approach is required[†].

14.5 Generalizations

Outline

In this section we deal briefly with two useful extensions of the hard-sphere approach: the construction of polymer chains and the way in which both rotational motion and inelasticity can be added (the latter intended for use in a macroscopic context only). In the first example the additions involve procedural details, so the program modifications are shown; in the second the changes are in the collision dynamics and so we concentrate on the mathematical details. Two further extensions, namely, the addition of a gravitational field and the use of hard-wall boundaries, are described in §15.3.

Hard-sphere polymer chains

Flexible polymer chains – reminiscent of a bead necklace – can be constructed by placing each pair of chain neighbors in a potential well with infinitely high walls and a width corresponding to the maximum bond elongation [rap79]. While this kind of model lacks the refinement of the alkane model considered in Chapter 10, it allows polymer studies to benefit from the advantages of event-driven MD.

We assume that the atoms still have unit diameter and define the maximum bond elongation to be *bondStretch*. Cell sizes will now be determined by this value, so that the collision event occurring when a bond becomes fully stretched will also be detected by the usual scan across cells. The initial state must be properly

---

† If there is a choice, the actual relative efficiency could well depend on the kind of processor used; for this reason we will forgo timing details.

constructed and we will also assume that all chain atoms are numbered consecutively. The only additional change needed is in `PredictEvent`, to ensure that when predicting collisions for chain neighbors, both normal and stretched bond collisions are examined, and the appropriate choice made.

```
int collCode;
real rr;
...
for (n = cellList[n]; n >= 0; n = cellList[n]) {
  if (n != na && n != nb && (nb >= -1 || n < na)) {             5
    tInt = timeNow - mol[n].time;
    VSub (dr, mol[na].r, mol[n].r);
    VVSAdd (dr, - tInt, mol[n].rv);
    VVSub (dr, shift);
    VSub (dv, mol[na].rv, mol[n].rv);                           10
    collCode = 0;
    if (abs (n - na) > 1 || n / chainLen != na / chainLen) {
      b = VDot (dr, dv);
      if (b < 0.) {
        vv = VLenSq (dv);                                       15
        d = Sqr (b) - vv * (VLenSq (dr) - 1.);
        if (d >= 0.) collCode = 1;
      }
    } else if (nb < MOL_LIMIT + 2 * NDIM) {
      collCode = 2;                                             20
      b = VDot (dr, dv);
      rr = VLenSq (dr);
      vv = VLenSq (dv);
      if (b < 0.) {
        d = Sqr (b) - vv * (rr - 1.);                           25
        if (d >= 0.) collCode = 1;
      }
    }
    if (collCode > 0) {
      if (collCode == 1) t = - (sqrt (d) + b) / vv;             30
      else {
        d = Sqr (b) - vv * (rr - Sqr (bondStretch));
        t = (sqrt (d) - b) / vv;
      }
      ScheduleEvent ...                                         35
    }
  }
}
```

The fact that chain neighbors are never more than one cell apart removes the need to search for new collisions of this type after a cell crossing. The role of `collCode` is to distinguish between the two types of collision event that are now possible; the formula used for the stretched bond collision is a particular instance of (14.2.5).

Rotation and inelasticity

Rotational motion is a feature that is readily added to the hard-sphere model [ber77]. The surfaces of the spheres are assumed to be rough, so that when a collision occurs not only is there a change in translational motion but the spins of the spheres also change.

It is a simple exercise in kinematics to show that the relative velocity at the point of impact (we now include the possibility that the spheres have different sizes and masses) is

$$g = v_{ij} - (\sigma_i \omega_i + \sigma_j \omega_j) \times r_{ij}/2\bar{\sigma} \tag{14.5.1}$$

where $\bar{\sigma} = (\sigma_i + \sigma_j)/2$. If the impulse is b, the velocity and angular velocity changes required to conserve linear and angular momentum are given by

$$m_i \Delta v_i = -m_j \Delta v_j = b \tag{14.5.2}$$

$$m_i \sigma_i \Delta \omega_i = m_j \sigma_j \Delta \omega_j = -r_{ij} \times b/2\kappa\bar{\sigma} \tag{14.5.3}$$

where κ is the numerical factor in the moment of inertia $I = \kappa m \sigma^2$; $\kappa = 1/10$ for solid spheres, $1/8$ for disks.

The impulse can be expressed in terms of the components of g parallel and perpendicular to r_{ij},

$$b = \bar{m} \big[g'^\| - g^\| + \kappa_1 \big(g'^\perp - g^\perp \big) \big] \tag{14.5.4}$$

where the primes denote values after the collision, $\kappa_1 = 1/(1 + 1/4\kappa)$, and the reduced mass is

$$\bar{m} = \frac{m_i m_j}{m_i + m_j} \tag{14.5.5}$$

The change in kinetic energy can also be written in terms of these components,

$$\Delta E_K = \tfrac{1}{2}\bar{m} \big[g'^{\|2} - g^{\|2} + \kappa_1 \big(g'^{\perp 2} - g^{\perp 2} \big) \big] \tag{14.5.6}$$

Since $\Delta E_K = 0$ for an elastic collision,

$$g'^\| = -g^\| , \qquad g'^\perp = \pm g^\perp \tag{14.5.7}$$

with the negative solution being applicable for rough surfaces; the positive solution is for smooth surfaces, in which case $r_{ij} \times b = 0$ and the spins do not change.

Another feature that can be incorporated into the hard-sphere model is inelasticity. While not relevant in the molecular context, the same general MD approach can be used for modeling granular materials, and here energy dissipation by means of highly inelastic collisions is a key element in the dynamics[†]. For inelastic collisions,

---

† Soft-particle granular models are discussed in Chapter 16.

a coefficient of restitution ϵ is introduced, with $0 \leq \epsilon \leq 1$, and the relative velocities before and after collision are related by $\boldsymbol{g}' = -\epsilon \boldsymbol{g}$, assuming, for convenience, that the same ϵ applies to both components. Expressions for $\boldsymbol{b}$ and ΔE_K follow immediately from (14.5.4) and (14.5.6).

14.6 Further study

14.1   Compare the thermodynamic properties of square-well and LJ fluids.

14.2   Study the equation of state of the hard-sphere fluid [erp84].

14.3   Examine trajectory sensitivity to small perturbations; here, unlike the corresponding soft-sphere treatment of this problem (§3.8), there is no numerical integration error.

14.4   Construct polymer chains using linked hard spheres and compare the effectiveness of the approach with the soft-sphere chains described in §9.2.

14.5   Measure the diffusion coefficient directly and by using the velocity autocorrelation function [erp85].

14.6   Add rotational motion and examine its effect on diffusion.

14.7   Nonspherical particles can be handled by checking for overlaps and then using an iterative method to determine the instant of collision [kus76, reb77, all89]; explore this approach.

15

Time-dependent phenomena

15.1 Introduction

Practically all the simulations described so far have involved systems that are either in equilibrium or in some time-independent stationary state; while individual results are subject to fluctuation, it is the well-defined averages over sufficiently long time intervals that are of interest. In this chapter we extend the MD approach to a class of problem in which the behavior is not only time dependent, but the properties themselves are also spatially dependent in ways that are not always predictable. The analysis of the behavior of such systems cannot be carried out following the methods described earlier, which were generally based on the evaluation of system-wide averages or correlations, and one is therefore compelled to resort to graphical methods. Here we focus on MD applications in fluid dynamics, a subject in which atomic matter is conventionally replaced by a continuous medium for practical purposes; recovering the atomic basis is part of trying to understand more complex fluid behavior of the type studied in rheology. For more on the microscopic approach to hydrodynamics see [mar92].

15.2 Open systems

Most current MD applications deal with closed systems; this implies either total isolation from the outside world, or coupling to the environment in a way described by one of the ensembles of statistical mechanics. The coupling can occur, for example, with the aid of a thermostat (§6.3), in which case the equilibrium properties are those of the canonical ensemble. Extending MD to open systems, where coupling to the external world is of a more general kind, introduces new problems, some of which will be encountered here. Not only are open systems out of thermodynamic equilibrium, but in many cases they are also spatially inhomogeneous and time dependent[†]. In some situations, the presence of physical walls – as opposed to periodic boundaries – is essential to obtain the desired behavior.

Two examples of open systems will be treated here, both from the realm of fluid dynamics. One of the problems is a study of convective flow driven by a temperature gradient, the other involves the flow of a fluid past a rigid obstacle. These represent examples of attempts to bridge the gap between the atomistic picture of the microscopic world, so ably captured by MD, and the more conventional world of fluid dynamics. Although the existence of atoms is generally irrelevant in the continuum picture, and is ignored by the equations of continuum fluid dynamics, in order to learn what the molecular constituents of a fluid really do while the fluid is flowing around the obstacle and exhibiting a range of quite complex behavior, it is necessary to return to the roots, and this implies MD simulation. These examples also benefit from the fact that the phenomena are primarily two-dimensional; three-dimensional simulations of the required size require a much heavier computational effort.

A flow problem of considerable importance is thermal convection [tri88]: a horizontal layer of fluid is heated from below and the resulting interplay between the upward flow produced by heating and the downward flow due to gravity leads to the formation of structured flow patterns in the shape of rolls and various other forms, either stationary or time dependent. In a wide variety of situations the entire description of a flow experiment can be condensed into a single dimensionless quantity; in this case it is the Rayleigh number

$$Ra = \frac{\alpha g d^3 \Delta T}{\nu \kappa} \tag{15.2.1}$$

that determines the behavior, where α is the thermal expansion coefficient of the fluid, ν $(= \eta/\rho)$ the kinematic viscosity, κ $(= \lambda/\rho C_P)$ the thermal diffusivity, d the layer height and ΔT the temperature difference. (To complete the story, a second dimensionless quantity, the Prandtl number $Pr = \nu/\kappa$, is also involved; this partly determines the nature of the convective motion.)

The problem of flow past a rigid obstacle [tri88] is another extensively studied problem in fluid dynamics. Here, the behavior is governed by the Reynolds number

$$Re = \frac{dv}{\nu} \tag{15.2.2}$$

where d is a characteristic length scale, namely, the diameter of a cylindrical obstacle oriented perpendicular to the flow, and v is the flow speed. Irrespective of the obstacle size, the flow speed and whether the fluid is a liquid or a gas, the flow

---

[†] The system considered in §12.3 was also in this category.

patterns depend only on Re; this kind of scaling behavior, as with the previous example, is known as dynamic similarity. At small Re values the flow is laminar (as opposed to turbulent), but the flow can be either stationary or time varying, with a pair of fixed eddies or a highly structured set of traveling vortices forming in the wake of the obstacle.

In setting up MD simulations of these systems it is important to ensure parameter combinations that produce the correct values of the dimensionless numbers such as Ra and Re; if they are too small nothing interesting should be expected, since in each case there exist threshold values for the onset of the instability responsible for the flow patterns. Even if the threshold is exceeded, there is no guarantee that a microscopically small MD system will resemble its macroscopic counterpart: there must exist a minimum region size – measurable in atomic diameters or mean free paths (whichever is the larger) – below which the characteristic fluid flow patterns cannot develop[†]. In addition to the size requirements, the duration of the simulation must be long enough to allow observation of any time-dependent behavior. The pessimist will also point out the use of highly exaggerated temperature gradients or shear rates – which must be many orders of magnitude larger than their real-world counterparts in order to overcome the inherent thermal fluctuations and compensate for the small system size – and question whether the concept of dynamic similarity has not been stretched a little too far.

15.3 Thermal convection

Motion in a gravitational field

Extending[♠] the hard-sphere method of §14.2 to include the effect of a uniform gravitational field is a relatively simple exercise. The collision prediction process is not affected at all because the same uniform acceleration is experienced by all atoms, so that the computation for determining the existence of a collision and when it occurs is independent of the field. Between collisions atoms follow parabolic rather than linear trajectories.

It is only the prediction of cell crossings that calls for special attention, because parabolic motion means that it is possible for an atom to leave a cell through the face (or edge) of entry. The resulting changes to `PredictEvent` follow, where the gravitational field `gravField` is assumed to act in the negative y direction; the algebraic problem solved here is locating the intersection of a parabola with a straight line. This version of the function is intended for use in two dimensions, but is readily extended to three.

---

† Remember that the edge length of a square region containing 10^5 atoms at liquid density is only a mere 1000 Å; in three dimensions the corresponding size is even smaller.

♠ pr_15_1

```
VecI signDir;
real h, h1, h2;
...
VCopy (w, mol[na].inCell);
if (mol[na].rv.x > 0.) ++ w.x;                                    5
signDir.x = (mol[na].rv.x < 0.);
if (gravField.y == 0. && mol[na].rv.y > 0.) ++ w.y;
VMul (w, w, region);
VDiv (w, w, cells);
VSAdd (rs, mol[na].r, 0.5, region);                              10
VVSub (w, rs);
WhenCross (x);
if (gravField.y != 0.) {
  h1 = Sqr (mol[na].rv.y) + 2. * gravField.y * w.y;
  if (mol[na].rv.y > 0.) {                                       15
    h2 = h1 + 2. * gravField.y * region.y / cells.y;
    if (h2 > 0.) {
      h = - sqrt (h2);
      signDir.y = 0;
    } else {                                                     20
      h = sqrt (h1);
      signDir.y = 1;
    }
  } else {
    h = sqrt (h1);                                               25
    signDir.y = 1;
  }
  tm.y = - (mol[na].rv.y + h) / gravField.y;
} else {
  WhenCross (y);                                                 30
  signDir.y = (mol[na].rv.y < 0.);
}
```

The role of `signDir` will become clear shortly when we continue with the discussion of changes to `PredictEvent`.

The function `UpdateMol` for updating positions must be modified to allow for parabolic trajectories, with the velocity also being updated,

```
void UpdateMol (int id)
{
  real tInt;

  tInt = timeNow - mol[id].time;                                 5
  VVSAdd (mol[id].r, tInt, mol[id].rv);
  VVSAdd (mol[id].r, 0.5 * Sqr (tInt), gravField);
  VVSAdd (mol[id].rv, tInt, gravField);
  mol[id].time = timeNow;
}                                                               10
```

Because local time variables are associated with each atom, a similar modification is required in *PredictEvent* for the collision prediction,

```
VVSAdd (dr, -0.5 * Sqr (tInt), gravField);
VVSAdd (dv, - tInt, gravField);
```

When checking for energy conservation, allowance should be made for the gravitational potential energy contribution.

Hard-wall boundaries

The boundaries used in this study are all rigid, although the side walls could be replaced by periodic boundaries. These hard walls can be smooth, in which case collisions with the walls are energy-conserving, specular collisions, but in most situations the walls will be rough, so that after undergoing a collision an atom loses all memory of its prior velocity. Simply randomizing the velocity direction (and possibly magnitude) will achieve this effect, as in §7.3, but here we will demonstrate an alternative approach in which the random element is absent.

Each wall is divided into a series of narrow strips with width typically equal to the atom diameter; when an atom collides with the wall the outcome alternates between specular collisions and velocity reversals, depending on the strip involved – a kind of corrugation effect. If the wall is also attached to a thermal reservoir held at constant temperature, as is the case for the top and bottom walls, the velocity magnitude can be changed to the required value (as in §7.3).

Modifications to the program described in §14.2 to incorporate these effects follow. All of the reference to periodic boundaries must be removed. Additional variables are required for the gravitational field (regarded as a vector quantity), the width of the wall corrugations, the wall temperatures and the selection of the pair of opposite walls that act as thermal reservoirs,

```
VecR gravField, roughWid;
real wallTemp[2];
int thermalWall[NDIM];
```

and there is additional input data,

```
NameR (gravField),
NameI (thermalWall),
NameR (wallTemp),
```

Collisions with the hard walls require a new form of processing, so an additional class of cell boundary event is introduced; in *SingleEvent*, all cell events are now identified by the modified test

```
if (evIdB < MOL_LIMIT + NDIM * 2 || evIdB >= MOL_LIMIT + 100)
```

The usual cell events are covered by the latter part of the test, while the former, together with $evIdB \geq MOL\_LIMIT$, accounts for the wall collisions.

The function *ProcessCellCrossing* is modified to handle wall collisions as well as regular cell crossings.

```
VecR rs;
VecI irs;
real vFac, vv;
int j, jj;
...                                                              5
j = evIdB - MOL_LIMIT;
if (j >= 100) {
  jj = j - 100;
  ... (same as before) ...
} else {                                                        10
  jj = j / 2;
  VComp (cellRange[j % 2], jj) = 0;
  VComp (mol[evIdA].rv, jj) *= -1.;
  VSAdd (rs, mol[evIdA].r, 0.5, region);
  VDiv (rs, rs, roughWid);                                      15
  VCopy (irs, rs);
  if (jj != 0 && irs.x % 2 == 0) mol[evIdA].rv.x *= -1.;
  if (jj != 1 && irs.y % 2 == 0) mol[evIdA].rv.y *= -1.;
  if (thermalWall[jj]) {
    vv = VLenSq (mol[evIdA].rv);                                20
    vFac = sqrt (NDIM * wallTemp[j % 2] / vv);
    VScale (mol[evIdA].rv, vFac);
  }
}
...                                                             25
```

In *PredictEvent*, changes are required in determining the identity of the cell event about to be scheduled and the range of cells that must be examined.

```
#define LimitCells(t)                                    \
    if (mol[na].inCell.t + cellRangeT[0].t == -1)        \
      cellRangeT[0].t = 0;                               \
    if (mol[na].inCell.t + cellRangeT[1].t == cells.t)   \
      cellRangeT[1].t = 0;                               5

VecI cellRangeT[2];
...
dir = (tm.x < tm.y) ? 0 : 1;
evCode = 100 + dir;                                        10
if (VComp (mol[na].inCell, dir) == 0 &&
```

```
        VComp (signDir, dir) == 1) evCode = 2 * dir;
    else if (VComp (mol[na].inCell, dir) == VComp (cells, dir) - 1 &&
        VComp (signDir, dir) == 0) evCode = 2 * dir + 1;
    ScheduleEvent (na, MOL_LIMIT + evCode, timeNow + VComp (tm, dir));    15
    cellRangeT[0] = cellRange[0];
    cellRangeT[1] = cellRange[1];
    LimitCells (x);
    LimitCells (y);
    for (m1y = cellRangeT[0].y; m1y <= cellRangeT[1].y; m1y ++) {          20
      ...
```

Here, $cellRangeT$ replaces $cellRange$ for the limits of the nested x and y cell loops.

As formulated above, an atom is deemed to collide with the wall when its center reaches the wall position. While not affecting the results here, it is more sensible (and necessary if atoms of mixed sizes are involved) if the collision occurs when the wall–atom distance equals the atom radius. In the present case the only changes needed are to enlarge the region slightly and shift the initial coordinates; in $SetParams$ add

```
    VAddCon (region, region, 1.);
```

and in $InitCoords$ (§2.4)

```
    VecR offset;
    ...
    VAddCon (gap, region, -1.);
    VDiv (gap, gap, initUcell);
    VSetAll (offset, 0.5);                                                5
    ...
      VSet (c, nx + 0.5, ny + 0.5);
      VMul (c, c, gap);
      VVSAdd (c, -0.5, region);
      VVAdd (c, offset);                                                  10
```

The width of the wall 'corrugation' is also specified in $SetParams$,

```
    VecI iw;
    ...
    VCopy (iw, region);
    VDiv (roughWid, region, iw);
```

Flow analysis

The flow is analyzed using the grid method of §7.3. Data collection uses the same function *GridAverage*, after removing any reference to the z component of velocity. The grid measurements are recorded at fixed time intervals *intervalGrid*; to do this, an additional event category similar to those used previously for other measurements is introduced.

Whenever the specified number of grid samples have been collected the results are appended to a file that stores the grid measurements, or 'snapshots', for the entire run. This is the task of the function *PutGridAverage*. Data are output in binary form, after scaling the values so that they can be stored as short (16-bit) integers in order to reduce file size. Enough information accompanies the grid data to permit reconstruction of the original values, albeit with reduced precision. The parameter *NHIST* has the value 4, with the four quantities treated in *GridAverage* being the density, the square of the velocity and the two velocity components; each quantity is computed for all atoms in every cell at a given instant, and the cell results are averaged over time. The serial number of each set of data is stored in *snapNumber*.

```
#define SCALE_FAC  32767.

void PutGridAverage ()
{
  real histMax[NHIST], w;                                          5
  int blockSize, fOk, hSize, j, n;
  short *hI;
  FILE *fp;

  hSize = VProd (sizeHistGrid);                                   10
  for (j = 0; j < NHIST; j ++) {
    histMax[j] = 0.;
    for (n = 0; n < hSize; n ++) {
      w = fabs (histGrid[j][n]);
      histMax[j] = Max (histMax[j], w);                           15
    }
    if (histMax[j] == 0.) histMax[j] = 1.;
  }
  fOk = 1;
  blockSize = (NHIST + 1) * sizeof (real) + sizeof (VecR) +       20
      (NDIM + 3) * sizeof (int) + NHIST * hSize * sizeof (short);
  if ((fp = fopen (fileName[FL_SNAP], "a")) != 0) {
    WriteF (blockSize);
    WriteF (histMax);
    WriteF (region);                                              25
    WriteF (runId);
    WriteF (sizeHistGrid);
```

```
      WriteF (snapNumber);
      WriteF (timeNow);
      AllocMem (hI, hSize, short);                                  30
      for (j = 0; j < NHIST; j ++) {
        for (n = 0; n < hSize; n ++)
            hI[n] = SCALE_FAC * histGrid[j][n] / histMax[j];
        WriteFN (hI, hSize);
      }                                                             35
      free (hI);
      if (ferror (fp)) fOk = 0;
      fclose (fp);
    } else fOk = 0;
    if (! fOk) ErrExit (ERR_SNAP_WRITE);                            40
  }
```

The file output operations have been simplified by defining

```
#define WriteF(x)       fwrite (&x, sizeof (x), 1, fp)
#define WriteFN(x, n)   fwrite (x, sizeof (x[0]), n, fp)
```

More on the subject of file usage appears in §18.6 and §18.7.

An analysis program♠, only part of which is shown here, would read this file, one snapshot at a time, using the function *GetGridAverage* (the complement of *PutGridAverage*). For the initial call, the variable *blockNum* is given the value −1 (as in *GetConfig* in §18.6) to ensure the necessary storage is allocated.

```
int GetGridAverage ()
{
  int fOk, n, j, k;
  short *hI;
  FILE *fp;                                                        5

  fOk = 1;
  fp = fopen (fName, "r");
  if (blockNum == -1) {
    if (! fp) fOk = 0;                                             10
  } else {
    fseek (fp, blockNum * blockSize, 0);
    ++ blockNum;
  }
  if (fOk) {                                                       15
    ReadF (blockSize);
    ...
    ReadF (timeNow);
    if (feof (fp)) return (0);
    if (blockNum == -1) {                                          20
      AllocMem2 (histGrid, NHIST, VProd (sizeHistGrid), real);
```

___
♠ pr_angridflow

```
      AllocMem (streamFun, VProd (sizeHistGrid), real);
      blockNum = 1;
    }
    AllocMem (hI, VProd (sizeHistGrid), short);                      25
    for (j = 0; j < NHIST; j ++) {
      ReadFN (hI, VProd (sizeHistGrid));
      for (n = 0; n < VProd (sizeHistGrid); n ++)
        histGrid[j][n] = hI[n] * histMax[j] / SCALE_FAC;
    }                                                               30
    free (hI);
    if (ferror (fp)) fOk = 0;
    fclose (fp);
  }
  if (! fOk) ErrExit (ERR_SNAP_READ);                               35
  return (1);
}
```

The array `streamFun` is to be used in evaluating (15.3.1) and

```
#define ReadF(x)       fread (&x, sizeof (x), 1, fp)
#define ReadFN(x, n)   fread (x, sizeof (x[0]), n, fp)
```

simplify the file input operations.

The size of the grid cells used for the analysis and the number of samples that contribute to a single time-averaged snapshot have yet to be specified. Ideally, both should allow the smallest spatial structures and the most rapid changes in the flow to be resolved. Opposing this goal is the sampling issue, since the smaller the number of atoms participating in the average for a single cell the larger the fluctuations. Compromise is necessary, and a typical (although very much a problem dependent) tradeoff for a square system with $N_m = 10^5$ might involve a 50×50 grid, with measurements collected at time intervals of between 0.1 and 1 (there is no benefit in having them too closely spaced) and a snapshot after every 100–500 measurements.

There is no shortage of methods for displaying the results of the flow analysis. Here we have chosen to plot the flow streamlines. This amounts to evaluating the stream function at each grid point and then constructing a contour plot of the function[†]. The stream function [tri88] is defined as the line integral

$$\psi (\boldsymbol{r}) = \int \rho (v_y \, dl_x - v_x \, dl_y) \tag{15.3.1}$$

evaluated along an arbitrary path from the origin to the point $\boldsymbol{r}$. The following approach to computing the stream function is based on (15.3.1), and could form

---

† An alternative way of exhibiting the flow results is to use arrow plots to show the grid-averaged flow direction.

the basis for an analysis or display program.

```
VecR w;
real *streamFun, sFirst;
...
VDiv (w, region, sizeHistGrid);
sFirst = 0.;                                                        5
for (iy = 0; iy < sizeHistGrid.y; iy ++) {
  for (ix = 0; ix < sizeHistGrid.x; ix ++) {
    n = iy * sizeHistGrid.x + ix;
    if (ix == 0) {
      sFirst -= histGrid[0][n] * histGrid[2][n] * w.y;            10
      streamFun[n] = sFirst;
    } else streamFun[n] = streamFun[n - 1] +
      histGrid[0][n] * histGrid[3][n] * w.x;
  }
}                                                                  15
```

Contour plots of local scalar properties such as density and temperature are also readily produced. Successive plots can be combined to produce an animated sequence that will clearly reveal any time-dependent behavior. We avoid discussing the details of how to produce different kinds of graphic output since standard software packages are generally available for this kind of work.

Results

The results shown here are for a hard-disk run using input data

```
density             0.4
eventMult           4
gravField           0.  -0.15
initUcell           400 100
intervalGrid        1.
intervalSum         1.
limitEventCount     600000000
limitGrid           100
runId               1
sizeHistGrid        120 30
temperature         1.
thermalWall         0 1
wallTemp            20. 1.
```

The system contains a total of $N_m = 40\,000$ disks, the walls are hard and rough, the lower boundary is hot and the upper cold, and the gravitational field acts in the downward direction. There is no temperature gradient in the initial state.

In Figure 15.1 we show the convective flow that has become firmly established by $t = 1800$, corresponding to approximately 2.4×10^8 collisions. Temperature

Fig. 15.1. Streamlines showing four convection rolls.

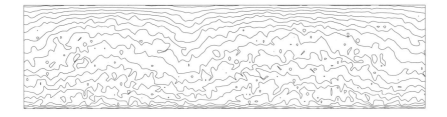

Fig. 15.2. Temperature contour plot.

is subject to both vertical and horizontal variations that are correlated with the flow, as can be seen in Figure 15.2; the simple vertical profile used in §7.3 would have concealed this information. Density (not shown) is found to be practically constant, except at the cold wall. The value of Ra can only be estimated very roughly [puh89, rap91c], but it turns out to be consistent with what is expected from continuum fluid dynamics based on the observed behavior.

In simulations of this kind there is normally an initial transient phase, during which the system gradually evolves into a final state appropriate to the choice of parameters [rap92]. How to identify the point at which the system may be said to have reached this final state is not always obvious. A steady state is easily detected, but slow periodic oscillations or completely random behavior are less readily identifiable as 'final states'. There is also the possibility of long-lived metastable states preceding the true final state. In short, there are few general rules.

15.4 Obstructed flow

Boundaries and driving force

There are two kinds of boundary that must be considered in this problem – the region boundaries and the obstacle perimeter. Region boundaries can be made periodic so that the flow recirculates, provided that any memory of nonuniform flow is erased by, for example, periodically randomizing the velocities of all atoms located very close to either end of the system. The obstacle boundary should be rough to ensure that the adjacent fluid layer is at rest, corresponding to the nonslip boundary that occurs experimentally. There are various ways of accomplishing this; a particularly simple approach for two-dimensional flow around a circular obstacle is to represent the obstacle as a ring of fixed atoms identical to those in the fluid.

The flow is driven by superimposing a velocity bias in the flow direction when carrying out the velocity randomization; a constant external field could have been used instead, exactly as in the case of Poiseuille flow (§7.3). Heat will be generated as a result of the sheared flow near the obstacle, but now there are no walls that can be used to remove the excess heat; a simple method of overcoming this problem is to cool the atoms while randomizing the velocities of atoms near the ends of the system.

Program details

The program♠ used in this case study is an extension of the basic soft-disk MD program using either cells or neighbor lists to assist the interaction calculations (because of the flow, some of the performance gain of the neighbor-list method might be lost). Leapfrog integration is used. Features that must be added to the program include the obstacle and the mechanism for driving the flow. The new variables appearing in this problem are

```
VecR obsPos;
real bdyStripWidth, flowSpeed, obsSize;
int nFixedMol, nFreeMol, stepDrive;
```

and the `Mol` structure contains an additional element `fixed`. The input data are

```
  NameR (bdyStripWidth),
  NameR (flowSpeed),
  NameR (obsPos),
  NameR (obsSize),
  NameI (stepDrive),
```

Given that the circular region occupied by the obstacle is devoid of atoms, initialization of the run begins by determining the precise number of atoms in the system. The atoms of the fluid initially occupy the sites of a square lattice; the atoms used to form the obstacle boundary are evenly spaced around the perimeter with separation $\approx r_c$. The obstacle size and position are specified relative to the

---

♠ pr_15_2

region size.

```
void EvalMolCount ()
{
  VecR c, gap;
  int nx, ny;

  nFixedMol = M_PI * obsSize * region.y / rCut;                          5
  VDiv (gap, region, initUcell);
  nFreeMol = 0;
  for (ny = 0; ny < initUcell.y; ny ++) {
    for (nx = 0; nx < initUcell.x; nx ++) {                              10
      VSet (c, nx + 0.5, ny + 0.5);
      VMul (c, c, gap);
      VVSAdd (c, -0.5, region);
      if (OutsideObs (&c)) ++ nFreeMol;
    }                                                                    15
  }
  nMol = nFixedMol + nFreeMol;
}

int OutsideObs (VecR *p)                                                 20
{
  int outside;

  outside = (Sqr (p->x - obsPos.x * region.x) +
    Sqr (p->y - obsPos.y * region.y) >                                  25
    Sqr (0.5 * obsSize * region.y + rCut));
  return (outside);
}
```

This test is easily modified for obstacles with other shapes. Atoms making up the obstacle boundary are distinguished from those of the fluid by `mol[].fixed`; these are set by

```
void SetMolType ()
{
  int n;

  for (n = 0; n < nFixedMol; n ++) mol[n].fixed = 1;                     5
  for (n = nFixedMol; n < nMol; n ++) mol[n].fixed = 0;
}
```

The initial state can now be prepared; flow will be in the positive *x* direction.

```
void InitCoords ()
{
  VecR c, gap, w;
  real ang;
```

```
  int n, nx, ny;                                                            5

  VDiv (gap, region, initUcell);
  VMul (w, obsPos, region);
  for (n = 0; n < nFixedMol; n ++) {
    ang = 2. * M_PI * n / nFixedMol;                                       10
    VSet (c, cos (ang), sin (ang));
    VSAdd (mol[n].r, w, 0.5 * obsSize * region.y, c);
  }
  n = nFixedMol;
  for (ny = 0; ny < initUcell.y; ny ++) {                                  15
    for (nx = 0; nx < initUcell.x; nx ++) {
      VSet (c, nx + 0.5, ny + 0.5);
      VMul (c, c, gap);
      VVSAdd (c, -0.5, region);
      if (OutsideObs (&c)) {                                               20
        mol[n].r = c;
        ++ n;
      }
    }
  }                                                                        25
}

void InitVels ()
{
  int n;                                                                   30

  VZero (vSum);
  DO_MOL {
    if (! mol[n].fixed) {
      VRand (&mol[n].rv);                                                  35
      VScale (mol[n].rv, velMag);
      VVAdd (vSum, mol[n].rv);
    } else VZero (mol[n].rv);
  }
  DO_MOL {                                                                 40
    if (! mol[n].fixed) {
      VVSAdd (mol[n].rv, - 1. / nFreeMol, vSum);
      mol[n].rv.x += flowSpeed;
    }
  }                                                                        45
}
```

The threefold task of maintaining fluid motion, removing excess heat and erasing the flow pattern is achieved by the following call from *SingleStep*,

```
  if (stepCount % stepDrive == 0) DriveFlow ();
```

and the function that does the work is

```
void DriveFlow ()
{
  int n;

  DO_MOL {
    if (! mol[n].fixed &&
        fabs (mol[n].r.x) > 0.5 * region.x - bdyStripWidth) {
      VRand (&mol[n].rv);
      VScale (mol[n].rv, velMag);
      mol[n].rv.x += flowSpeed;
    }
  }
}
```

The following test must be added to the loops of *LeapfrogStep* to prevent atoms belonging to the obstacle boundary from moving,

```
if (mol[n].fixed) continue;
```

Data collection is again based on the grid method, with the fixed atoms forming the obstacle boundary excluded.

Results

The results demonstrated here use a soft-disk simulation with the following input data; there are $N_m = 125\,000$ atoms – relatively large systems are required in order to allow the flow patterns to develop properly [rap87].

bdyStripWidth	3.
deltaT	0.005
density	0.8
flowSpeed	2.
initUcell	500 250
limitGrid	200
nebrTabFac	8
obsPos	-0.25 0.
obsSize	0.2
rNebrShell	0.4
runId	1
sizeHistGrid	120 60
stepAvg	100
stepDrive	40
stepEquil	0
stepGrid	20
stepLimit	100000
temperature	1.

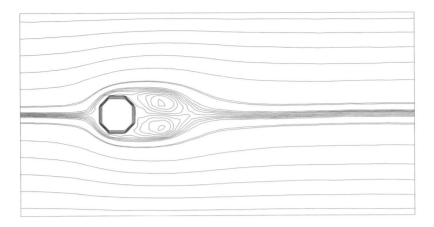

Fig. 15.3. Streamline plot showing the vortices that form early in the run.

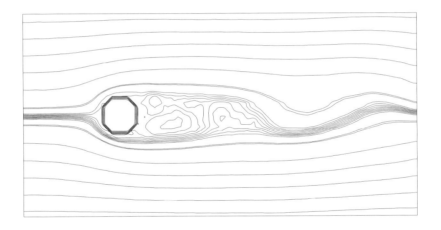

Fig. 15.4. An example of the oscillating flow pattern.

Two examples of the flow streamlines are shown here. Figure 15.3 illustrates
the transient phase, here at $t = 100$, in which a pair of counter-rotating eddies
have formed at the downstream edge of the obstacle. The flow eventually develops
oscillatory behavior (which corresponds to vortex shedding, although the present
system is barely large enough to see this effect); Figure 15.4 shows a snapshot at
$t = 400$ that is typical of the behavior observed. Note that the contours are not
evenly spaced; they are closer together in the area that is most strongly affected by
the obstacle in order to reveal the flow patterns in greater detail.

15.5 Further study

15.1 The thermal convection problem can also be treated using a soft-disk fluid (an extension of the case study in §7.3); explore.

15.2 Flow studies of this kind can also be carried out in three dimensions. In the case of Taylor–Couette flow in a rotating cylinder, a problem thoroughly explored by continuum methods, MD simulation leads to quantitatively correct behavior [hir98, hir00]. More extensive MD study of this and other rotating fluid problems could be carried out.

15.3 Other examples of flow problems exhibiting space and/or time dependence that call for more detailed analysis include pipe flow of immiscible fluids [kop89, tho89], stick–slip flow [tho91] and the flow of chain molecules [edb87, rap94]; these are examples of fluids that are not adequately described by continuum means, so that there is considerable scope for research using the MD approach.

16

Granular dynamics

16.1 Introduction

The importance of understanding the processes governing the transport of granular materials [jae96] has long been recognized, particularly because of its industrial relevance. Methods analogous to MD modeling turn out to be appropriate for the study of granular matter, although the constituent particles are, of course, no longer the atoms and molecules of MD.

Mere inspection reveals the complexity of granular matter. The grains themselves are irregularly shaped, often covered with asperities, and are normally polydisperse. Grain collisions are highly inelastic and friction is important for forming heaps. The wear and tear of collisions can alter the shape of the grains to some extent; electrostatic forces, moisture, adhesion and the presence of air can all affect the behavior. Which of these, and other, characteristics must be incorporated into the model to reproduce the key features of the behavior can only be established empirically.

The goal of this brief departure from simulation at the molecular scale is to demonstrate the wider applicability of the approach, but not to provide a survey of either granular dynamics simulation techniques or applications; reviews of the subject include [bar94, her95]. The discussion of this chapter deals with methods based on soft-particle MD and, while there are many fascinating granular systems to choose from, the examples here deal with vibrating layers [biz98, rap98], mainly because of the visual impact of the results. The methods are readily extended to other kinds of problem.

16.2 Granular models

Background

Unlike the molecular scale applications of MD, where the interactions are based on well-founded experimental and theoretical considerations, much remains to be done in order to achieve a detailed understanding of the nature of the interactions between granular particles. In particular, a simplified representation of the salient features of the grain structure and interactions is a prerequisite for simulation, with detailed comparisons between simulation and experiment used to establish how good an approximation has been achieved [sch96].

The models presently in widespread use tend to be based on inelastically colliding hard or soft spheres (or disks in the case of two-dimensional simulations), more often the latter [cun79, wal83, wal92]; rotational motion of the particles may or may not be included in the dynamics. Normal and tangential velocity-dependent ('viscous') damping forces are used to represent the inelasticity of the interactions. Static friction is difficult to model, but its effect can at least be partially emulated by a force that opposes any sliding motion while particles are in contact (which, for a soft-particle model, means within interaction range). In all cases energy is no longer a conserved quantity.

More complex models utilize nonspherical particles; examples include rigid [gal93] and flexible [pos93] assemblies of spheres, as well as more complex structures formed of elastic triangles and deformable damped beams [pos95]. As the level of detail increases so does the computational effort, and it is important to have some way of assessing which of the features of the models are actually responsible for the observed behavior. An example of this is the relative importance of friction and particle geometry [pos93] where, to some extent, nonconvex particles can compensate for a lack of static friction since they can mutually interlock.

Particle interactions

Consider a pair of spherical, granular particles i and j, with diameters σ_i and σ_j. The repulsive force between the particles can be expressed as

$$
\boldsymbol{f}_v = \begin{cases} \dfrac{48}{r_{ij}} \left[\left(\dfrac{\sigma_{ij}}{r_{ij}} \right)^{12} - \dfrac{1}{2} \left(\dfrac{\sigma_{ij}}{r_{ij}} \right)^6 \right] \hat{\boldsymbol{r}}_{ij} & r_{ij} < 2^{1/6}\sigma_{ij} \\ 0 & r_{ij} \geq 2^{1/6}\sigma_{ij} \end{cases} \tag{16.2.1}
$$

where $\sigma_{ij} = (\sigma_i + \sigma_j)/2$ is the effective mean particle diameter; this is a slightly generalized form of the soft-sphere interaction (2.2.4) and it acts whenever overlap occurs, namely, when $r_{ij} < 2^{1/6}\sigma_{ij}$. Alternatives to the overlap force (16.2.1) that are also in routine use include functions that depend on the overlap $\sigma_{ij} - r_{ij}$, either linearly or to the 3/2-power, the latter a result due to Hertz, but the choice of function is often of little consequence as far as the overall behavior is concerned. Note that because of the slight degree of softness the particle diameter is not precisely defined in these models.

What distinguishes the interactions used for granular media from those in MD studies of molecular systems is the presence of dissipative forces that act over the duration of each collision. The first of these is a normal damping force

$$f_d = -\gamma_n (\hat{r}_{ij} \cdot v_{ij}) \hat{r}_{ij} \tag{16.2.2}$$

which depends on the component of the relative velocity of the particles $v_{ij} = v_i - v_j$ along the direction between particle centers. The factor γ_n is the normal damping coefficient and it is assumed to be the same for all particles. The total force in the direction of $\hat{r}_{ij}$ is

$$f_n = f_v + f_d \tag{16.2.3}$$

Frictional damping acts in a plane normal to $\hat{r}_{ij}$ at the point of contact of the particles. The relative transverse velocity of the particle surfaces at this point, allowing for particle rotation, is

$$v_{ij}^s = v_{ij} - (\hat{r}_{ij} \cdot v_{ij}) \hat{r}_{ij} - \left(\frac{\sigma_i \omega_i + \sigma_j \omega_j}{\sigma_i + \sigma_j} \right) \times r_{ij} \tag{16.2.4}$$

Sliding friction has the form

$$f_s = -\min \left(\gamma_s | v_{ij}^s |, \; \mu | f_n | \right) \hat{v}_{ij}^s \tag{16.2.5}$$

where γ_s is the sliding friction coefficient and the static friction coefficient μ sets an upper bound proportional to $|f_n|$. Despite the simplifications, empirical models similar to this – possibly with some minor variations – have proved quite successful for a wide range of problems. The need for an easily described model is essential, because it is often necessary to simulate relatively large systems over long periods of time to capture both the space and time dependence of the behavior.

The translational and rotational accelerations of the particles, a_i and α_i, depend on sums of the above forces for all interacting pairs. The contribution of each pair is included by adding its value of $f_n + f_s$ to the total $m_i a_i$, subtracting the same quantity from $m_j a_j$, and also subtracting $\hat{r}_{ij} \times f_s / 2\kappa$ from both $m_i \sigma_i \alpha_i$ and $m_j \sigma_j \alpha_j$, where $m_i = \sigma_i^3$ is the particle mass and κ the numerical factor in the moment of inertia.

16.3 Vibrating granular layer

Two-dimensional version

The model described here represents a horizontal layer of grains in a container whose base oscillates sinusoidally in the vertical direction. The container sides are assumed periodic (hard walls are an alternative) while the top boundary – which plays a very minor role in the calculation – is elastically reflecting. The computations♠ use the neighbor-list method for interaction processing and leapfrog integration.

The details of each particle are contained in the structure

```
typedef struct {
  VecR r, rv, ra;
  real wv, wa, diam;
} Mol;
```

where `wv` and `wa` are the (scalar) angular velocity and acceleration, and `diam` the particle diameter (*NDIM* must be set to 2 for this two-dimensional simulation).

The force computations are carried out by the following version of the function `ComputeForces` which includes evaluation of the velocity-dependent forces and makes provision for particles of different sizes.

```
void ComputeForces ()
{
  VecR dr, dv;
  real aMass, dFac, drv, fcVal, ft, ftLim, rr, rri, rri3, rSep, vRel,
     ws, wt;                                                               5
  int j1, j2, n;

  DO_MOL {
    VZero (mol[n].ra);
    mol[n].wa = 0.;                                                        10
  }
  for (n = 0; n < nebrTabLen; n ++) {
    j1 = nebrTab[2 * n];
    j2 = nebrTab[2 * n + 1];
    dFac = 0.5 * (mol[j1].diam + mol[j2].diam);                            15
    VSub (dr, mol[j1].r, mol[j2].r);
    VWrap (dr, x);
    rr = VLenSq (dr);
    if (rr < Sqr (rCut * dFac)) {
      rSep = sqrt (rr);                                                    20
      rri = Sqr (dFac) / rr;
      rri3 = Cube (rri);
      fcVal = 48. * rri3 * (rri3 - 0.5) / rr;
```

```
      VSub (dv, mol[j1].rv, mol[j2].rv);
      drv = VDot (dr, dv);                                               25
      fcVal -= fricDyn * drv / rr;
      VVSAdd (mol[j1].ra, fcVal, dr);
      VVSAdd (mol[j2].ra, - fcVal, dr);
      VVSAdd (dv, - drv / rr, dr);
      ws = (mol[j1].diam * mol[j1].wv + mol[j2].diam * mol[j2].wv) /     30
         (mol[j1].diam + mol[j2].diam);
      dv.x += ws * dr.y;
      dv.y -= ws * dr.x;
      vRel = VLen (dv);
      ftLim = fricStat * fabs (fcVal) * rSep / vRel;                    35
      ft = - Min (ftLim, fricDyn);
      VVSAdd (mol[j1].ra, ft, dv);
      VVSAdd (mol[j2].ra, - ft, dv);
      wt = ft * vRel;
      if (VCross (dr, dv) > 0.) wt = - wt;                              40
      mol[j1].wa += wt;
      mol[j2].wa += wt;
    }
  }
  ComputeBdyForces ();                                                  45
  DO_MOL {
    aMass = Sqr (mol[n].diam);
    VScale (mol[n].ra, 1. / aMass);
    mol[n].wa /= 2. * inertiaK * aMass * mol[n].diam;
    mol[n].ra.y -= gravField;                                          50
  }
}
```

The damping coefficient γ_s and the static friction constant μ are represented by `fricDyn` and `fricStat`, respectively. Gravitational acceleration is also taken into account; the boundary interactions are handled by `ComputeBdyForces` below.

Construction of the neighbor list (`BuildNebrList`) is done in the usual manner, with a few minor changes since the cells must be sized to accommodate the largest of the particles (here this is not a problem since $\sigma_i \le 1$) and the system is only periodic in the x direction. The following modification allows for different particle sizes when examining pairs:

```
  dFac = 0.5 * (mol[j1].diam + mol[j2].diam);
  if (VLenSq (dr) < Sqr (rCut * dFac + rNebrShell)) {
```

Interactions with the vibrating base and the top of the container are treated as follows; interaction with the base involves the normal force component only (the tangential part, as above, is optional).

```
void ComputeBdyForces ()
```

```
{
  VecR dr;
  real dFac, drv, fcVal, rr, rri, rri3;
  int n;                                                            5

  DO_MOL {
    dFac = 0.5 * (mol[n].diam + 1.);
    dr.y = mol[n].r.y - basePos;
    if (fabs (dr.y) < rCut * dFac) {                                10
      rr = Sqr (dr.y);
      rri = Sqr (dFac) / rr;
      rri3 = Cube (rri);
      fcVal = 48. * rri3 * (rri3 - 0.5) / rr;
      drv = dr.y * (mol[n].rv.y - baseVel);                         15
      fcVal -= fricDyn * drv / rr;
      mol[n].ra.y += fcVal * dr.y;
    }
    dr.y = mol[n].r.y - 0.5 * region.y;
    ... (as above, but without velocity damping) ...               20
  }
}
```

The base vibration is governed by two parameters, the amplitude *vibAmp* and the frequency *vibFreq*. The current base position *basePos* and velocity *baseVel* are determined at each timestep by

```
void SetBase ()
{
  nBaseCycle = vibFreq * stepCount * deltaT;
  curPhase = vibFreq * stepCount * deltaT - nBaseCycle;
  basePos = - 0.5 * region.y + vibAmp *                            5
    (1. - cos (2. * M_PI * curPhase));
  baseVel = 2. * M_PI * vibFreq * vibAmp * sin (2. * M_PI * curPhase);
}
```

The function *LeapfrogStep* includes half-timestep updates of the angular velocities; there is no need to evaluate angular coordinates as they are not required by the simulation (since they do not appear explicitly in the interactions they can be omitted entirely, unless there is interest in examining the rotational motion); the fact that the velocities used in the interactions are evaluated at times shifted by a half timestep from the coordinates should not cause problems (alternatively, a tentative full-timestep update could be used, as in §11.6). In *EvalProps* the rotational motion must be included in the kinetic energy calculation,

```
vvSum += Sqr (mol[n].diam) * (vv + inertiaK * Sqr (mol[n].diam) *
  Sqr (mol[n].wv));
```

The initial particle positions form a closely spaced array positioned just above the base,

```
void InitCoords ()
{
  VecR c, gap;
  int n, nx, ny;

  SetBase ();                                                          5
  VDiv (gap, region, initUcell);
  gap.y = rCut;
  n = 0;
  for (ny = 0; ny < initUcell.y; ny ++) {                            10
    c.y = ny * gap.y + basePos + 1.;
    for (nx = 0; nx < initUcell.x; nx ++) {
      c.x = (nx + 0.5) * gap.x - 0.5 * region.x;
      mol[n].r = c;
      ++ n;                                                          15
    }
  }
}
```

with randomly directed initial velocities of magnitude *velMag* and zero angular velocities. Particle sizes are chosen randomly over a narrow range, bounded above by unity (the reason for not using a single size is simply to avoid lattice-like packing artifacts),

```
void SetMolSizes ()
{
  int n;

  DO_MOL mol[n].diam = 1. - 0.15 * RandR ();                          5
}
```

New variables and input data items are

```
real basePos, baseVel, curPhase, fricDyn, fricStat, gravField,
  inertiaK, vibAmp, vibFreq;
int nBaseCycle;

  NameR (fricDyn),                                                    5
  NameR (gravField),
  NameR (vibAmp),
  NameR (vibFreq),
```

and the values that are set in *SetParams*

```
  region.x = initUcell.x / 0.95;
  region.y = region.x;
```

```
fricStat = 0.5;
inertiaK = 0.125;
velMag = 1.;                                                          5
```

Three-dimensional version

The extension to three dimensions[●] requires only a minimum of changes. The granular particles have become spheres instead of disks, `mol[].wv` and `mol[].wa` are of type *VecR*, and the vertical direction is now z. Modification of the various functions to handle these changes is straightforward. Changes to *ComputeForces* reflect the fact that rotation now involves vectors rather than scalars:

```
DO_MOL {
  ...
  VZero (mol[n].wa);
}
for (n = 0; n < nebrTabLen; n ++) {                                   5
  ...
  VWrap (dr, y);
  ...
  if (rr < Sqr (rCut * dFac)) {
    ...                                                               10
    VSSAdd (ws, mol[j1].diam, mol[j1].wv, mol[j2].diam, mol[j2].wv);
    VScale (ws, 1. / (mol[j1].diam + mol[j2].diam));
    VCross (wt, ws, dr);
    VVSub (dv, wt);
    ...                                                               15
    VCross (wt, dr, dv);
    VScale (wt, - ft / rSep);
    VVAdd (mol[j1].wa, wt);
    VVAdd (mol[j2].wa, wt);
  }                                                                   20
}
...
DO_MOL {
  aMass = Cube (mol[n].diam);
  ...                                                                 25
  VScale (mol[n].wa, ...);
  ...
}
```

16.4 Wave patterns

The analysis of this category of problem, as with several other of the case studies, relies heavily on the ability to directly visualize the system as the simulation progresses. We will not deal with quantitative issues but will simply show two

● pr_16_3

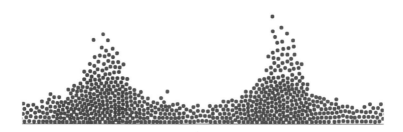

Fig. 16.1. Standing waves in a two-dimensional vibrated granular layer.

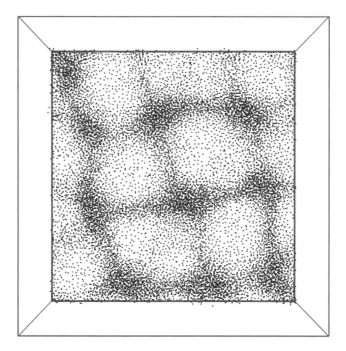

Fig. 16.2. A typical wave pattern occurring in a thin, three-dimensional vibrated layer. The view is from above (the container outline is shown in perspective); darker regions correspond to the wave peaks.

computer-generated images for particular parameter choices that display the kind of behavior typical of the two- and three-dimensional simulations. Snapshots of the particle coordinates are stored during the run, using suitably modified versions of the function *PutConfig* (§18.6); graphics programs for generating the imagery are not discussed here.

Both examples involve periodic side boundaries. In two dimensions, the data for the comparatively small system shown here include

`deltaT`	`0.004`
`fricDyn`	`5.`
`gravField`	`8.`
`initUcell`	`70 10`
`stepSnap`	`25`
`vibAmp`	`2.`
`vibFreq`	`0.4`

and the changes for the three-dimensional case, involving a container with a square base and a layer of less than half the thickness, are

`gravField`	`6.`
`initUcell`	`60 60 4`

Snapshots of typical standing wave patterns are shown in Figures 16.1 and 16.2. In both cases there appear to be preferred wavelengths; this can be established with greater certainty by considering larger systems. In three dimensions the waves are seen to form a roughly square pattern; since the system is periodic the alignment of the entire pattern can rotate to help achieve the preferred wavelength (other patterns, such as hexagons, can also appear).

16.5 Further study

16.1 Examine which characteristics of the particle interactions are needed to obtain the waves and which play only a minor role.

16.2 Explore how the wave patterns depend on the parameters defining the problem.

16.3 Study particle motion within the waves.

16.4 Instead of using continuous potentials, model these systems with the event-driven approach of Chapter 14 [hon92, biz98].

16.5 Phenomena involving granular segregation are especially fascinating and simulations of the kind described here have succeeded in reproducing some of these effects, for example [rap02b]; investigate.

17

Algorithms for supercomputers

17.1 Introduction

The previous chapters concentrated on translating physical problems into practical simulations. Computational efficiency, beyond the use of cells and neighbor lists (as well as hierarchical subdivision when appropriate), received little attention. For 'conventional' computers, there is not a great deal more that the average user can do in this respect, assuming that a reasonably effective programming style has been adopted. This attitude is no longer adequate when modern, high performance, multiprocessor machines are to serve as the platforms for large-scale simulation.

In this chapter we focus on ways of adapting the basic MD approach to take advantage of advanced computer architectures; since enhanced performance comes not only from a faster processor clock cycle, but also from a number of fundamental changes in the way computers process data, this is a subject that cannot be ignored. The subject is also a relatively complex one and, at best, only peripheral to the goals of the practicing simulator. We will therefore not delve too deeply into the issues involved, but will merely focus on three examples, all of which can be valuable for large-scale MD simulation; the first employs message-passing parallelism, the second involves parallelism achieved by the use of computational threads and shared memory, and the third demonstrates how to rearrange data to achieve effective vector processing[†].

17.2 The quest for performance

It comes as no surprise to learn that spreading a computational task over several processors is a way to complete the job sooner. Multiprocessor systems benefit from an economy of scale, and high performance computers now almost always

---

[†] The vectorizeable program can be run on a nonvector machine in its present form (although this would only be done only for development purposes), and the parallel programs can also be run on a uniprocessor system (also just for testing) if the necessary supporting software is available.

consist of at least a few processing units, if not more; the number of processors per machine begins at two and extends into the thousands, and there is apparently no limit in sight. The ideal kind of problem for such a machine – assuming that each processor, or each small group of processors, has its own private memory – is one that can be partitioned into a number of smaller computations that are carried out in parallel on all the processors, without too much data having to be shared between them. Molecular dynamics systems with short-range forces fall comfortably into this category.

Another means of extracting higher performance from computer hardware is to resort to vector processing. This entails pipelining the computations in assembly-line fashion. The constraint placed on a computation, if it is to be effectively vectorized, is that data items should be organized into relatively long vectors in such a way that all items can be processed independently, without fear that the processing of one item will affect a later one. This condition is not always easily satisfied.

Needless to say, both these architectural features are increasingly likely to be encountered[†]. The price of utilizing them effectively is increased algorithm and software complexity. Unlike 'simple' computers, where optimizing compilers can take a typical program and massage it to achieve reasonable performance, the needs of parallel and vector processing cannot always be resolved in this way, because it is not always obvious – assuming that it is possible at all – how to automate the process of optimally mapping an intrinsically serial computation onto the more 'complex' computer architectures. In addition to actually producing a working parallel and/or vectorized program, the efficiency of the end result must be considered: a parallel computation with large interprocessor communication overheads is doomed to failure, as is a vectorized computation that either uses the vector capability inefficiently because the vectors are too short, or spends a disproportionate amount of time rearranging data into vectorizeable form.

17.3 Techniques for parallel processing

Living with multiprocessor computers

The taxonomy of multiprocessing is far from simple. Among the factors to be taken into account are whether individual processors all carry out the same operation during each cycle, or whether they are able to act independently; whether each

---

[†] Beyond these two most visible features of many modern computers, there are a number of more subtle and less adequately documented processor features that can have a significant impact on performance. Just to name some of them: the internal processor registers, primary and secondary (and sometimes even tertiary) caches, address space mapping and memory interleaving. An algorithm that in some way conflicts with certain engineering design assumptions, for example, in its pattern of memory accesses, can experience a drastic performance drop. Beyond drawing the attention of the reader to the existence of such potential pitfalls, there is little more that can be said without addressing each computer model individually.

processor has its own private memory, or all share a common memory, or both; whether processors communicate with one another by passing messages across a communication network, or through common memory; the nature and topology of the communication network, if any. Some of these features can influence the way in which software ought to be organized.

We will avoid becoming involved in these issues here by assuming a particular generic architecture, one that is in fact widespread because it is the simplest to implement. The assumption is that there are several independent processors, each with private memory, communicating over a network – the message-passing approach. Systems of this kind are easily assembled by simply linking modest personal computers using standard (or above standard) network hardware; depending on the nature of the computation, the performance may or may not be satisfactory, but the same general approach is often found embedded in more customized hardware designs.

In order for the message-passing scheme to work efficiently, it is vital that the communication overheads be kept low compared with the time a processor spends computing; the ideal application consists of a lot of calculation with small amounts of data being transferred from time to time. The communication overhead is composed of two parts: the time to initiate a message transfer, typically a constant value, and a transfer time that is roughly proportional to the message length. There is also the issue of load balancing; obviously if all processors can be kept busy doing useful computing the overall system utilization will be optimal, but if some processors have more work to do than others, overall effectiveness is reduced.

Algorithm organization

There are different ways to partition an MD computation among multiple processors. The act of partitioning can focus on the computations, on the atoms involved, or on the simulation region. While all three elements are part of every scheme, the emphasis differs; this has a considerable impact on the memory and communication requirements of each method.

If it is just the computations that are partitioned, then all information about the system resides in the memory of each processor, but each only carries out the interaction computations for certain atom pairs. The information about the forces on each atom is then combined. This approach is extremely wasteful in terms of memory and is best suited to small computations only (and, perhaps, shared-memory computers).

The second partitioning scheme is based on the atoms themselves, and assigns each atom to a particular processor for the duration of the simulation, irrespective of its spatial location [rai89]. While conceptually simple, large amounts of

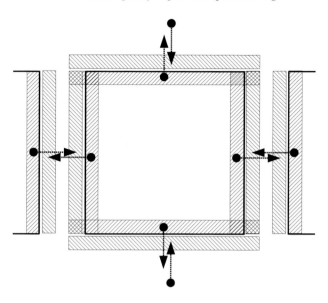

Fig. 17.1. The portion of the simulation region (for a two-dimensional subdivision) represented by the square outline is handled by a single processor; it contains shaded areas denoting subregions whose atoms interact with atoms in adjacent processors, and is surrounded by shaded areas denoting subregions from adjacent processors whose interactions must be taken into account; arrows indicate the flow of data between processors.

communication are required to handle interactions between atoms assigned to different processors. For long-range interactions this may not be a problem, but for the short-range case the third choice turns out to be far more efficient.

The third scheme subdivides space and assigns each processor a particular subregion [rap91b]. All the atoms that are in a given subregion at some moment in time reside in the processor responsible and when an atom moves between subregions all the associated variables are explicitly transferred from one processor to another. Thus there is economy insofar as memory is concerned, and also in the communication required to allow atoms to transfer between processors, since comparatively few atoms make such a move during a single timestep. More importantly, assuming there are some 10^4 or more atoms per subregion (in three dimensions) and a relatively short-ranged potential, most of the interactions will occur among atoms in the subregion and relatively few between atoms in adjacent subregions – see Figure 17.1. In order to accommodate the latter, copies of the coordinates of atoms close to any subregion boundary are transferred to the processor handling the adjacent subregion prior to the interaction computation. This transfer also involves only a small fraction of the atoms.

It is this third scheme that will be described here. The only requirement is that communication be reasonably efficient, with the associated system overheads –

both hardware and software – kept to a low level in comparison with the amount of computation involved. Under these circumstances, both computation speed and memory requirements scale in the expected linear way with the numbers of atoms and processors[†].

17.4 Distributed computation

Overview

The program described here is based on the three-dimensional soft-sphere computation described in §3.4 that uses both cells and neighbor lists. Some parts of the original program can be used unchanged, but wherever atoms become aware of the existence of subregions the computations must take this into account. While conceptually straightforward, the distributed computation involves numerous details that must be treated with care.

The functions handling tasks related to communications are referenced in a generic form that does not assume any particular message-passing software package. The text describes what each of these functions is supposed to do and, in practice, since all such software must provide similar functionality, a simple series of macro substitutions may be all that is required to produce a working program; an actual implementation concludes the section.

In addition to the interaction calculations, integration of the equations of motion, initialization and measurements, each of which is modified to a lesser or greater degree for the distributed implementation, it is also necessary to specify which processor is responsible for each spatial subregion, to identify the atoms participating in each data transfer and to carry out the transfers. Several kinds of data transfer are used:

- for interaction calculations it is necessary to copy information about the coordinates of atoms close to subregion boundaries;
- when atoms move between subregions their entire descriptions are transferred;
- while making measurements, the values computed separately in each processor must be combined to produce an overall result.

Basic computations

The program description[♦] begins with the neighbor-list construction, on the assumption that the atoms are already in the correct processors and that copies of the coordinates of atoms in adjacent subregions have been made available. These tasks

---

[†] In practice, the performance depends on the details of the processor architecture, communication infrastructure and operating system; performance that grows linearly with the number of processors is not always achievable.

[♦] pr_17_1

will be described later. Two new variables play an important role here; *nMolMe* is the number of atoms currently in the subregion and *nMolCopy* is the number of additional atoms from adjacent subregions (processors) whose coordinates have been copied to this processor because they are potential interaction candidates. Given the additional information, this version of *BuildNebrList* (and the other functions that follow) can be executed independently on each processor.

```
#define OFFSET_LIST                                                    \
   {{7}, {7,8,9}, {9}, {6,7,13}, {5,6,7,8,9,10,11,12,13},             \
    {9,10,11}, {13}, {11,12,13}, {11}, {2,7},                         \
    {2,3,4,7,8,9}, {4,9}, {1,2,6,7,13},                               \
    {0,1,2,3,4,5,6,7,8,9,10,11,12,13}, {4,9,10,11}, {13},             \
    {11,12,13}, {11}}
#define OFFSET_LEN                                                     \
   {1,3,1,3,9,3,1,3,1,2,6,2,5,14,4,1,3,1}

void BuildNebrList ()
{
  VecR cellBase, dr, invWid, rs, t1, t2;
  VecI cc, m1v, m2v, vOff[] = OFFSET_VALS;
  real rrNebr;
  int c, indx, j1, j2, m1, m1x, m1y, m1z, m2, n, offset, tOffset,
    vOffList[][N_OFFSET] = OFFSET_LIST, vOffTableLen[] = OFFSET_LEN;

  VAddCon (t1, cells, -2.);
  VSub (t2, subRegionHi, subRegionLo);
  VDiv (invWid, t1, t2);
  VSetAll (t1, 1.);
  VDiv (t1, t1, invWid);
  VSub (cellBase, subRegionLo, t1);
  rrNebr = Sqr (rCut + rNebrShell);
  for (n = nMolMe + nMolCopy; n < nMolMe + nMolCopy + VProd (cells);
    n ++) cellList[n] = -1;
  for (n = 0; n < nMolMe + nMolCopy; n ++) {
    VSub (rs, mol[n].r, cellBase);
    VMul (cc, rs, invWid);
    c = VLinear (cc, cells) + nMolMe + nMolCopy;
    cellList[n] = cellList[c];
    cellList[c] = n;
  }
  nebrTabLen = 0;
  for (m1z = 0; m1z < cells.z - 1; m1z ++) {
    for (m1y = 0; m1y < cells.y; m1y ++) {
      for (m1x = 0; m1x < cells.x; m1x ++) {
        VSet (m1v, m1x, m1y, m1z);
        tOffset = 13;
        if (m1z == 0) tOffset -= 9;
        if (m1y == 0) tOffset -= 3;
        else if (m1y == cells.y - 1) tOffset += 3;
```

```
      if (m1x == 0) tOffset -= 1;
      else if (m1x == cells.x - 1) tOffset += 1;
      m1 = VLinear (m1v, cells) + nMolMe + nMolCopy;                    45
      for (offset = 0; offset < vOffTableLen[tOffset]; offset ++) {
        indx = vOffList[tOffset][offset];
        VAdd (m2v, m1v, vOff[indx]);
        m2 = VLinear (m2v, cells) + nMolMe + nMolCopy;
        DO_CELL (j1, m1) {                                              50
          DO_CELL (j2, m2) {
            if (m1 != m2 || j2 < j1) {
              VSub (dr, mol[j1].r, mol[j2].r);
              ... (identical to standard version) ...
    }                                                                   55
```

Several points should be noted[†]. The cell array is defined separately for each subregion and includes an additional layer of cells that completely surrounds the subregion; it is here that all the atoms copied from adjacent subregions are located. Periodic boundaries are not mentioned at this stage of the computation because, as will be shown later, they are treated during the copying operation. The vector variables *subRegionLo* and *subRegionHi* contain the spatial limits of the subregion handled by each processor. Because periodic boundaries are handled by other means, the range of adjacent cells scanned during neighbor-list construction depends on the cell location; for cells that are located on a subregion face, edge or corner, fewer adjacent cells need be examined – the data in *vOffList* makes provision for all 18 distinct cases and the way it is used together with *vOffTableLen* should be apparent from the listing.

Only minor changes are needed in the force calculation. The principal reason for the changes is in order to evaluate accumulated properties such as the potential energy and virial sum. The force computation does not distinguish between atoms that really belong to the subregion and those that are merely copies from an adjacent subregion, since the force contributions associated with the latter are simply discarded afterwards. Energy and virial sums are treated in a manner that ensures the correct contributions from atom pairs that interact across a subregion boundary. Once again, no mention of periodic wraparound is needed.

```
  void ComputeForces ()
  {
    ...
    for (n = 0; n < nMolMe + nMolCopy; n ++) VZero (mol[n].ra);
    for (n = 0; n < nebrTabLen; n ++) {                                5
      j1 = nebrTab[2 * n];
      j2 = nebrTab[2 * n + 1];
      VSub (dr, mol[j1].r, mol[j2].r);
```

[†] The reader new to distributed processing should bear in mind that all variables are local to each processor; the concept of a global variable does not exist in a message-passing environment.

```
    rr = VLenSq (dr);
    if (rr < rrCut) {                                              10
      ... (identical to standard version) ...
      if (j1 < nMolMe) {
        uSum += uVal;
        virSum += fcVal * rr;
      }                                                            15
      if (j2 < nMolMe) {
        uSum += uVal;
        virSum += fcVal * rr;
      }
    }                                                              20
  }
  uSum *= 0.5;
  virSum *= 0.5;
}
```

Integration uses the leapfrog method. The only change to *LeapfrogStep* is the
use of *nMolMe* for the number of atoms to be processed. The atoms whose coordinates were copied from adjacent subregions are readily excluded from this calculation since they appear in the *mol* array after the first *nMolMe* entries corresponding
to atoms in the subregion.

In establishing the initial state we encounter communication[†] for the first time,
albeit still in a very minor role. All the initialization has been combined into a
single function *InitState* that is executed concurrently on all processors, with
each processor determining which atoms belong to its subregion. Since a particular
atom is no longer associated with a fixed memory location, unlike the uniprocessor
version of the program, each atom is labeled with a unique identifier *mol[].id*.
Although not really needed in this particular example, it is sometimes necessary to
be able to distinguish individual atoms (a polymer fluid would be one such example). A simple cubic lattice is used for the initial state. The macros *VGe* and *VLt*
(§18.2) compare vector components.

```
void InitState ()
{
  VecR vSumL;
  VecR c, gap;
  int n, np, nx, ny, nz;                                          5
  ValList msg1[] = {
    ValR (vSumL)
  };
  ValList msg2[] = {
```

[†] The distributed environment makes its presence felt both during algorithm development and at the programming stage; any features that are added to the simulation must take this extra software overhead into account.
Much of the communication processing is hidden away from the application, as indeed it should be. It is a pity
that even the few details included here have to be mentioned at all, but this situation will persist until some
standardized method (or language) for programming parallel computers achieves widespread acceptance.

```
      ValR (vSum)                                                    10
    };

    VDiv (gap, region, initUcell);
    VZero (vSumL);
    nMol = 0;                                                        15
    nMolMe = 0;
    for (nz = 0; nz < initUcell.z; nz ++) {
      for (ny = 0; ny < initUcell.y; ny ++) {
        for (nx = 0; nx < initUcell.x; nx ++) {
          VSet (c, nx + 0.25, ny + 0.25, nz + 0.25);                 20
          VMul (c, c, gap);
          VVSAdd (c, -0.5, region);
          VRand (&mol[nMolMe].rv);
          if (VGe (c, subRegionLo) && VLt (c, subRegionHi)) {
            mol[nMolMe].r = c;                                       25
            VScale (mol[nMolMe].rv, velMag);
            VVAdd (vSumL, mol[nMolMe].rv);
            mol[nMolMe].id = nMol;
            ++ nMolMe;
            if (nMolMe > nMolMeMax) {                                30
              errCode = ERR_TOO_MANY_MOLS;
              -- nMolMe;
            }
          }
          ++ nMol;                                                   35
        }
      }
    }
    if (ME_BOSS) {
      vSum = vSumL;                                                  40
      DO_SLAVES {
        MsgRecvUnpack (np, 121, msg1);
        VVAdd (vSum, vSumL);
      }
      VScale (vSum, 1. / nMol);                                      45
      MsgBcPackSend (122, msg2);
    } else {
      MsgPackSend (0, 121, msg1);
      MsgBcRecvUnpack (122, msg2);
    }                                                                50
    DO_MOL_ME {
      VVSub (mol[n].rv, vSum);
      VZero (mol[n].ra);
    }
  }                                                                  55
```

The latter part of *InitState* includes several communication operations and
a summary of what occurs here is as follows. There is a macro *ME_BOSS* that is
able to distinguish one processor from all the others; the mission of the processor

having the status of 'boss' is to collect the values of the array *vSum* from all the other 'slave' processors after each has computed its local values, evaluate the total sums, compute the values that each processor must subtract from all its atoms' velocities to ensure a zero center of mass velocity and, finally, broadcast these values to each of the slave processors. The slaves perform the complementary task; they accumulate *vSum*, send it to the boss and wait for the necessary values to be returned. Although one particular processor has been designated the boss and the remainder slaves, this should not detract from the fact that, with a couple of minor exceptions, all processors perform completely equivalent tasks and are mutually synchronized by the data transfers that occur throughout the calculation.

The function *MsgRecvUnpack* waits for, and accepts, the message specified in the *ValList* argument (see §18.5) from a designated processor; the numerical argument appearing here and in other communication functions is an arbitrary value used by the message software to distinguish between different kinds of message[†]. This function also processes the received message, by storing its contents sequentially in the indicated locations. *MsgPackSend* performs the complementary operation, by collecting the specified values into the message body and then sending it to the designated processor. There is also a broadcast capability; the function *MsgBcPackSend* is used by the boss processor to broadcast an identical message to all the slaves, and each slave processor will use *MsgBcRecvUnpack* to receive this message. Other communication functions will appear in due course as needed.

Another function that must retrieve a small amount of information from each processor is *EvalProps*. The copy of *EvalProps* running on the boss processor produces the same final results as the uniprocessor version, but it must first collect the partial results from the slaves.

```
void EvalProps ()
{
  VecR vSumL;
  real uSumL, virSumL, vv, vvMax, vvMaxL, vvSumL;
  int errCodeL, n, np;                                            5
  ValList msg[] = {
    ValI (errCodeL),
    ValR (uSumL),
    ValR (virSumL),
    ValR (vSumL),                                                10
    ValR (vvMaxL),
    ValR (vvSumL)
  };

  VZero (vSumL);                                                  15
```

[†] This feature is not used here, but it is especially helpful during development for ensuring that messages sent and received correspond to one another.

```
vvSumL = 0.;
vvMaxL = 0.;
DO_MOL_ME {
  VVAdd (vSumL, mol[n].rv);
  vv = VLenSq (mol[n].rv);                                      20
  vvSumL += vv;
  vvMaxL = Max (vvMaxL, vv);
}
if (ME_BOSS) {
  vSum = vSumL;                                                 25
  vvSum = vvSumL;
  vvMax = vvMaxL;
  DO_SLAVES {
    MsgRecvUnpack (np, 161, msg);
    if (errCodeL != ERR_NONE) errCode = errCodeL;              30
    vvMax = Max (vvMax, vvMaxL);
    vvSum += vvSumL;
    VVAdd (vSum, vSumL);
    uSum += uSumL;
    virSum += virSumL;                                         35
  }
  dispHi += sqrt (vvMax) * deltaT;
  if (dispHi > 0.5 * rNebrShell) nebrNow = 1;
  kinEnergy.val = 0.5 * vvSum / nMol;
  totEnergy.val = kinEnergy.val + uSum / nMol;                 40
  pressure.val = density * (vvSum + virSum) / (nMol * NDIM);
} else {
  errCodeL = errCode;
  uSumL = uSum;
  virSumL = virSum;                                            45
  MsgPackSend (0, 161, msg);
}
}
```

Finally, the appropriately modified version of the function `SingleStep` is

```
void SingleStep ()
{
  ValList msg[] = {
    ValI (moreCycles),
    ValI (nebrNow)                                              5
  };

  ++ stepCount;
  timeNow = stepCount * deltaT;
  LeapfrogStep (1);                                            10
  if (nebrNow > 0) DoParlMove ();
  DoParlCopy ();
  if (nebrNow > 0) {
    nebrNow = 0;
    if (ME_BOSS) dispHi = 0.;                                  15
```

```
    BuildNebrList ();
  }
  ComputeForces ();
  LeapfrogStep (2);
  EvalProps ();                                            20
  if (ME_BOSS) {
    MsgBcPackSend (151, msg);
  } else MsgBcRecvUnpack (151, msg);
  if (ME_BOSS) {
    if (stepCount >= stepEquil) {                          25
      AccumProps (1);
      if (stepCount > stepEquil &&
        (stepCount - stepEquil) % stepAvg == 0) {
        AccumProps (2);
        PrintSummary (stdout);                             30
        AccumProps (0);
      }
    }
  }
}                                                          35
```

The boss processor is responsible for collecting the results and handling the output. The communication operations appearing here provide each processor with the current value of *nebrNow* informing it whether an update of the neighbor list is due. The functions *DoParlMove* and *DoParlCopy*, described below, deal with the interprocessor data transfers needed for interaction calculations and atom movements.

An important detail should be apparent from the way communications in the above functions are organized. Obviously, there must be a one-to-one correspondence between messages sent and messages received. Equally significant, however, is the fact that these message transfer operations are used to synchronize the processors[†]. When writing parallel software based on a message-passing paradigm, it is important to plan the communications carefully, otherwise deadlock and race conditions can occur that are difficult to diagnose.

Message-passing operations

We now turn to the functions where the majority of the interprocessor communication occurs, namely, the movement of atoms between subregions and the copying of coordinate data prior to the interaction calculations.

The functions responsible for deciding which atoms should be moved, and then actually doing the work, including coordinate adjustment for periodic wraparound, are as follows. There are six directions (in three dimensions) to be considered,

---

[†] None of the additional capabilities found in parallel software systems for explicit synchronization, such as creating a barrier that no processor can pass until it receives permission, are required here.

and each is treated in turn; atoms can of course participate in more than one such transfer. The special case that a processor is its own neighbor, which occurs when no subdivision of the region is made in a particular direction, is also taken into account. Once all the moves are complete, each processor compresses its own data to eliminate gaps in the arrays.

```
#define OutsideProc(b)                                          \
   (sDir == 0 && VComp (mol[n].r, dir) <                        \
   VComp (subRegionLo, dir) + b ||                              \
   sDir == 1 && VComp (mol[n].r, dir) >=                        \
   VComp (subRegionHi, dir) - b)
#define NWORD_MOVE  (2 * NDIM + 1)

void DoParlMove ()
{
  int dir, n, nIn, nt, sDir;

  for (dir = 0; dir < NDIM; dir ++) {
    for (sDir = 0; sDir < 2; sDir ++) {
      nt = 0;
      DO_MOL_ME {
        if (mol[n].id >= 0) {
          if (OutsideProc (0.)) {
            trPtr[sDir][trBuffMax * dir + nt] = n;
            ++ nt;
            if (NWORD_MOVE * nt > NDIM * trBuffMax) {
              errCode = ERR_COPY_BUFF_FULL;
              -- nt;
            }
          }
        }
      }
      nOut[sDir][dir] = nt;
    }
    for (sDir = 0; sDir < 2; sDir ++) {
      nt = nOut[sDir][dir];
      PackMovedData (dir, sDir, &trPtr[sDir][trBuffMax * dir], nt);
      if (VComp (procArraySize, dir) > 1) {
        MsgSendInit ();
        MsgPackI (&nt, 1);
        MsgPackR (trBuff, NWORD_MOVE * nt);
        if (sDir == 1) MsgSendRecv (VComp (procNebrLo, dir),
          VComp (procNebrHi, dir), 140 + 2 * dir + 1);
        else MsgSendRecv (VComp (procNebrHi, dir),
          VComp (procNebrLo, dir), 140 + 2 * dir);
        MsgRecvInit ();
        MsgUnpackI (&nIn, 1);
        MsgUnpackR (trBuff, NWORD_MOVE * nIn);
      } else nIn = nt;
      if (nMolMe + nIn > nMolMeMax) {
```

```
      errCode = ERR_TOO_MANY_MOVES;                                    45
      nIn = 0;
    }
    UnpackMovedData (nIn);
  }
}                                                                      50
RepackMolArray ();
}
```

The only new communication function appearing here is *MsgSendRecv*, which both transmits data to a neighboring processor and receives data from the opposite neighbor; in some message-passing systems a function of this kind can be used to achieve overlapped data transfers. Several functions for packing and unpacking messages also make an appearance here (such as *MsgPackR* and *MsgUnpackI*) and these will be discussed later. The size of the multiprocessor configuration and the way the simulation region is subdivided – for example, into slices spanning the entire region, or into smaller boxes as in Figure 17.1 – are specified by *procArraySize*, while the location of each processor in the (up to three-dimensional) multiprocessor array is specified by *procArrayMe*[†].

Message packing and unpacking are two-stage processes; the work that is specific to the application data (shown below) is kept separate from the actual filling or emptying of message buffers (in *DoParlMove* above). The periodic boundaries are addressed at this stage.

```
void PackMovedData (int dir, int sDir, int *trPtr, int nt)
{
  real rShift;
  int j;
                                                                       5
  rShift = 0.;
  if (sDir == 1 &&
      VComp (procArrayMe, dir) == VComp (procArraySize, dir) - 1)
    rShift = - VComp (region, dir);
  else if (sDir == 0 && VComp (procArrayMe, dir) == 0)                 10
    rShift = VComp (region, dir);
  for (j = 0; j < nt; j ++) {
    VToLin (trBuff, NWORD_MOVE * j, mol[trPtr[j]].r);
    trBuff[NWORD_MOVE * j + dir] += rShift;
    VToLin (trBuff, NWORD_MOVE * j + NDIM, mol[trPtr[j]].rv);          15
    trBuff[NWORD_MOVE * j + 2 * NDIM] = mol[trPtr[j]].id;
    mol[trPtr[j]].id = -1;
  }
}
                                                                       20
```

---

† Note the checks – here and subsequently – to ensure storage arrays are not overfilled; such safety measures should always be present whenever unpredictable amounts of data are involved.

```
void UnpackMovedData (int nIn)
{
  int j;

  for (j = 0; j < nIn; j ++) {                                        25
    VFromLin (mol[nMolMe + j].r, trBuff, NWORD_MOVE * j);
    VFromLin (mol[nMolMe + j].rv, trBuff, NWORD_MOVE * j + NDIM);
    mol[nMolMe + j].id = trBuff[NWORD_MOVE * j + 2 * NDIM];
  }
  nMolMe += nIn;                                                      30
}
```

Repacking is required to remove gaps due to atoms that have moved out,

```
void RepackMolArray ()
{
  int j, n;

  j = 0;                                                              5
  DO_MOL_ME {
    if (mol[n].id >= 0) {
      mol[j] = mol[n];
      ++ j;
    }                                                                 10
  }
  nMolMe = j;
}
```

The functions for copying the coordinates of atoms close to subregion bound-aries to adjacent processors (prior to the interaction calculations) are very similar; the sets of atoms involved are updated only when the neighbor list is about to be rebuilt.

```
#define NWORD_COPY  (NDIM + 1)

void DoParlCopy ()
{
  real rCutExt;                                                      5
  int dir, n, nIn, nt, sDir;

  rCutExt = rCut + rNebrShell;
  nMolCopy = 0;
  for (dir = 0; dir < NDIM; dir ++) {                                10
    if (nebrNow > 0) {
      for (sDir = 0; sDir < 2; sDir ++) {
        nt = 0;
        for (n = 0; n < nMolMe + nMolCopy; n ++) {
          if (OutsideProc (rCutExt)) {                               15
            trPtr[sDir][trBuffMax * dir + nt] = n;
```

```
          ++ nt;
          if (NWORD_COPY * nt > NDIM * trBuffMax) {
            errCode = ERR_COPY_BUFF_FULL;
            -- nt;                                                   20
          }
        }
      }
      nOut[sDir][dir] = nt;
    }                                                                25
  }
  for (sDir = 0; sDir < 2; sDir ++) {
    nt = nOut[sDir][dir];
    PackCopiedData (dir, sDir, &trPtr[sDir][trBuffMax * dir], nt);
    if (VComp (procArraySize, dir) > 1) {                            30
      MsgSendInit ();
      MsgPackI (&nt, 1);
      MsgPackR (trBuff, NWORD_COPY * nt);
      if (sDir == 1) MsgSendRecv (VComp (procNebrLo, dir),
          VComp (procNebrHi, dir), 130 + 2 * dir + 1);               35
      else MsgSendRecv (VComp (procNebrHi, dir),
          VComp (procNebrLo, dir), 130 + 2 * dir);
      MsgRecvInit ();
      MsgUnpackI (&nIn, 1);
      MsgUnpackR (trBuff, NWORD_COPY * nIn);                         40
    } else nIn = nt;
    if (nMolMe + nMolCopy + nIn > nMolMeMax) {
      errCode = ERR_TOO_MANY_COPIES;
      nIn = 0;
    }                                                                45
    UnpackCopiedData (nIn);
  }
}
}
                                                                     50
void PackCopiedData (int dir, int sDir, int *trPtr, int nt)
{
  real rShift;
  int j;
                                                                     55
  rShift = 0.;
  if (sDir == 1 &&
      VComp (procArrayMe, dir) == VComp (procArraySize, dir) - 1)
      rShift = - VComp (region, dir);
  else if (sDir == 0 && VComp (procArrayMe, dir) == 0)              60
      rShift = VComp (region, dir);
  for (j = 0; j < nt; j ++) {
    VToLin (trBuff, NWORD_COPY * j, mol[trPtr[j]].r);
    trBuff[NWORD_COPY * j + dir] += rShift;
    trBuff[NWORD_COPY * j + NDIM] = mol[trPtr[j]].id;                65
  }
}
```

```
void UnpackCopiedData (int nIn)
{                                                                        70
  int j;

  for (j = 0; j < nIn; j ++) {
    VFromLin (mol[nMolMe + nMolCopy + j].r, trBuff, NWORD_COPY * j);
    mol[nMolMe + nMolCopy + j].id = trBuff[NWORD_COPY * j + NDIM];        75
  }
  nMolCopy += nIn;
}
```

The main program and initialization function for the distributed computation are as follows.

```
int main (int argc, char **argv)
{
  MsgStartup ();
  if (ME_BOSS) {
    GetNameList (argc, argv);                                            5
    PrintNameList (stdout);
  }
  InitSlaves ();
  NebrParlProcs ();
  SetParams ();                                                          10
  SetupJob ();
  moreCycles = 1;
  while (moreCycles) {
    SingleStep ();
    if (stepCount == stepLimit) moreCycles = 0;                          15
  }
  MsgExit ();
}

void SetupJob ()                                                         20
{
  AllocArrays ();
  InitRand (randSeed);
  stepCount = 0;
  InitState ();                                                          25
  nebrNow = 1;
  if (ME_BOSS) AccumProps (0);
}
```

We have glossed over two details that are intimately associated with the message-passing software, namely, how to ensure that all the processors run copies of the same program and how each processor obtains its distinct identity *procMe* (which determines, among other things, who is the boss). The former may be automatic, or require some user action either before or while running the program; the latter may be as simple as a function call[†].

New variables introduced in this program are

```
VecR subRegionHi, subRegionLo;
VecI procArrayMe, procArraySize, procNebrHi, procNebrLo;
real *trBuff;
int **trPtr, nOut[2][NDIM], errCode, nMolCopy, nMolMe, nMolMeMax,
   nProc, procMe, trBuffMax;
```

There are new input data items specifying the maximum number of atoms a processor can hold (including copies), the number of processors and the way the simulation region is subdivided, and the size of the buffers used for collecting data to be transferred,

```
NameI (nMolMeMax),
NameI (procArraySize),
NameI (trBuffMax),
```

In *SetParams*, the subregion limits are established and the cell array size (with an extra cell at each end) is determined from the size of the subregion,

```
VecR w;
...
nebrTabMax = nebrTabFac * nMolMeMax;
VDiv (w, region, procArraySize);
VMul (subRegionLo, procArrayMe, w);
VVSAdd (subRegionLo, -0.5, region);
VAdd (subRegionHi, subRegionLo, w);
VScale (w, 1. / (rCut + rNebrShell));
VAddCon (cells, w, 2.);
```

Memory allocation differs from the standard case,

```
void AllocArrays ()
{
  int k;

  AllocMem (mol, nMolMeMax, Mol);
  AllocMem (cellList, VProd (cells) + nMolMeMax, int);
  AllocMem (nebrTab, 2 * nebrTabMax, int);
  AllocMem (trBuff, NDIM * trBuffMax, real);
  AllocMem2 (trPtr, 2, NDIM * trBuffMax, int);
}
```

[†] Examination of the software documentation will resolve these questions.

Finally, each processor discovers who its neighbors are, based on the value of its own individual copy of *procMe*,

```
void NebrParlProcs ()
{
  VecI t;
  int k;

  nProc = VProd (procArraySize);                                            5
  procArrayMe.x = procMe % procArraySize.x;
  procArrayMe.y = (procMe / procArraySize.x) % procArraySize.y;
  procArrayMe.z = procMe / (procArraySize.x * procArraySize.y);
  for (k = 0; k < NDIM; k ++) {                                             10
    t = procArrayMe;
    VComp (t, k) = (VComp (t, k) +
        VComp (procArraySize, k) - 1) % VComp (procArraySize, k);
    VComp (procNebrLo, k) = VLinear (t, procArraySize);
    t = procArrayMe;                                                       15
    VComp (t, k) = (VComp (t, k) + 1) % VComp (procArraySize, k);
    VComp (procNebrHi, k) = VLinear (t, procArraySize);
  }
}
```

and the boss processor, the only one with access to the input data file, distributes its contents to all the other processors,

```
void InitSlaves ()
{
  ValList initVals[] = {
    ValR (deltaT),
    ValR (density),                                                         5
    ValI (initUcell),
    ValI (nebrTabFac),
    ValI (nMolMeMax),
    ValI (procArraySize),
    ValI (randSeed),                                                       10
    ValR (rNebrShell),
    ValI (stepEquil),
    ValI (stepLimit),
    ValR (temperature),
    ValI (trBuffMax)                                                       15
  };

  if (ME_BOSS) {
    MsgBcPackSend (111, initVals);
  } else MsgBcRecvUnpack (111, initVals);                                  20
}
```

Additional details

In order to complete the distributed MD demonstration, we provide the extra information needed to produce a working program for use with MPI, a message passing software standard [gro96] on which several widely available software packages are based. The two missing details are:

- initialize the computation and let each processor know its identity;
- express the generic communication functions as actual MPI functions.

The following is required by `main` to initialize the MPI system, allocate buffer storage, and obtain the values of `nProc` and `procMe`. We assume the reader is familiar with the MPI functions (all prefixed with `MPI_`) used here[†].

```
#define MsgStartup()                                        \
   MPI_Init (&argc, &argv),                                 \
   MPI_Comm_size (MPI_COMM_WORLD, &nProc),                  \
   MPI_Comm_rank (MPI_COMM_WORLD, &procMe),                 \
   AllocMem (buffSend, BUFF_LEN, real),                     \       5
   AllocMem (buffRecv,BUFF_LEN, real)
```

We also require

```
#define BUFF_LEN   64000

MPI_Status mpiStatus;
real *buffRecv, *buffSend;
int buffWords, mpiNp;                                               5
```

The replacement of the generic communication functions, either by calls to their MPI equivalents, or by other function calls or operations, is accomplished with the following macro definitions:

```
#define MsgSend(to, id)                                     \
   MPI_Send (buffSend, buffWords, MPI_MY_REAL, to, id,      \
   MPI_COMM_WORLD)
#define MsgRecv(from, id)                                   \
   MPI_Recv (buffRecv, BUFF_LEN, MPI_MY_REAL, from, id,     \       5
   MPI_COMM_WORLD, &mpiStatus)
#define MsgSendRecv(from, to, id)                           \
   MPI_Sendrecv (buffSend, buffWords, MPI_MY_REAL, to, id,  \
   buffRecv, BUFF_LEN, MPI_MY_REAL, from, id,               \
   MPI_COMM_WORLD, &mpiStatus)                                       10
```

---

† MPI typically requires the value of `nProc` to be entered on the command line, so it must agree with the value obtained from `procArraySize` which specifies the 'shape' of the subdivision. The appropriate MPI runtime software, with the necessary include files and libraries, must of course be installed on the computer in order to use this program. Information on how to compile and run programs based on MPI is to be found in the appropriate user documentation. (The first edition used the alternative PVM [gei94] package.)

```
#define MsgBcSend(id)                                               \
   for (mpiNp = 1; mpiNp < nProc; mpiNp ++)                          \
      MsgSend (mpiNp, id)
#define MsgBcRecv(id)        MsgRecv (0, id)
#define MsgExit()           MPI_Finalize (); exit (0)                      15
#define MsgSendInit()       buffWords = 0
#define MsgRecvInit()       buffWords = 0
#define MsgPackR(v, nv)     DoPackReal (v, nv)
#define MsgPackI(v, nv)     DoPackInt (v, nv)
#define MsgUnpackR(v, nv)   DoUnpackReal (v, nv)                           20
#define MsgUnpackI(v, nv)   DoUnpackInt (v, nv)
```

Further useful definitions are

```
#define MsgRecvUnpack(from, id, ms)                                 \
   {MsgRecv (from, id);                                             \
    MsgRecvInit ();                                                 \
    UnpackValList (ms, sizeof (ms));}
#define MsgPackSend(to, id, ms)                                     \      5
   {MsgSendInit ();                                                 \
    PackValList (ms, sizeof (ms));                                  \
    MsgSend (to, id);}
#define MsgBcRecvUnpack(id, ms)                                     \
   {MsgBcRecv (id);                                                 \      10
    MsgRecvInit ();                                                 \
    UnpackValList (ms, sizeof (ms));}
#define MsgBcPackSend(id, ms)                                       \
   if (nProc > 1) {                                                 \
      MsgSendInit ();                                               \      15
      PackValList (ms, sizeof (ms));                                \
      MsgBcSend (id);                                               \
   }
#define ME_BOSS      (procMe == 0)
#define MPI_MY_REAL  MPI_DOUBLE                                            20
#define DO_MOL_ME    for (n = 0; n < nMolMe; n ++)
#define DO_SLAVES    for (np = 1; np < nProc; np ++)
```

The packing and unpacking operations for blocks of data are

```
void DoPackReal (real *w, int nw)
{
  int n;

  if (buffWords + nw >= BUFF_LEN) errCode = ERR_MSG_BUFF_FULL;             5
  else {
    for (n = 0; n < nw; n ++) buffSend[buffWords + n] = w[n];
    buffWords += nw;
  }
}                                                                         10
```

```
void DoUnpackReal (real *w, int nw)
{
  int n;

  for (n = 0; n < nw; n ++) w[n] = buffRecv[buffWords + n];      15
  buffWords += nw;
}
```

together with analogous functions *DoPackInt* and *DoUnpackInt*. Finally, packing
of *ValList* arrays is carried out by

```
void PackValList (ValList *list, int size)
{
  int k;

  for (k = 0; k < size / sizeof (ValList); k ++) {               5
    switch (list[k].vType) {
      case N_I:
        MsgPackI (list[k].vPtr, list[k].vLen);
        break;
      case N_R:                                                  10
        MsgPackR (list[k].vPtr, list[k].vLen);
        break;
    }
  }
}                                                                15
```

and there is a complementary function *UnpackValList*.

This approach is preferable to embedding actual MPI calls in the body of the
program from the point of view of portability; note also that only a subset of the
available MPI functionality is utilized, for the same reason.

17.5 Shared-memory parallelism

Overview

In the case of several processors sharing access to a common memory there are two
approaches available. One is to use the same scheme for message passing described
previously; depending on the computer and operating system, it is possible that this
will run faster than if the processors have to communicate across a network – even
a fast one – because of specialized functions for internal communication. The alternative is to use an approach based on threads (or 'lightweight' processes); these
are basically replica copies of the computational process that access a common
region of memory and are supposed to have a relatively low computational overhead associated with their use. This turns out to be a much simpler approach than
message passing; it does, however, require care to ensure that data are accessed
in a consistent manner by the different threads and, of course, it cannot be used

when processors do not share common memory. The overall efficiency depends on the processor architecture and the nature of the problem, and it is quite possible that performance will not scale as efficiently with the number of processors as the message-passing approach[†].

Use of computational threads

The example used here[♠] is, once again, a simple soft-sphere MD simulation. We will show how threads can be introduced into two key parts of the program. The first is the leapfrog integrator, included on account of its simplicity rather than because of its heavy computational requirements. The second involves the neighbor-list construction and force evaluation procedures, which is where most of the computation time is spent and which, therefore, stand to benefit most from thread usage. Several macro definitions are used to conceal programming details.

The leapfrog integration function is changed so that now it calls another function `LeapfrogStepT`. There will be one such call for each thread and it is this new function that does the real work.

```
void LeapfrogStep (int part)
{
  int ip;

  THREAD_PROC_LOOP (LeapfrogStepT, part);                              5
}

void *LeapfrogStepT (void *tr)
{
  int ip, n;                                                          10

  QUERY_THREAD ();
  switch (QUERY_STAGE) {
    case 1:
      THREAD_SPLIT_LOOP (n, nMol) {                                   15
        VVSAdd (mol[n].rv, 0.5 * deltaT, mol[n].ra);
        VVSAdd (mol[n].r, deltaT, mol[n].rv);
      }
      break;
    case 2:                                                          20
      THREAD_SPLIT_LOOP (n, nMol)
        VVSAdd (mol[n].rv, 0.5 * deltaT, mol[n].ra);
      break;
  }
  return (NULL);                                                     25
}
```

[†] Parallel compilers can, in principle, produce the kind of code described here, although it is often simpler and more efficient (and sometimes essential) to introduce the changes into the MD code manually.

[♠] pr_17_2

The above functions need explanation. When *LeapfrogStep* is called, it spawns several threads that can execute concurrently on separate processors. It does this by running through a loop that starts each one of *nThread* processes separately. This is carried out by *THREAD_PROC_LOOP*. The first argument is the name of the function (here *LeapfrogStepT*) to be executed by the thread, the second (*part*) is a value to be passed across to the thread. The function *LeapfrogStep* returns only after all the threads have completed their work.

The function *LeapfrogStepT* is executed by each of the threads and, since the threads execute in parallel and in a totally unsynchronized manner, care is required to ensure that data are handled properly; in particular, any data written by one thread should not be accessed (either for reading or for writing) by another concurrent thread. The form of the function header and the *return* statement are a requirement of the thread functions.

The reference to the macro *QUERY_THREAD* produces the serial number of the thread, a value between 0 and *nThread-1*, which is placed in *ip*. It is used in *THREAD_SPLIT_LOOP*, which is a loop over a subset of the atoms defined as

```
#define THREAD_SPLIT_LOOP(j, jMax)                              \
   for (j = ip * jMax / nThread;                                \
        j < (ip + 1) * jMax / nThread; j ++)
```

The value of the argument *part* is supplied by the macro *QUERY_STAGE*. The outcome of executing all the threads is that the leapfrog update is applied to all atoms. A similar approach can be used, for example, in *ApplyBoundaryCond*, although functions that are responsible for only a small fraction of the overall workload may not warrant conversion to use threads.

The force computation is a little more intricate. It requires multiple neighbor lists, one for each thread, containing distinct subsets of atom pairs; individual atoms will generally appear in more than one of these lists. It also requires additional storage for acceleration and potential energy values that are computed separately for each of the lists and subsequently combined to produce the correct values. Note that allowing the different threads to update a common array of acceleration values would violate the restrictions on what threads are permitted to do with data and would be guaranteed to produce incorrect results.

Replacing the function *BuildNebrList* of §3.4 involves the following alterations:

```
void BuildNebrList ()
{
  int ip, n;

  for (n = nMol; n < nMol + VProd (cells); n ++) cellList[n] = -1;    s
  THREAD_PROC_LOOP (BuildNebrListT, 1);
```

```
  THREAD_PROC_LOOP (BuildNebrListT, 2);
}

void *BuildNebrListT (void *tr)                                          10
{
  ...
  int ip;

  QUERY_THREAD ();                                                      15
  switch (QUERY_STAGE) {
    case 1:
      VDiv (invWid, cells, region);
      DO_MOL {
        VSAdd (rs, mol[n].r, 0.5, region);                              20
        VMul (cc, rs, invWid);
        if (cc.z % nThread == ip) {
          c = VLinear (cc, cells) + nMol;
          ...
        }                                                              25
      }
      break;
    case 2:
      rrNebr = Sqr (rCut + rNebrShell);
      nebrTabLenP[ip] = 0;                                             30
      THREAD_SPLIT_LOOP (m1z, cells.z) {
        for (m1y = 0 ...
          ...
            if (VLenSq (dr) < rrNebr) {
              if (nebrTabLenP[ip] * nThread >= nebrTabMax)             35
                ErrExit (ERR_TOO_MANY_NEBRS);
              nebrTabP[ip][2 * nebrTabLenP[ip]] = j1;
              nebrTabP[ip][2 * nebrTabLenP[ip] + 1] = j2;
              ++ nebrTabLenP[ip];
            }                                                          40
          ...
        }
      break;
  }
  return (NULL);                                                       45
}
```

Here, during the first call to *BuildNebrListT* (the first stage), each thread is responsible for constructing the linked lists of cell occupants for only a fraction of the cells, based on their *z* coordinates. During the second call, each thread produces a separate portion of the neighbor list, stored in the array *nebrTabP[ip][]*, involving pairs of atoms for which the first member of the pair lies in a cell being handled by that thread. Thus, while the atoms themselves typically appear in more than one of the neighbor lists, the pairs themselves appear only once.

The corresponding function *ComputeForces* becomes

```
void ComputeForces ()
{
  int ip, iq;

  THREAD_PROC_LOOP (ComputeForcesT, 1);
  THREAD_PROC_LOOP (ComputeForcesT, 2);
  uSum = 0.;
  THREAD_LOOP uSum += uSumP[iq];
}


void *ComputeForcesT (void *tr)
{
  ...
  int ip, iq;

  QUERY_THREAD ();
  switch (QUERY_STAGE) {
    case 1:
      rrCut = Sqr (rCut);
      DO_MOL VZero (raP[ip][n]);
      uSumP[ip] = 0.;
      for (n = 0; n < nebrTabLenP[ip]; n ++) {
        j1 = nebrTabP[ip][2 * n];
        j2 = nebrTabP[ip][2 * n + 1];
        ...
        if (rr < rrCut) {

          ...
          VVSAdd (raP[ip][j1], fcVal, dr);
          VVSAdd (raP[ip][j2], - fcVal, dr);
          uSumP[ip] += uVal;
        }
      }
      break;
    case 2:
      THREAD_SPLIT_LOOP (n, nMol) {
        VZero (mol[n].ra);
        THREAD_LOOP VVAdd (mol[n].ra, raP[iq][n]);
      }
      break;
  }
  return (NULL);
}
```

In the first call to *ComputeForcesT*, the particular neighbor list associated with thread *ip* is processed, with acceleration values being stored in array *raP[ip][]* and potential energies in *uSumP[ip]*. The second call accumulates these separate contributions, where, for brevity,

```
#define THREAD_LOOP                                          \
   for (iq = 0; iq < nThread; iq ++)
```

Additional quantities declared here are

```
pthread_t *pThread;
VecR **raP;
real *uSumP;
int **nebrTabP, *nebrTabLenP, funcStage, nThread;
```

where the elements *pThread* are required by the thread processing; the necessary array allocations (in *AllocArrays* – the usual *nebrTab* is not required, having been replaced by corresponding arrays *nebrTabP* private to each thread) and input data item are

```
   AllocMem (pThread, nThread, pthread_t);
   AllocMem (uSumP, nThread, real);
   AllocMem (nebrTabLenP, nThread, int);
   AllocMem2 (raP, nThread, nMol, VecR);
   AllocMem2 (nebrTabP, nThread, 2 * nebrTabMax / nThread, int);       5

   NameI (nThread),
```

Finally, the remaining definitions used in the program are

```
   #define QUERY_THREAD()   ip = (int) tr
   #define QUERY_STAGE      funcStage
   #define THREAD_PROC_LOOP(tProc, fStage)                     \
      funcStage = fStage;                                      \
      for (ip = 1; ip < nThread; ip ++)                        \      5
         pthread_create (&pThread[ip], NULL, tProc,            \
         (void *) ip);                                         \
      tProc ((void *) 0);                                      \
      for (ip = 1; ip < nThread; ip ++)                        \
         pthread_join (pThread[ip], NULL);                            10
```

Further information about the functions prefixed with *pthread_* can be found in the documentation of the thread library functions. It is left to the reader to explore the performance benefits that can be obtained by this approach (a prerequisite being a computer with multiple processors and shared memory).

17.6 Techniques for vector processing

Pipelined computation

When supercomputers first appeared, their performance derived from two principal features: a fast processor clock and the use of pipelined vector processing. With the appearance of comparatively cheap microprocessors, and their employment in parallel computers, the role of vector processing temporarily diminished. However, the reader may rest assured that it is returning once again to boost microprocessor performance. Some pipelining is already to be found in these chips (as well as multiple instruction units), but the need for application software to take its existence into account is less of an issue than when 'serious' vector processing is crucial to performance. There are still mainframe supercomputers that rely heavily on vector processing.

We will again concentrate on the soft-sphere simulation; this turns out to be the most difficult to vectorize effectively. For longer-range forces the number of interacting neighbors of each atom is large, and since a slightly modified form of the all-pairs version of the interaction function can then be used (provided there are not too many atoms) the compiler is able to handle the vectorization automatically [bro86]. Likewise for small systems with only short-range forces, where the loss of efficiency due to using the all-pairs method is adequately compensated by the vectorization speedup. But for bigger systems (beyond several hundred atoms) with short-range forces, for which cells are essential and neighbor lists advisable if the memory is available, compilers are unable to rearrange the conventional MD program so that it may be vectorized effectively. The problem must be solved the hard way – by rearranging the computation into a more suitable form [rap91a].

A specific computer architecture will not be invoked here. Instead, the computations will be expressed in a form that any processor with certain common hardware features – to be enumerated later – and an effective compiler will be able to execute relatively efficiently. The approach differs from previous sections, where we spelled out the communication and thread functions to be used; a similar approach could have been used here as well, but since the result of reformulating the algorithm is a program that can be vectorized automatically, there is little point in doing the work manually (by calling an assortment of vector functions directly).

Effective vector processing requires long sequences of data items (the vectors) that can be processed independently in pipeline fashion. Because of the overhead in filling these pipes initially, there is an advantage to using long vectors, although the minimum useful length depends on the specific machine. Vector computers emphasize floating-point performance, but in recognition of the fact that most problems are not organized in precisely the form needed, the machines are also able to carry out certain kinds of data rearrangement at high speed. More precisely, the capability exists to gather randomly distributed data items speedily into a single array based on a vector of addresses, as well as the converse scatter operation; the MD code below relies heavily on such operations.

Layer method

The problem with the cell method, which makes it impossible to vectorize effectively, is its use of linked lists to identify the atoms in each cell. The method we describe here takes the same cell occupancy data and rearranges it into a form more suitable for vector processing. The approach can also be extended to produce neighbor lists [gre89b] organized in a way that can be automatically vectorized.

In the original, cell-based version of `ComputeForces` in §3.4 there are several nested loops; the outermost is over cells and within this there is a loop over the offsets to neighboring cells; two further inner loops consider all pairs of cell occupants. We will now proceed to rearrange these loops. If the cell occupants are assumed to be ordered in some (arbitrary) way, then the role of the two outermost loops in the revised version is to produce all valid pairings of the ith and jth occupants of whatever pair of cells happens to be under consideration by the inner loops (the case where both cells are the same is also covered). Within these two outer loops there is a loop over possible relative offsets between cell pairs. Finally, the innermost loop is over all cells, with the second cell of the pair obtained using the known cell offset. If the number of cells is close to the number of atoms, the innermost loop will fulfill the principal requirement of effective vectorization – a large repetition count. Of course this is not the whole story, because it is the details of the processing carried out by the innermost loop that determine performance and, in addition, the work needed to rearrange the data for this computation must be taken into account.

The reordered cell occupancy data are stored in 'layers' [rap91a], as shown in Figure 17.2. Each layer consists of one storage element per cell, and there are as many layers as the maximum cell occupancy. The first atom in a cell (the order has no special significance) is placed in the first layer, and so on; the first layer will be practically full, but later layers will be less densely populated. Since the two outer loops scan all pairs of layers and the number of cycles of the inner loops is independent of cell occupancy, it follows that computation time varies quadratically with the number of layers. Thus the method is most effective for relatively dense systems, where cell occupancy does not vary too widely.

The presence of periodic boundaries complicates the algorithm, so they are eliminated by a process of replication (§3.4), reminiscent of the copying operation used earlier in the distributed approach. Before beginning the interaction computation (`ComputeForces`), all atoms close to any boundary are duplicated so that they appear just beyond the opposite boundary. These replicas are used for all

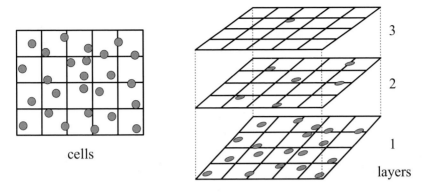

Fig. 17.2. The reorganization of the cell contents into layers for efficient vector processing.

interactions that would otherwise extend across one or more boundaries. The cell array must be enlarged to provide a shell of thickness r_c (the interaction cutoff range) surrounding the region.

The replication function♠ follows. Each spatial dimension is treated in turn and atoms near edges or corners may be replicated more than once. A check for array overflow is included[†].

```
void ReplicateMols ()
{
  int k, n, na;

  nMolRep = nMol;                                                    5
  for (k = 0; k < NDIM; k ++) {
    na = nMolRep;
    for (n = 0; n < nMolRep; n ++) {
      if (fabs (VComp (mol[n].r, k)) >= 0.5 * VComp (region, k) -
          rCut) {                                                    10
        mol[na].r = mol[n].r;
        if (VComp (mol[na].r, k) > 0.)
          VComp (mol[na].r, k) -= VComp (region, k);
        else VComp (mol[na].r, k) += VComp (region, k);
        ++ na;                                                       15
      }
    }
    nMolRep = na;
    if (nMolRep >= nMolMax) ErrExit (ERR_TOO_MANY_REPLICAS);
  }                                                                  20
}
```

♠ pr_17_3
† To vectorize the loops in this function it may be necessary to divide each loop over atoms into two parts: the first loop identifies the atoms to be replicated and stores their indices in a separate array; the second does the replication using this index array.

The interactions are computed in several stages. The first stage determines the cell occupied by each atom. Next the layers are constructed by a method to be discussed below. Then come the multiply-nested loops: the outer two loops select all possible layer pairs, within them is the loop over cell offsets and inside this are four successive loops where all the interactions are computed. The first of the innermost loops considers all cell pairings involving the chosen layers and the specified offset, determines whether a valid pair of atoms is to be found there and, if so, adds this information to a list. The second loop processes the listed atom pairs and computes their interactions. The final two loops add these newly computed terms to the accumulated interactions of the respective atoms.

The structure *Mol* used here has the form

```
typedef struct {
  VecR r, rv, ra;
  real u;
  int inCell;
} Mol;                                                               5
```

where u is the potential energy of the individual atom and is required in order to achieve vectorization; the new variables appearing in the program are

```
Mol *mol;
VecR *raL;
real *uL;
int **layerMol, **molPtr, *inside, *molId, bdyOffset, nLayerMax,
    nMolMax, nMolRep;                                                5
```

The arrays are allocated by

```
void AllocArrays ()
{
  int k;

  AllocMem (mol, nMolMax, Mol);                                      5
  AllocMem (molId, nMolMax, int);
  AllocMem (raL, VProd (cells), VecR);
  AllocMem (uL, VProd (cells), real);
  AllocMem (inside, 2 * bdyOffset + VProd (cells), int);
  AllocMem2 (molPtr, 2, VProd (cells), int);                         10
  AllocMem2 (layerMol, nLayerMax, 2 * bdyOffset + VProd (cells), int);
}
```

The variable *nMolMax* is the maximum number of atoms, including replicas, for which storage is available, *nLayerMax* is the number of available layers and the constant *bdyOffset* is used to avoid any problems with negative array indices when shifted layers are paired (see later). The meaning of the other variables will become clear in due course. If more than a few layers are needed the array `layerMol` will dominate the storage requirements of the program.

The interaction calculation follows.

```
#define OFFSET_VALS                                                   \
   {{-1,-1,-1}, {0,-1,-1}, {1,-1,-1}, {-1,0,-1}, {0,0,-1},    \
    {1,0,-1}, {-1,1,-1}, {0,1,-1}, {1,1,-1}, {-1,-1,0},       \
    {0,-1,0}, {1,-1,0}, {-1,0,0}, {0,0,0}, {1,0,0},           \
    {-1,1,0}, {0,1,0}, {1,1,0}, {-1,-1,1}, {0,-1,1},          \
    {1,-1,1}, {-1,0,1}, {0,0,1}, {1,0,1}, {-1,1,1},           \
    {0,1,1}, {1,1,1}}

void ComputeForces ()
{
  VecR dr, invWid, regionEx, rs, t;
  VecI cc, vOff[] = OFFSET_VALS;
  real fcVal, rr, rrCut, rri, rri3;
  int ic, layer1, layer2, m1, m2, n, na, nat, nLayer, nPair,
      offset, offsetLo;

  VCopy (t, cells);
  VAddCon (t, t, -2.);
  VDiv (invWid, t, region);
  VSetAll (t, 2.);
  VDiv (t, t, invWid);
  VAdd (regionEx, region, t);
  for (n = 0; n < nMolRep; n ++) {
    VSAdd (rs, mol[n].r, 0.5, regionEx);
    VMul (cc, rs, invWid);
    mol[n].inCell = VLinear (cc, cells);
    molId[n] = n;
  }
  nLayer = 0;
  for (na = nMolRep; na > 0; na = nat) {
    if (nLayer >= nLayerMax) ErrExit (ERR_TOO_MANY_LAYERS);
    for (ic = 0; ic < VProd (cells); ic ++)
      layerMol[nLayer][bdyOffset + ic] = -1;
    for (n = 0; n < na; n ++)
      layerMol[nLayer][bdyOffset + mol[molId[n]].inCell] = molId[n];
    for (ic = 0; ic < VProd (cells); ic ++) {
      n = layerMol[nLayer][bdyOffset + ic];
      if (n >= 0) mol[n].inCell = -1;
    }
    nat = 0;
    for (n = 0; n < na; n ++) {
      if (mol[molId[n]].inCell >= 0) molId[nat ++] = molId[n];
    }
    ++ nLayer;
  }
```

```
for (n = 0; n < nMolRep; n ++) {
  VZero (mol[n].ra);
  mol[n].u = 0.;
}
rrCut = Sqr (rCut);                                                        50
for (layer1 = 0; layer1 < nLayer; layer1 ++) {
  for (layer2 = layer1; layer2 < nLayer; layer2 ++) {
    offsetLo = (layer2 == layer1) ? 14 : 0;
    for (offset = offsetLo; offset < 27; offset ++) {
      nPair = 0;                                                           55
      m2 = bdyOffset + VLinear (vOff[offset], cells);
      for (m1 = bdyOffset; m1 < bdyOffset + VProd (cells); m1 ++) {
        if ((inside[m1] || inside[m2]) &&
            layerMol[layer1][m1] >= 0 && layerMol[layer2][m2] >= 0) {
          molPtr[0][nPair] = layerMol[layer1][m1];                         60
          molPtr[1][nPair] = layerMol[layer2][m2];
          ++ nPair;
        }
        ++ m2;
      }                                                                    65
      for (n = 0; n < nPair; n ++) {
        VSub (dr, mol[molPtr[0][n]].r, mol[molPtr[1][n]].r);
        rr = VLenSq (dr);
        if (rr < rrCut) {
          rri = 1. / rr;                                                   70
          rri3 = Cube (rri);
          fcVal = 48. * rri3 * (rri3 - 0.5) * rri;
          VSCopy (raL[n], fcVal, dr);
          uL[n] = 4. * rri3 * (rri3 - 1.) + 1.;
        } else {                                                          75
          VZero (raL[n]);
          uL[n] = 0.;
        }
      }
      for (n = 0; n < nPair; n ++) {                                      80
        VVAdd (mol[molPtr[0][n]].ra, raL[n]);
        mol[molPtr[0][n]].u += uL[n];
      }
      for (n = 0; n < nPair; n ++) {
        VVSub (mol[molPtr[1][n]].ra, raL[n]);                             85
        mol[molPtr[1][n]].u += uL[n];
      }
    }
  }
}                                                                         90
uSum = 0.;
DO_MOL uSum += mol[n].u;
uSum *= 0.5;
}
```

Each of the innermost loops can be shown to satisfy the basic requirement of vectorization, namely, that each item processed is independent of all others[†]. The other characteristic of the loops is that, at least for the earliest layers, the vectors processed are of length proportional to N_m.

The one slightly subtle detail in this computation is the assignment of atoms to layers. The way this is done is to make several iterations over the set of atoms, one such cycle for each layer, each iteration assigning whichever atoms happen to be in each cell to the layer being filled. Thus several atoms may be assigned to the same position in a layer (in the array `layerMol`), but all atoms except the last one encountered in each cell will be overwritten. A check is made after each cycle to see which atoms were recorded in the layer; these are removed from the set before proceeding to the next layer, by zeroing the values of `inCell` for these atoms and compressing the identities of the remaining atoms held in `molId`. This process is repeated until no unassigned atoms remain. Clearly such a scheme would be wasteful in terms of computation, were it not for the fact that it can be vectorized.

Other comments about the above function are the following. Computation of the cell size ignores the outermost cells, because these are outside the simulation region. The array `inside` is used to distinguish quickly cells that are within the simulation region from those adjoining the boundary; this is preferable to a test based on the three indices needed to specify cell position. When a layer is paired with itself, only positive cell offsets need be considered.

Changes to *SetParams* are

```
VecI t;
...
VAddCon (cells, cells, 2);
t = initUcell;
VAddCon (t, t, 2);                                              5
nMolMax = 4 * VProd (t);
bdyOffset = cells.x * (cells.y + 1) + 1;
```

and *nLayerMax* must be added to the input data. Initialization also requires

```
void SetupLayers ()
{
  int n, nLayer, nx, ny, nz;

  for (n = 0; n < 2 * bdyOffset + VProd (cells); n ++) inside[n] = 0;   5
  for (nz = 1; nz < cells.z - 1; nz ++) {
    for (ny = 1; ny < cells.y - 1; ny ++) {
```

[†] Supplementary, system-specific compiler directives may have to be used to inform the compiler of this fact, since it is not at all obvious from just reading the program without understanding the underlying organization.

```
    for (nx = 1; nx < cells.x - 1; nx ++) {
      inside[bdyOffset + (nz * cells.y + ny) * cells.x + nx] = 1;
    }
                                                                    10
  }
}
for (nLayer = 0; nLayer < nLayerMax; nLayer ++) {
  for (n = 0; n < 2 * bdyOffset + VProd (cells); n ++)
    layerMol[nLayer][n] = -1;
                                                                    15
}
}
```

The rest of the program is unchanged.

17.7 Further study

17.1 Implement the distributed MD computation on your favorite multiprocessor computer and measure the communication overheads. How does performance vary with system size and number of processors?

17.2 Examine the performance of the threaded approach on a shared-memory computer.

17.3 Extend the vectorized layer method to include neighbor lists.

18

More about software

18.1 Introduction

This chapter includes a summary of the definitions of various structures and macros used in the programs listed in the book, descriptions of some general-purpose mathematical functions and discussion of software topics such as file and data handling needed in the case studies.

18.2 Structures and macro definitions

The use of structures to represent cartesian vectors in two and three dimensions, together with a set of macro definitions that describe frequently used vector operations, leads to more concise programs that not only are less susceptible to typing and similar errors, but that are also more readable. Here, for the convenience of the reader, all the definitions associated with vectors, and other quantities, that are to be found scattered throughout the text, are gathered together in one place.

We begin with the structures and definitions required for three-dimensional vectors. The floating-point and integer vector structures are

```
typedef struct {
  real x, y, z;
} VecR;

typedef struct {                                          5
  int x, y, z;
} VecI;
```

A series of basic vector arithmetic definitions follow.

```
#define VSet(v, sx, sy, sz)                    \
    (v).x = sx,                                \
    (v).y = sy,                                \
```

481

```
    (v).z = sz
#define VCopy(v1, v2)                                                    \
    (v1).x = (v2).x,                                                     \
    ... (ditto for other components) ...
#define VScale(v, s)                                                     \
    (v).x *= s,                                                          \
    ...
#define VSCopy(v2, s1, v1)                                              \
    (v2).x = (s1) * (v1).x,                                             \
    ...
#define VAdd(v1, v2, v3)                                                \
    (v1).x = (v2).x + (v3).x,                                           \
    ...
#define VSub(v1, v2, v3)                                                \
    (v1).x = (v2).x - (v3).x,                                           \
    ...
#define VSAdd(v1, v2, s3, v3)                                          \
    (v1).x = (v2).x + (s3) * (v3).x,                                   \
    ...
#define VSSAdd(v1, s2, v2, s3, v3)                                     \
    (v1).x = (s2) * (v2).x + (s3) * (v3).x,                           \
    ...
#define VMul(v1, v2, v3)                                                \
    (v1).x = (v2).x * (v3).x,                                           \
    ...
#define VDiv(v1, v2, v3)                                                \
    (v1).x = (v2).x / (v3).x,                                           \
    ...
```

Product operations, including the scalar and vector products, are

```
#define VDot(v1, v2)                                                    \
    ((v1).x * (v2).x + (v1).y * (v2).y + (v1).z * (v2).z)
#define VCross(v1, v2, v3)                                              \
    (v1).x = (v2).y * (v3).z - (v2).z * (v3).y,                        \
    (v1).y = (v2).z * (v3).x - (v2).x * (v3).z,                        \
    (v1).z = (v2).x * (v3).y - (v2).y * (v3).x
#define VWDot(v1, v2, v3)                                               \
    ((v1).x * (v2).x * (v3).x + (v1).y * (v2).y * (v3).y +           \
    (v1).z * (v2).z * (v3).z)
```

Matrix–vector products (for both the matrix and its transpose) are

```
#define MVMul(v1, m, v2)                                                         \
    (v1).x = (m)[0] * (v2).x + (m)[3] * (v2).y + (m)[6] * (v2).z,             \
    (v1).y = (m)[1] * (v2).x + (m)[4] * (v2).y + (m)[7] * (v2).z,             \
    (v1).z = (m)[2] * (v2).x + (m)[5] * (v2).y + (m)[8] * (v2).z
#define MVMulT(v1, m, v2)                                                        \
    (v1).x = (m)[0] * (v2).x + (m)[1] * (v2).y + (m)[2] * (v2).z,             \
```

```
(v1).y = (m)[3] * (v2).x + (m)[4] * (v2).y + (m)[5] * (v2).z,   \
(v1).z = (m)[6] * (v2).x + (m)[7] * (v2).y + (m)[8] * (v2).z
```

Other useful operations include

```
#define VSetAll(v, s)                                          \
   VSet (v, s, s, s)
#define VAddCon(v1, v2, s)                                     \
   (v1).x = (v2).x + (s),                                      \
   ...                                                                    5
#define VProd(v)                                               \
   ((v).x * (v).y * (v).z)
#define VGe(v1, v2)                                            \
   ((v1).x >= (v2).x && (v1).y >= (v2).y &&                    \
    (v1).z >= (v2).z)                                                     10
#define VLt(v1, v2)                                            \
   ((v1).x < (v2).x && (v1).y < (v2).y &&                      \
    (v1).z < (v2).z)
#define VLinear(p, s)                                          \
   (((p).z * (s).y + (p).y) * (s).x + (p).x)                             15
#define VCSum(v)                                               \
   ((v).x + (v).y + (v).z)
#define VComp(v, k)                                            \
   *((k == 0) ? &(v).x : ((k == 1) ? &(v).y : &(v).z))
```

together with two operations for converting between vectors and array elements,

```
#define VToLin(a, n, v)                                        \
   a[(n) + 0] = (v).x,                                         \
   a[(n) + 1] = (v).y,                                         \
   a[(n) + 2] = (v).z
#define VFromLin(v, a, n)                                      \
   VSet (v, a[(n) + 0], a[(n) + 1], a[(n) + 2])                          5
```

The two-dimensional equivalents are not shown, except for the cross product, which is regarded as a scalar and defined as

```
#define VCross(v1, v2)                                         \
   ((v1).x * (v2).y - (v1).y * (v2).x)
```

The choice of whether to use two- or three-dimensional vectors in a program can be made to depend on the value of *NDIM* by using conditional compilation,

```
#if NDIM == 3
#define VSet...
...
#elif NDIM == 2
```

```
#define VSet...                                                    5
...
#endif
```

Additional vector operations that do not explicitly depend on *NDIM* are

```
#define VZero(v)          VSetAll (v, 0)
#define VLenSq(v)         VDot (v, v)
#define VWLenSq(v1, v2)   VWDot(v1, v2, v2)
#define VLen(v)           sqrt (VDot (v, v))
#define VVAdd(v1, v2)     VAdd (v1, v1, v2)                        5
#define VVSub(v1, v2)     VSub (v1, v1, v2)
#define VVSAdd(v1, s2, v2)  VSAdd (v1, v1, s2, v2)
#define VInterp(v1, s2, v2, v3)                             \
   VSSAdd (v1, s2, v2, 1. - (s2), v3)
```

Quaternions are handled in a corresponding manner,

```
typedef struct {
  real u1, u2, u3, u4;
} Quat;

#define QSet(q, s1, s2, s3, s4)                             \    5
   (q).u1 = s1,                                             \
   (q).u2 = s2,                                             \
   (q).u3 = s3,                                             \
   (q).u4 = s4
#define QZero(q)                                            \   10
   QSet (q, 0, 0, 0, 0)
#define QScale(q, s)                                        \
   (q).u1 *= s,                                             \
   ...
#define QSAdd(q1, q2, s3, q3)                               \   15
   (q1).u1 = (q2).u1 + (s3) * (q3).u1,                      \
   ...
#define QLenSq(q)                                           \
   (Sqr ((q).u1) + Sqr ((q).u2) + Sqr ((q).u3) +
   Sqr ((q).u4))                                                 20
#define QMul(q1, q2, q3)                                    \
   (q1).u1 =   (q2).u4 * (q3).u1 - (q2).u3 * (q3).u2 +      \
               (q2).u2 * (q3).u3 + (q2).u1 * (q3).u4,       \
   (q1).u2 =   (q2).u3 * (q3).u1 + (q2).u4 * (q3).u2 -      \
               (q2).u1 * (q3).u3 + (q2).u2 * (q3).u4,       \   25
   (q1).u3 = - (q2).u2 * (q3).u1 + (q2).u1 * (q3).u2 +      \
               (q2).u4 * (q3).u3 + (q2).u3 * (q3).u4,       \
   (q1).u4 = - (q2).u1 * (q3).u1 - (q2).u2 * (q3).u2 -      \
               (q2).u3 * (q3).u3 + (q2).u4 * (q3).u4
```

as are complex variables,

```
typedef struct {
  real R, I;
} Cmplx;

#define CSet(a, x, y)                                    \     5
    a.R = x,                                             \
    a.I = y
#define CAdd(a, b, c)                                    \
    a.R = b.R + c.R,                                     \
    a.I = b.I + c.I                                            10
#define CSub(a, b, c)                                    \
    a.R = b.R - c.R,                                     \
    a.I = b.I - c.I
#define CMul(a, b, c)                                    \
   a.R = b.R * c.R - b.I * c.I,                          \     15
   a.I = b.R * c.I + b.I * c.R
```

Other macro definitions used are

```
#define Sqr(x)      ((x) * (x))
#define Cube(x)     ((x) * (x) * (x))
#define Sgn(x, y)   (((y) >= 0) ? (x) : (- (x)))
#define IsEven(x)   ((x) & ~1)
#define IsOdd(x)    ((x) & 1)                                  5
#define Nint(x)                                          \
    (((x) < 0.) ? (- (int) (0.5 - (x))): ((int) (0.5 + (x))))
#define Min(x1, x2)                                      \
    (((x1) < (x2)) ? (x1) : (x2))
#define Max(x1, x2)                                      \    10
    (((x1) > (x2)) ? (x1) : (x2))
#define Min3(x1, x2, x3)                                 \
    (((x1) < (x2)) ? (((x1) < (x3)) ? (x1) : (x3)) :     \
                     (((x2) < (x3)) ? (x2) : (x3)))
#define Max3(x1, x2, x3)                                 \    15
    (((x1) > (x2)) ? (((x1) > (x3)) ? (x1) : (x3)) :     \
                     (((x2) > (x3)) ? (x2) : (x3)))
#define Clamp(x, lo, hi)                                 \
    (((x) >= (lo) && (x) <= (hi)) ? (x) : (((x) < (lo)) ? \
    (lo) : (hi)))                                              20
```

Macros are introduced to help deal with periodic boundary conditions,

```
#define VWrap(v, t)                                      \
    if (v.t >= 0.5 * region.t) v.t -= region.t;          \
    else if (v.t < -0.5 * region.t) v.t += region.t
#define VShift(v, t)                                     \
    if (v.t >= 0.5 * region.t) shift.t -= region.t;      \     5
    else if (v.t < -0.5 * region.t) shift.t += region.t
```

```
#define VShiftWrap(v, t)                                    \
   if (v.t >= 0.5 * region.t) {                             \
     shift.t -= region.t;                                   \
     v.t -= region.t;                                       \    10
   } else if (v.t < -0.5 * region.t) {                      \
     shift.t += region.t;                                   \
     v.t += region.t;                                       \
   }
#define VCellWrap(t)                                        \    15
   if (m2v.t >= cells.t) {                                  \
     m2v.t = 0;                                             \
     shift.t = region.t;                                    \
   } else if (m2v.t < 0) {                                  \
     m2v.t = cells.t - 1;                                   \    20
     shift.t = - region.t;                                  \
   }
```

and, since these tend to be repeated for each of the vector components, the following are also used (here in three-dimensional form),

```
#define VWrapAll(v)                                         \
   {VWrap (v, x);                                           \
    VWrap (v, y);                                           \
    VWrap (v, z);}
#define VShiftAll(v)                                        \    5
   {VShift (v, x);                                          \
    VShift (v, y);                                          \
    VShift (v, z);}
#define VCellWrapAll()                                      \
   {VCellWrap (x);                                          \    10
    VCellWrap (y);                                          \
    VCellWrap (z);}
```

The cell offsets used for interaction and neighbor-list computations (with standard periodic boundaries) are, in three dimensions,

```
#define OFFSET_VALS                                         \
   {{0,0,0}, {1,0,0}, {1,1,0}, {0,1,0}, {-1,1,0},           \
    {0,0,1}, {1,0,1}, {1,1,1}, {0,1,1}, {-1,1,1},           \
    {-1,0,1}, {-1,-1,1}, {0,-1,1}, {1,-1,1}}
#define N_OFFSET  14                                             5
```

and in two dimensions,

```
#define OFFSET_VALS                                         \
   {{0,0}, {1,0}, {1,1}, {0,1}, {-1,1}}
#define N_OFFSET  5
```

Some of the measurements make use of

```
typedef struct {
  real val, sum, sum2;
} Prop;

#define PropZero(v)                                           \           5
    v.sum = 0.,                                               \
    v.sum2 = 0.
#define PropAccum(v)                                          \
    v.sum += v.val,                                           \
    v.sum2 += Sqr (v.val)                                                 10
#define PropAvg(v, n)                                         \
    v.sum /= n,                                               \
    v.sum2 = sqrt (Max (v.sum2 / n - Sqr (v.sum), 0.))
#define PropEst(v)                                            \
    v.sum, v.sum2                                                         15
```

Several macros are introduced to help with reading and writing files (by concealing the standard Unix file functions),

```
#define ReadF(x)        fread (&x, sizeof (x), 1, fp)
#define ReadFN(x, n)    fread (x, sizeof (x[0]), n, fp)
#define WriteF(x)       fwrite (&x, sizeof (x), 1, fp)
#define WriteFN(x, n)   fwrite (x, sizeof (x[0]), n, fp)
```

Text characters used in some programs are defined as

```
#define CHAR_MINUS   '-'
#define CHAR_ZERO    '0'
```

Finally, the following shorthand is employed for frequently appearing loops:

```
#define DO_MOL                                                \
    for (n = 0; n < nMol; n ++)
#define DO_CELL(j, m)                                         \
    for (j = cellList[m]; j >= 0; j = cellList[j])
```

18.3 Allocating arrays

One particularly useful feature of C and other modern programming languages is the ability to allocate storage for arrays dynamically, with the amount of storage being determined at runtime (as opposed to fixed-size arrays whose dimensions are specified in the program source). All arrays used in the programs whose sizes depend on the input parameters are allocated dynamically. As a consequence, there are no preset size limits built into the C implementations of the programs.

The following pair of macro definitions handles the allocations of one- and two-dimensional arrays of any kind of variable or structure,

```
#define AllocMem(a, n, t)                                    \
   a = (t *) malloc ((n) * sizeof (t))
#define AllocMem2(a, n1, n2, t)                              \
   AllocMem (a, n1, t *);                                    \
   AllocMem (a[0], (n1) * (n2), t);                          \
   for (k = 1; k < n1; k ++) a[k] = a[k - 1] + n2;
```

If reducing computation time is a serious issue, then there may be more to memory usage than is apparent here, and the concerned user will have to take into account various hardware architectural features when going about the task. Considerations such as localizing memory access, the actual number of array dimensions, minimizing address computation and avoiding cache conflicts are typical examples of the factors influencing efficiency. Disregarding such details can sometimes lead to computation times several times (very occasionally, even an order of magnitude) longer than is really necessary. Determining whether near-optimal performance has been achieved is not always a simple task.

18.4 Utility functions

Mathematical functions

Several standard computational functions are required in a few of the case studies. They are all to be found in [pre92], as well as in various widely available mathematical function libraries. In the interest of completeness, we include listings of the customized versions of these functions used in the case studies.

Matrix multiplication (§8.5, §11.5) uses the function

```
void MulMat (real *a, real *b, real *c, int n)
{
  int i, j, k;

  for (i = 0; i < n; i ++) {
    for (j = 0; j < n; j ++) {
      Mn (a, i, j) = 0.;
      for (k = 0; k < n; k ++)
        Mn (a, i, j) += Mn (b, i, k) * Mn (c, k, j);
    }
  }
}
```

where

```
#define MAT(a, n, i, j)  (a)[(i) + n * (j)]
#define Mn(a, i, j)       MAT (a, n, i, j)
```

are introduced to aid readability by hiding the fact that matrices are stored as singly-indexed arrays. The function for multiplying a matrix by a vector (§11.5) is

```
void MulMatVec (real *a, real *b, real *c, int n)
{
  int i, k;

  DO (i, n) {                                                          5
    a[i] = 0.;
    DO (k, n) a[i] += Mn (b, i, k) * c[k];
  }
}
```

The function needed for solving a set of linear equations efficiently (§10.3, §11.5), based on the Crout version of the LU method, is as follows [ral78]; the matrix a is stored as a one-dimensional array in column order, and the vector on the right-hand side of the equation is stored in x and overwritten by the solution.

```
#define A(i, j)     MAT (a, n, i, j)
#define Swap(a, b)  v = a, a = b, b = v
#define N_MAX       100

void SolveLineq (real *a, real *x, int n)                             5
{
  real vMaxI[N_MAX], v, vMax;
  int ptrMax[N_MAX], i, j, k, m;

  if (n > N_MAX) exit (0);                                            10
  for (i = 0; i < n; i ++) {
    vMax = 0.;
    for (j = 0; j < n; j ++) {
      if ((v = fabs (A(i, j))) > vMax) vMax = v;
    }                                                                 15
    vMaxI[i] = 1. / vMax;
  }
  for (m = 0; m < n; m ++) {
    vMax = 0.;
    for (i = m; i < n; i ++) {                                       20
      for (k = 0; k < m; k ++) A(i, m) -= A(i, k) * A(k, m);
      if ((v = fabs (A(i, m)) * vMaxI[i]) > vMax) {
        vMax = v;
        ptrMax[m] = i;
      }                                                               25
    }
    if (m != ptrMax[m]) {
      for (k = 0; k < n ; k ++) Swap (A(m, k), A(ptrMax[m], k));
      vMaxI[ptrMax[m]] = vMaxI[m];
    }                                                                 30
    for (j = m + 1; j < n; j ++) {
      for (k = 0; k < m; k ++) A(m, j) -= A(m, k) * A(k, j);
```

```
    A(m, j) /= A(m, m);
  }
}
for (i = 0; i < n; i ++) {                                           35
  Swap (x[ptrMax[i]], x[i]);
  for (j = 0; j < i; j ++) x[i] -= A(i, j) * x[j];
  x[i] /= A(i, i);
}                                                                    40
for (i = n - 2; i >= 0; i --) {
  for (j = i + 1; j < n; j ++) x[i] -= A(i, j) * x[j];
}
}
```

The function (§5.4) for fast (in place, forward direction) Fourier transformation [hig76] of complex data arrays of size 2^n is

```
void FftComplex (Cmplx *a, int size)
{
  Cmplx t, w, wo;
  real theta;
  int i, j, k, n;                                                    5

  k = 0;
  for (i = 0; i < size; i ++) {
    if (i < k) {
      t = a[i];                                                      10
      a[i] = a[k];
      a[k] = t;
    }
    n = size / 2;
    while (n >= 1 && k >= n) {                                       15
      k -= n;
      n /= 2;
    }
    k += n;
  }                                                                  20
  for (n = 1; n < size; n *= 2) {
    theta = M_PI / n;
    CSet (wo, cos (theta) - 1., sin (theta));
    CSet (w, 1., 0.);
    for (k = 0; k < n; k ++) {                                       25
      for (i = k; i < size; i += 2 * n) {
        j = i + n;
        CMul (t, w, a[j]);
        CSub (a[j], a[i], t);
        CAdd (a[i], a[i], t);                                        30
      }
      CMul (t, w, wo);
      CAdd (w, w, t);
    }
```

```
  }                                                                35
}
```

Numerical integration (§5.3) is based on the simplest trapezoidal rule and assumes unit spacing of the points,

```
real Integrate (real *f, int nf)
{
  real s;
  int i;
                                                                  5
  s = 0.5 * (f[0] + f[nf - 1]);
  for (i = 1; i < nf - 1; i ++) s += f[i];
  return (s);
}
```

The solution of a cubic equation (§9.4), assuming all roots are real and the cubic term has a coefficient of unity, is given by

```
void SolveCubic (real *g, real *a)
{
  real q1, q2, t;

  q1 = sqrt (Sqr (a[0]) - 3. * a[1]) / 3.;                        5
  q2 = (a[0] * (2. * Sqr (a[0]) - 9. * a[1]) + 27. * a[2]) / 54.;
  t = acos (q2 / (q1 * q1 * q1));
  g[0] = -2. * q1 * cos (t / 3.) - a[0] / 3.;
  g[1] = -2. * q1 * cos ((t + 2. * M_PI) / 3.) - a[0] / 3.;
  g[2] = -2. * q1 * cos ((t + 4. * M_PI) / 3.) - a[0] / 3.;       10
}
```

with the roots being returned in the array *g*.

Random number generation

Random numbers generally play only a minor part in MD simulations and the need to ensure 'high-quality' values is far less important than in Monte Carlo work. Molecular chaos will tend to eradicate all but the most egregious irregularities in whatever random numbers are generated. We have therefore (out of habit) used a simple, but adequate, method[†] for producing uniformly distributed values in (0, 1), the origins of which are shrouded in antiquity!

```
#define IADD    453806245
#define IMUL    314159269
```

---

† Readers wanting well-documented procedures should check what is available on the local computer, or turn to the literature [knu69, jam90, pre92].

```
#define MASK    2147483647
#define SCALE   0.4656612873e-9
                                                                    5
int randSeedP = 17;

real RandR ()
{
  randSeedP = (randSeedP * IMUL + IADD) & MASK;                    10
  return (randSeedP * SCALE);
}
```

The variable *randSeed* is optionally supplied by the user as input data. Initialization of the random number sequence uses this value to set *randSeedP* (in which case the run will be reproducible), or, if zero, a (difficult to reproduce) value based on the system clock is chosen,

```
void InitRand (int randSeedI)
{
  struct timeval tv;

  if (randSeedI != 0) randSeedP = randSeedI;                       5
  else {
    gettimeofday (&tv, 0);
    randSeedP = tv.tv_usec;
  }
}                                                                  10
```

Random vectors are also required. The following function produces unit vectors in three dimensions with uniformly distributed random orientation, using a rejection method [mar72].

```
void VRand (VecR *p)
{
  real s, x, y;

  s = 2.;                                                         5
  while (s > 1.) {
    x = 2. * RandR () - 1.;
    y = 2. * RandR () - 1.;
    s = Sqr (x) + Sqr (y);
  }                                                               10
  p->z = 1. - 2. * s;
  s = 2. * sqrt (1. - s);
  p->x = s * x;
  p->y = s * y;
}                                                                 15
```

The average success rate at each attempt is $\pi/4$ (76%). If the vectors are two-dimensional use

```
void VRand (VecR *p)
{
  real s;

  s = 2. * M_PI * RandR ();                              5
  p->x = cos (s);
  p->y = sin (s);
}
```

Other functions

A sorting function is required (§4.4). This version shows the Heapsort method [knu73]; the data are not rearranged, but the correctly ordered (ascending) indices are returned in *seq*.

```
void Sort (real *a, int *seq, int n)
{
  real q;
  int i, ir, ixt, j, k;
                                                          5
  for (j = 0; j < n; j ++) seq[j] = j;
  if (n > 1) {
    k = n / 2;
    ir = n - 1;
    while (1) {                                          10
      if (k > 0) {
        -- k;
        ixt = seq[k];
        q = a[ixt];
      } else {                                           15
        ixt = seq[ir];
        q = a[ixt];
        seq[ir] = seq[0];
        -- ir;
        if (ir == 0) {                                   20
          seq[0] = ixt;
          break;
        }
      }
      i = k;                                             25
      j = 2 * k + 1;
      while (j <= ir) {
        if (j < ir && a[seq[j]] < a[seq[j + 1]]) ++ j;
        if (q < a[seq[j]]) {
          seq[i] = seq[j];                               30
```

```
        i = j;
        j = 2 * j + 1;
      } else j = ir + 1;
    }
    seq[i] = ixt;                                                    35
  }
 }
}
```

Error reporting is done with a minimum of fuss and errors will generally just terminate the job:

```
void ErrExit (int code)
{
  printf ("Error: %s\n", errorMsg[code]);
  exit (0);
}                                                                     5
```

where the error codes and messages used by the programs are

```
enum {ERR_NONE, ERR_BOND_SNAPPED, ERR_CHECKPT_READ, ERR_CHECKPT_WRITE,
   ERR_COPY_BUFF_FULL, ERR_EMPTY_EVPOOL, ERR_MSG_BUFF_FULL,
   ERR_OUTSIDE_REGION, ERR_SNAP_READ, ERR_SNAP_WRITE,
   ERR_SUBDIV_UNFIN, ERR_TOO_MANY_CELLS, ERR_TOO_MANY_COPIES,
   ERR_TOO_MANY_LAYERS, ERR_TOO_MANY_LEVELS, ERR_TOO_MANY_MOLS,        5
   ERR_TOO_MANY_MOVES, ERR_TOO_MANY_NEBRS, ERR_TOO_MANY_REPLICAS};

char *errorMsg[] = {"", "bond snapped", "read checkpoint data",
   "write checkpoint data", "copy buffer full", "empty event pool",
   "message buffer full", "outside region", "read snap data",          10
   "write snap data", "subdivision unfinished", "too many cells",
   "too many copied mols", "too many layers", "too many levels",
   "too many mols", "too many moved mols", "too many neighbors",
   "too many replicas"};
```

The variable *moreCycles* determines when a run should terminate. Orderly termination can occur as a result of the number of timesteps reaching a limit, the execution time exceeding some preset value, or a decision by the user. It is possible (in some environments) to send a signal to the program manually telling it to stop (after writing its checkpoint file); in Unix, making provision for such a signal[†] requires inserting the call

```
SetupInterrupt ();
```

[†] There is a user command *kill* for sending any desired signal (here the signal used is *SIGUSR1*) to an executing job.

near the start of `main` and adding the functions

```
void ProcInterrupt ()
{
  moreCycles = 0;
}

void SetupInterrupt ()
{
  signal (SIGUSR1, ProcInterrupt);
}
```

18.5 Organizing input data

Very little attention has been paid to the subject of input data for the programs, beyond simply listing the necessary data items. The reason we were able to do this is because all that a program requires is the file containing a list of variable names and their input values; the processing, completeness checks and even an annotated printout of the values are all handled transparently. While such a service is standard in Fortran – the 'namelist' feature – there is no corresponding facility in C. We have therefore had to roll our own; here it is for the curious, starting with the data structure and macro definitions that must be included in any program using the feature.

```
typedef enum {N_I, N_R} VType;

typedef struct {
  char *vName;
  void *vPtr;
  VType vType;
  int vLen, vStatus;
} NameList;

#define NameI(x)   {#x, &x, N_I, sizeof (x) / sizeof (int)}
#define NameR(x)   {#x, &x, N_R, sizeof (x) / sizeof (real)}
```

The way the `NameList` structure is used is typically

```
NameList nameList[] = {
  NameI (intVariable),
  NameR (realVector),
};
```

and the data input function `GetNameList` that is called from `main` employs this information to process the input data file; the return code can be used to check for

errors. The name of the input file is taken from the name of the program and the
extension '.`in`' is appended[†].

```c
#define NP_I   ((int *)   (nameList[k].vPtr) + j)
#define NP_R   ((real *)  (nameList[k].vPtr) + j)

int GetNameList (int argc, char **argv)
{                                                               5
  int id, j, k, match, ok;
  char buff[80], *token;
  FILE *fp;

  strcpy (buff, argv[0]);                                       10
  strcat (buff, ".in");
  if ((fp = fopen (buff, "r")) == 0) return (0);
  for (k = 0; k < sizeof (nameList) / sizeof (NameList); k ++)
    nameList[k].vStatus = 0;
  ok = 1;                                                       15
  while (1) {
    fgets (buff, 80, fp);
    if (feof (fp)) break;
    token = strtok (buff, " \t\n");
    if (! token) break;                                         20
    match = 0;
    for (k = 0; k < sizeof (nameList) / sizeof (NameList); k ++) {
      if (strcmp (token, nameList[k].vName) == 0) {
        match = 1;
        if (nameList[k].vStatus == 0) {                         25
          nameList[k].vStatus = 1;
          for (j = 0; j < nameList[k].vLen; j ++) {
            token = strtok (NULL, ", \t\n");
            if (token) {
              switch (nameList[k].vType) {                      30
                case N_I:
                  *NP_I = atol (token);
                  break;
                case N_R:
                  *NP_R = atof (token);                         35
                  break;
              }
            } else {
              nameList[k].vStatus = 2;
              ok = 0;                                           40
            }
          }
          token = strtok (NULL, ", \t\n");
          if (token) {
            nameList[k].vStatus = 3;                            45
```

---

† Several standard file and character-string functions from the C library are used here; details are in the appropriate documentation.

```
                ok = 0;
              }
              break;
          } else {
              nameList[k].vStatus = 4;                                    50
              ok = 0;
          }
        }
      }
    }
    if (! match) ok = 0;                                                  55
  }
  fclose (fp);
  for (k = 0; k < sizeof (nameList) / sizeof (NameList); k ++) {
    if (nameList[k].vStatus != 1) ok = 0;
  }                                                                       60
  return (ok);
}
```

The function for printing an annotated record of the input data is

```
void PrintNameList (FILE *fp)
{
  int j, k;

  fprintf (fp, "NameList -- data\n");                                     5
  for (k = 0; k < sizeof (nameList) / sizeof (NameList); k ++) {
    fprintf (fp, "%s\t", nameList[k].vName);
    if (strlen (nameList[k].vName) < 8) fprintf (fp, "\t");
    if (nameList[k].vStatus > 0) {
      for (j = 0; j < nameList[k].vLen; j ++) {                           10
        switch (nameList[k].vType) {
          case N_I:
            fprintf (fp, "%d ", *NP_I);
            break;
          case N_R:                                                      15
            fprintf (fp, "%#g ", *NP_R);
            break;
        }
      }
    }                                                                    20
    switch (nameList[k].vStatus) {
      case 0:
        fprintf (fp, "** no data");
        break;
      case 1:                                                            25
        break;
      case 2:
        fprintf (fp, "** missing data");
        break;
      case 3:                                                            30
        fprintf (fp, "** extra data");
```

```
        break;
      case 4:
        fprintf (fp, "** multiply defined");
        break;                                                        35
    }
    fprintf (fp, "\n");
  }
  fprintf (fp, "----\n");
}                                                                     40
```

The macro *NameVal* is used by analysis programs (an example is shown in §5.4) to locate individual data items at the start of the output file written by *PrintNameList*,

```
#define NameVal(x)                                        \
  if (! strncmp (bp, #x, strlen (#x))) {                  \
    bp += strlen (#x);                                    \
    x = strtod (bp, &bp);                                 \
  }                                                            5
```

The *ValList* structure used in §17.4 is a simplified version of *NameList*. The structure and its associated macros are

```
typedef struct {
  void *vPtr;
  VType vType;
  int vLen;
} ValList;                                                       5

#define ValI(x)   {&x, N_I, sizeof (x) / sizeof (int)}
#define ValR(x)   {&x, N_R, sizeof (x) / sizeof (real)}
```

18.6 Configuration snapshot files

Snapshots of the system configuration are stored on disk by the following function; the details will vary, depending on the problem.

```
#define SCALE_FAC 32767.

void PutConfig ()
{
  VecR w;                                                          5
  int blockSize, fOk, n;
  short *rI;
  FILE *fp;

  fOk = 1;                                                        10
  blockSize = (NDIM + 1) * sizeof (real) + 3 * sizeof (int) +
```

```
          nMol * NDIM * sizeof (short);
   if ((fp = fopen (fileName[FL_SNAP], "a")) != 0) {
     WriteF (blockSize);
     WriteF (nMol);                                              15
     WriteF (region);
     WriteF (stepCount);
     WriteF (timeNow);
     AllocMem (rI, NDIM * nMol, short);
     DO_MOL {                                                    20
       VDiv (w, mol[n].r, region);
       VAddCon (w, w, 0.5);
       VScale (w, SCALE_FAC);
       VToLin (rI, NDIM * n, w);
     }                                                           25
     WriteFN (rI, NDIM * nMol);
     free (rI);
     if (ferror (fp)) fOk = 0;
     fclose (fp);
   } else fOk = 0;                                               30
   if (! fOk) ErrExit (ERR_SNAP_WRITE);
}
```

Note the use of binary data as opposed to human-readable text; this reduces
the storage requirements substantially although, when in this form, data files may
not be readily transportable between different kinds of computers and operating
systems. To reduce storage further, the coordinate data are scaled to fit into a two-
byte word, rather than the eight (or four) bytes used for floating-point variables
in the MD code; the consequent loss of precision is acceptable for the kinds of
analysis that we carry out on the data. The variable *blockSize* is set to the total
number of bytes written per snapshot, allowing more flexibility in the applications
that process the data later. The name of the output file is stored as one of the entries
in the character array *fileName* and the function *SetupFiles* must be called at
the start of the job to handle file related matters – details appear later.

Reading the data into an analysis program uses the following function; the variable *blockNum* indicates which set of configuration data from the file is to be read,
in effect allowing random access.

```
int GetConfig ()
{
  VecR w;
  int fOk, n;
  short *rI;                                                     5

  fOk = 1;
  if (blockNum == -1) {
    if ((fp = fopen (fileName[FL_SNAP], "r")) == 0) fOk = 0;
  } else {                                                       10
```

```
      fseek (fp, blockNum * blockSize, 0);
      ++ blockNum;
    }
    if (fOk) {
      ReadF (blockSize);                                    15
      if (feof (fp)) return (0);
      ReadF (nMol);
      ReadF (region);
      ReadF (stepCount);
      ReadF (timeNow);                                      20
      if (blockNum == -1) {
        SetCellSize ();
        AllocArrays ();
        blockNum = 1;
      }                                                     25
      AllocMem (rI, NDIM * nMol, short);
      ReadFN (rI, NDIM * nMol);
      DO_MOL {
        VFromLin (w, rI, NDIM * n);
        VScale (w, 1. / SCALE_FAC);                         30
        VAddCon (w, w, -0.5);
        VMul (mol[n].r, w, region);
      }
      free (rI);
      if (ferror (fp)) fOk = 0;                             35
    }
    if (! fOk) ErrExit (ERR_SNAP_READ);
    return (1);
  }
```

The first time *GetConfig* is called, *blockNum* should have the value –1; during
the initial call *AllocArrays* is used to allocate the necessary storage and, if cells
are needed in the analysis, *SetCellSize* must be called just prior to this.

18.7 Managing extensive computations

Computations short enough to run to completion without fear of interruption are
often inadequate for serious work. A production program may need to be equipped
with the means to save its present state on disk and be able to resume computation
from this saved state – the checkpoint/restart mechanism. Since there is no generally available procedure for this task, responsibility falls on the user. Here we
provide a demonstration[♠] of how this can be accomplished.

Everything that the program is unable to reconstruct quickly from available information must be included in the checkpoint file. We will assume that the program
has access to the original input data, so that this need not be included (a minor

----

♠ pr_18_1

detail). For added security, two copies of the checkpoint file will be maintained and they will be updated alternately; thus if the job aborts while writing the file (the usual reason being a lack of file space) the previous copy will still be available. The newer version can be identified in various ways, one of which is to have yet another file just for this purpose (if file modification times are accessible they could be used instead).

The file used for recording either grid averages or sets of atomic configurations (snapshots) for later analysis or graphics work is also introduced at this point. Here is a sample set of files that might be used; the *xxnn* prefix in each file name is replaced by the two-letter program mnemonic *progId* and a two-digit serial number *runId* identifying the run[†].

```
enum {FL_CHECKA, FL_CHECKB, FL_CKLAST, FL_SNAP};
char *fileNameR[] = {"xxnnchecka.data", "xxnncheckb.data",
  "xxnncklast.data", "xxnnsnap.data"}, fileName[5][20];
char *progId = "md";
int runId;                                                     5
```

All necessary files are created at the beginning of what is generally a series of continued runs by the following function (*CHAR_ZERO* denotes the character '0'). The variable *doCheckpoint* indicates whether checkpointing is activated, *newRun* is used to determine whether this is the first run of the series and another input variable *recordSnap* specifies whether configuration snapshot files are required; alternatively, the unwanted parts of this function could be omitted.

```
void SetupFiles ()
{
  FILE *fp;
  int k;

  for (k = 0; k < sizeof (fileNameR) / sizeof (fileNameR[0]); k ++) {   5
    strcpy (fileName[k], fileNameR[k]);
    fileName[k][0] = progId[0];
    fileName[k][1] = progId[1];
    fileName[k][2] = runId / 10 + CHAR_ZERO;                            10
    fileName[k][3] = runId % 10 + CHAR_ZERO;
  }
  if (! doCheckpoint) {
    newRun = 1;
  } else if ((fp = fopen (fileName[FL_CKLAST], "r")) != 0) {            15
    newRun = 0;
    fclose (fp);
  } else {
    newRun = 1;
```

---

† Note the distinction between *fileNameR*, which is usually represented by read-only character strings, and *fileName*, into which these strings are copied and then modified.

```
    fp = fopen (fileName[FL_CHECKA], "w");                        20
    fclose (fp);
    fp = fopen (fileName[FL_CHECKB], "w");
    fclose (fp);
    fp = fopen (fileName[FL_CKLAST], "w");
    fputc (CHAR_ZERO + FL_CHECKA, fp);                           25
    fclose (fp);
  }
  if (newRun && recordSnap) {
    fp = fopen (fileName[FL_SNAP], "w");
    fclose (fp);                                                 30
  }
}
```

A simpler form of this function, for use in programs that read configuration files
(by calls to *GetConfig*), is

```
void SetupFiles ()
{
  strcpy (fileName[FL_SNAP], fileNameR[FL_SNAP]);
  fileName[FL_SNAP][0] = progId[0];
  fileName[FL_SNAP][1] = progId[1];                               5
  fileName[FL_SNAP][2] = runId / 10 + CHAR_ZERO;
  fileName[FL_SNAP][3] = runId % 10 + CHAR_ZERO;
}
```

The functions used to access the checkpoint files follow; the data are stored on
disk in binary form. The details of the actual data to be read or written are specific
to the problem; here, as an example, the data for a basic soft-sphere simulation
using the leapfrog method are recorded; acceleration values are included (they are
part of the *mol* structures) because of the two-step nature of the leapfrog method.
Two copies of the checkpoint file are kept and they are written alternately.

```
void PutCheckpoint ()
{
  int fOk, fVal;
  FILE *fp;
                                                                 5
  fOk = 0;
  if ((fp = fopen (fileName[FL_CKLAST], "r+")) != 0 ) {
    fVal = FL_CHECKA + FL_CHECKB - (fgetc (fp) - CHAR_ZERO);
    rewind (fp);
    fputc (CHAR_ZERO + fVal, fp);                               10
    fclose (fp);
    fOk = 1;
  }
  if (fOk && (fp = fopen (fileName[fVal], "w")) != 0) {
    WriteF (kinEnergy);                                         15
```

```
      WriteF (stepCount);
      WriteF (timeNow);
      WriteF (totEnergy);
      WriteFN (mol, nMol);
      if (ferror (fp)) fOk = 0;
      fclose (fp);
    } else fOk = 0;
    if (! fOk) ErrExit (ERR_CHECKPT_WRITE);
}

void GetCheckpoint ()
{
    int fOk, fVal;
    FILE *fp;

    fOk = 0;
    if ((fp = fopen (fileName[FL_CKLAST], "r")) != 0) {
      fVal = fgetc (fp) - CHAR_ZERO;
      fclose (fp);
      fOk = 1;
    }
    if (fOk && (fp = fopen (fileName[fVal], "r")) != 0) {
      ReadF (kinEnergy);
      ReadF (stepCount);
      ReadF (timeNow);
      ReadF (totEnergy);
      ReadFN (mol, nMol);
      if (ferror (fp)) fOk = 0;
      fclose (fp);
    } else fOk = 0;
    if (! fOk) ErrExit (ERR_CHECKPT_READ);
}
```

To use these capabilities add the following to `main`,

```
while (moreCycles) {
   ...
   if (doCheckpoint && stepCount % stepCheckpoint == 0)
      PutCheckpoint ();
}
if (doCheckpoint) PutCheckpoint ();
```

and *SetupJob* must include

```
SetupFiles ();
if (newRun) {
  printf ("new run\n");
  InitCoords ();
   ...
   if (doCheckpoint) {
```

```
        PutCheckpoint ();
        PutCheckpoint ();
    }
  } else {
    printf ("continued run\n");
    GetCheckpoint ();
  }
```
10

18.8 Header files

The simulation programs all share a common header file *mddefs.h* that is included during compilation. This file contains the definitions appearing in this chapter as well as all the function prototypes (these are essential for functions returning non-integer results and optional – though recommended – in other cases). It also ensures that several standard C header files – the actual list may vary slightly depending on the operating system – are included and defines the floating-point type. The file typically begins with

```
#include <stdlib.h>
#include <stdio.h>
#include <math.h>
#include <string.h>
#include <signal.h>
#include <sys/time.h>

typedef double real;
```
5

19

The future

19.1 Role of simulation

Computer simulation in general, and molecular dynamics in particular, represent a new scientific methodology. Instead of adopting the traditional theoretical practice of constructing layer upon layer of assumption and approximation, this modern alternative attacks the original problem in all its detail. Unfortunately, phenomena that are primarily quantum mechanical in nature still present conceptual and technical obstacles, but, insofar as classical problems are concerned, the simulational approach is advancing as rapidly as computer technology permits. For this class of problem, the limits of what can be achieved remain well beyond the horizon.

Theoretical breakthroughs involve both new concepts and the mathematical tools with which to develop them. Most of the major theoretical advances of the just-finished twentieth century rest upon mathematical foundations developed during the preceding century, if not earlier. Whether still undeveloped mathematical tools and new concepts will ever replace the information presently only obtainable by computer simulation, or whether the simulation is the solution, is something only the future can tell. Whether computer modeling will become an integral part of theoretical science, or whether it will continue to exist independently, is also a big unknown. After all, theory, as we know it, has not been around for very long.

To what extent can simulation replace experiment? In the more general sense, this is already happening in engineering fields, where models are routinely constructed from well-established foundations. Many MD applications are still at the stage of attempting to solve the inverse problem, working backwards from experimental data to elucidate the invisible microscopic details; this is not done directly, however, but over the course of time the models gradually evolve until they are capable of yielding quantitatively correct behavior. To use MD as a predictive tool demands a high degree of confidence in the model and the methodology; this implies a thorough understanding of the effects of finite system size, limited timescales, numerical integration and the consequences of using approximate (classical) potential functions.

The systems studied by MD are normally many orders of magnitude smaller than the corresponding systems in nature. With increasing computer speed and memory this gap can be reduced, but never eliminated – only in two dimensions are the largest MD calculations similar in size to the mesoscopic systems studied experimentally. The same is true for timescales as well. The great success of the MD approach is due to the fact that these two limitations are, in many instances, irrelevant: the phenomena MD is used to study can often be exhaustively explored despite these limitations, because the lengths and times involved fall within the regime that MD can handle.

While one of the aims of MD is reproducing experiment, it also offers the opportunity for probing behavior at a much more detailed level than is obtainable by thermodynamic, mechanical, or spectroscopic techniques. While such 'numerical' measurements cannot be confirmed in the laboratory in a direct fashion, they sometimes have implications that can be tested experimentally; if these predictions are substantiated they can lead to improved understanding of the systems under study. The wealth of detail potentially available is what makes MD such a useful and important method, the question only being what data to examine and how to convert them into manageable form. Complete trajectories represent one extreme, thermodynamic averages the other.

19.2 Limits of growth

The MD field is still comparatively young; it has grown but remains tied to the advance of computer technology. Remarkable results have been obtained from what could be thought of as extremely small systems: that a few hundred to a few thousand model atoms not only permit studies of structure and dynamics, but also yield quantitative results in good agreement with experiment, is now a familiar fact of life. Of course life is not always so idyllic, and there are phenomena requiring length and time scales that exceed the capabilities of even the most powerful of computers. But if one is permitted to extrapolate from past rates of computer performance growth, the severity of these limitations should gradually diminish.

MD simulation has been a direct beneficiary of the rapid growth in computing power and the even greater improvement in the cost–performance ratio. Efficient compilers can help to a certain extent, but there is no substitute for a well-tuned algorithm, especially as the system size increases. Even when the performance of individual processors stops growing at the present rate and distributed computing becomes unavoidable, the nature of MD, with its computations often based on highly localized information, makes the distributed environment ideal for large-scale problem solving.

It is important to be aware of the way MD computations scale in order to appreciate the kinds of problems that might be approachable in the foreseeable future. The amount of computation grows at least linearly with the number of particles and so too will the processing time (for a fixed number of processing nodes in one's parallel computer). But this is not the whole story, because the time over which a simulated system must be observed in order to examine a particular class of phenomenon can also increase with system size. Propagating disturbances such as sound waves cross a system in a time proportional to its linear size L, but any process governed by diffusion requires a time of order L^2. Processes involving, for example, large polymer molecules, occur on timescales that are truly macroscopic, representing an extreme situation beyond the capability of any (presently) conceivable computer. Thus the prognosis is mixed.

19.3 Visualization and interactivity

Computation is no longer merely number crunching. Along with ready access to high performance computing comes the ability to observe the system being simulated. The MD practitioner need no longer be content with graphs of P plotted as a function of T, or some correlation as a function of time, but is now able to observe a system as it freezes, allowing the eye to capture some of the more subtle cooperative effects as the molecules reorganize. The systems that can be examined in this fashion are already of a size that, only a few years ago, were considered clients for the supercomputers of the day. Given that the human eye is without peer for many kinds of information reduction, it is obvious that the visual approach is an important one – the computational equivalent of the optical microscope.

Visualization takes many forms. In representing the results of discrete-particle simulations one has the choice of directly observing the particles themselves, or a display, typically involving scalar or vector fields, of suitably averaged quantities such as velocity, vorticity, temperature, concentration and stress, to name but a few. Data can be represented by means of arrow, contour and surface plots, as well as in less conventional, but visually rich forms, including the extensive use of color and animation. A number of examples (regrettably without the color) appeared at various points in the book.

Once it becomes possible to observe the system as it develops, the next step is to introduce a certain amount of interactivity into the simulation by allowing the user to control both the parameters of the simulation and the way the results are displayed. Realtime visualization can prove invaluable while developing models and computational techniques, in debugging, demonstration, comparison of models and selecting parameters. There are few general rules for adding an interactive capability to a simulation since the approach is at least partly determined by the software environment; the one critical requirement is responsiveness – if the user fails to receive rapid feedback interactivity is of little value. An example of a simple user interface of this kind was shown in Figure 2.6.

19.4 Coda

In retrospect, one can do no better than to borrow from Anatole France: *'Si les plats que je vous offre sont mal préparés, c'est moins la faute de mon cuisinier que celle de la chimie, qui est encore dans l'enfance.'* Quoted apologetically in a text on quantum mechanics [mes64], it is no less appropriate for this volume of recipes, despite the entirely classical foundations.

Appendix

List of variables

An alphabetical listing of the global variables used in the MD programs follows; variables used in the separate analysis programs are omitted (as are local variables).

The first part of the listing shows the elements that can belong to the various structures used in the programs (excluding those listed in §18.2); in some cases, especially `Mol`, only subsets of these elements are required.

`Mol` – atom or molecule:

`chg`	charge
`diam`	σ_i
`en`	e_i
`fixed`	fixed atom
`id`	atom label
`inCell`	cell occupied by atom
`inChain`	part of chain
`inClust`	cluster membership
`inObj`	part of object
`logRho`	$\log \rho_i$
`next`	link between atom data
`nBond`	number of bonds formed
`q,qv,qa,...`	quaternion components, derivatives, etc.
`rCol`	coordinates of latest collision
`rf`	$\sum_j r_{ij\,x} f_{ij\,y}$
`rMatT`	rotation matrix
`r,rv,ra,...`	coordinates, velocities, etc.
`s,sv,sa,...`	s_i for linear molecule, derivatives, etc.
`time`	last coordinate update
`torq`	n_i
`typeA`	atom type
`u`	interaction energy
`wv,wa`	ω_i, $\dot{\omega}_i$ (vector or scalar)

Site – site in molecule:

    f   force
    r   coordinates

MSite – site in molecule (reference state):

    r   site coordinates
    typeF   site type
    typeRdf   site type for RDF

Cons – constraint:

    bLenSq $r^2_{i_k j_k}$
    distSq $b^2_{i_k j_k}$
    site1/2   atoms i_k and j_k
    vec   constraint vector s_k

TBuf – buffer for time-dependent measurement:

    acf...   autocorrelation values
    count   number of measurements in set
    ddDiffuse   squared dipole displacements
    org...   origin values
    rrDiffuse   squared displacements
    rTrue   'true' coordinates

Clust – cluster:

    head   first in cluster
    next   next in cluster
    size   cluster size

Poly – linked chain:

    L   link array
    nLink   number of links
    ra,wa   acceleration of site zero

Link – link in chain:

    bV,cV,hV $b_k, c_k, \hat{h}_k$
    fV,gV,xV,yV F^e_k, G_k, X_k, Y_k
    inertiaM $\mathcal{I}_k$
    mass m_k
    omega,omegah ω_k
    r,rv   site coordinates, velocity
    rMatT R^T_k
    s,sv,svh,sa,... θ_k and derivatives
    torq n_k

TLevel – tree-code subdivision level:

    cCur,cLast   cell indices used in scanning tree
    fCell,lCell   cell range
    coordSum   coordinate sum

`TCell` – tree-code cell:

`cm,midPt`	center of charge, midpoint
`atomPtr,subPtr`	pointers to atoms, cells
`nOcc,nSub`	numbers of atoms, occupied descendants

`MpCell` – multipole cell:

`le,me`	multipole and local expansions
`occ`	cell occupied

`MpTerms` – multipole or local expansion:

`c,s`	sets of coefficients

`EvTree` – node in event tree:

`circAL/R,circBL/R`	circular list pointers
`idA/B`	event identifiers
`left,right,up`	tree pointers
`time`	event time

Global variables are as follows:

`alpha`	α
`avAcf...`	autocorrelation averages
`basePos/Vel`	base position, velocity
`bbDistSq`	average bond length squared
`bdyOffset`	array offset
`bdySlide`	offset across sliding boundary
`bdyStripWidth`	size of flow adjustment region
`blockNum/Size`	used with snapshot files
`bondAng/Len`	bond angle, length
`bondLim`	limit of bond stretch
`boundPairEng`	energy threshold
`cellList`	pointers used by cell method
`cellRange`	cells examined after collision
`cells`	cell array size
`cellWid`	cell width
`chainHead`	monomers in chain head
`chainLen`	monomers in chain
`chargeMag`	charge magnitude
`coll/crossCount`	event counters
`consDevA/L`	constraint deviations
`consPrec`	constraint accuracy
`consVec`	used in solving constraint equations
`count...`	measurement counters
`cumRdf`	cumulative RDF
`curCellsEdge`	cells per region edge
`curLevel`	level in cell hierarchy
`curPhase`	vibration phase
`ddDiffuseAv`	angular diffusion mean
`deltaT`	Δt
`density`	ρ

dihedAngCorr/Org	used in dihedral angle correlations
dilateRate/1/2	current and previous γ
dipoleInt	μ^2
dipoleOrder	$\langle M \rangle$ (and average)
diskInitPos/Vel	initial projectile position, speed
dispHi	accumulated maximum displacement
distFac	minimum distance criterion
dvirSum1/2	used in computing *dilateRate*
eeDistSq	$\langle R^2 \rangle$
embedWt	χ
enTransSum	accumulated kinetic energy transfer
errCode	error code
errorMsg	error messages
eventCount	event counter
eventMult	used to specify *poolSize*
evIdA/B	event identifiers
extPressure	external P
farSiteDist	distance to furthest site in molecule
fileName	file names
flowSpeed	nominal flow speed
fricDyn/Stat	friction coefficients
fSpaceLimit	n_c
g1Sum/g2Sum	G_1, G_2
gMomRatio1/2	$\langle g_2/g_1 \rangle$, $\langle g_3/g_1 \rangle$
gravField	force driving flow, gravity
heatForce	fictitious force
helixOrder	order parameter S
helixPeriod	atoms in single helix turn
hFunction	H-function
histGrid	accumulated coarse-grained measurements
hist...	histograms used in measuring distributions
inertiaK	κ
initSep	initial spacing
initUcell/chain	size of unit cell array for initial state
inside	identifies boundary and interior cells
intAcf...	integrated autocorrelation function
interval...	measurement interval
intType	interaction type
kinEnergy	E_K (and average)
kinEnInitSum	accumulated E_K for setting initial T
kinEnVal	current E_K
latticeCorr	$|\rho(\boldsymbol{k})|$
layerMol	storage for layer contents
limit...	upper limit of counter
lMat	constraint matrix $\boldsymbol{L}$
massS/V	M_s, M_v
maxCells	maximum number of cells
maxEdgeCells	maximum size of cell array
maxLevel	levels in cell hierarchy
maxOrd	multipole expansion order

`max/minPairEng`	range of pair energies
`mInert`	I, or I_x, etc.
`mMat`	constraint matrix $\boldsymbol{M}$
`molId`	atom labels
`molPtr`	atom pointers
`moreCycles`	controls program execution
`mpCells`	size of cell array
`mpCellList`	pointers used with cells
`nBaseCycle`	base cycle count
`nBuff...`	number of data collection buffers
`nCellEdge`	number of cells in each direction
`nChain`	number of chains
`nCons`	number of constraints
`nCycleR/V`	constraint iteration count
`nDihedAng`	number of dihedral angles
`nDof`	number of degrees of freedom
`nebrNow`	neighbor-list refresh due
`nebrTab`	storage for neighbor list
`nebrTabFac`	used to determine `nebrTabMax`
`nebrTabLen`	neighbor-list length
`nebrTabMax`	maximum neighbor-list length
`nebrTabPtr`	pointers used with neighbor list
`next...Time`	time of next measurement event
`nFixed/FreeMol`	numbers of fixed and mobile atoms
`nFunCorr`	number of k values used for correlation functions
`nLayerMax`	maximum number of layers
`nMol`	N_m
`nMolCopy`	number of copied atoms
`nMolDisk/Wall`	object sizes
`nMolMax`	maximum number of atoms
`nMolMe`	actual number of atoms in processor
`nMolMeMax`	maximum number of atoms per processor
`nMolRep`	number of replica atoms
`nOut`	number of atoms requiring transfer
`nPressCycle`	pressure correction cycle count
`nProc`	number of processors
`nSite`	total number of interaction sites
`nThread`	number of threads
`nVal...`	number of values in measurement
`obsPos/Size`	obstacle position and size
`pertTrajDev`	velocity perturbation
`poolSize`	size of event pool
`pressure`	P (and average)
`procArrayMe`	location in parallel processor array
`procArraySize`	size of parallel processor array
`procMe`	processor identity
`procNebrHi/Lo`	neighbor processors
`profileT/V`	T, v profiles
`progId`	program identifier
`pTensorXZ`	P_{xz}

`pThread`	used in thread processing
`radGyrSq`	$\langle S^2 \rangle$
`raL/uL`	acceleration and energy for each atom
`randSeed`	random number seed
`range...`	upper limit of distribution measurement
`rCut/A`	r_c
`region`	region size
`regionVol`	region volume
`rNebrShell`	r_n
`roughWid`	wall corrugation width
`rrDiffuseAv`	diffusion mean
`rSwitch`	r_s
`runId`	job identifier
`shearRate`	γ
`shearVisc`	η (and average)
`sitesMol`	number of interaction sites in molecule
`sizeHistGrid`	size of `histGrid`
`sizeHist...`	size of `hist...`
`siteSep`	tetrahedral molecule size
`snapNumber`	serial number of snapshot data
`solConc`	solvent concentration
`splineA2/3`	a_2, a_3
`stepCount`	timestep counter
`stepLimit`	total run length
`step...`	timesteps between measurements or other activities
`streamFun`	stream function evaluated on grid
`subRegionHi/Lo`	subregion limits
`tCos/tSin`	tabulated trigonometric functions
`temperature`	T
`tempFinal/Init`	temperature ranges
`tempReduceFac`	temperature factor
`thermalCond`	λ (and average)
`thermalWall`	specifies wall attached to heat bath
`timeNow`	current time
`tolPressure`	tolerance for pressure correction
`totEnergy`	E (and average)
`totEnVal`	current E
`trBuff`	buffer for interprocessor transfers
`trBuffMax`	size of `trBuff`
`trPtr`	pointers to atoms requiring transfer
`twistAng`	θ'
`uCon`	torsion parameter
`uSum`	total interaction energy
`valST`	used for space–time correlation measurements
`valTrajDev`	trajectory deviation
`varL,...`	region size variables
`varS,...`	temperature feedback variables
`varV,...`	volume feedback variables
`velMag`	initial velocity value
`vibAmp/Freq`	vibration amplitude, frequency

`virSum`	virial sum
`vSum`	velocity sum
`vv...Sum`	velocity-squared sums
`wallTemp/Hi/Lo`	thermal wall temperatures
`wellSep`	n_w

Case study software

The programs in the software package available for use with the book are listed below. The simulation programs are labeled numerically, corresponding to the chapters in which they are introduced. Programs used to analyze results (and do other things) are also listed here; the more complex of these programs are described in the book, while other, simpler programs are included for completeness but are not discussed.

2. Basic molecular dynamics

`pr_02_1`	all pairs, two dimensions
`pr_02_2`	velocity distribution

3. Simulating simple systems

`pr_03_1`	cells and leapfrog
`pr_03_2`	neighbor list and leapfrog
`pr_03_3`	cells and PC
`pr_03_4`	neighbor list and PC
`pr_03_5`	trajectory separation

4. Equilibrium properties of simple fluids

`pr_04_1`	thermodynamics, soft spheres
`pr_04_2`	thermodynamics, LJ
`pr_04_3`	RDF, soft spheres
`pr_04_4`	long-range order
`pr_04_5`	configuration snapshots
`pr_anblockavg`	block-averaged variance
`pr_anclust`	cluster analysis
`pr_anvorpol`	Voronoi polyhedra analysis

5. Dynamical properties of simple fluids

`pr_05_1`	diffusion
`pr_05_2`	velocity autocorrelation function
`pr_05_3`	transport coefficients
`pr_05_4`	space–time correlations
`pr_andiffus`	diffusion analysis
`pr_anspcor`	space–time correlation analysis
`pr_antransp`	transport coefficient analysis

6. Alternative ensembles

`pr_06_1`	feedback PT

pr_15_2	soft-disk obstructed flow
pr_angridflow	process flow snapshots

16. Granular dynamics

pr_16_1	two-dimensional vibrating layer
pr_16_2	three-dimensional vibrating layer

17. Algorithms for supercomputers

pr_17_1	distributed processing
pr_17_2	parallel processing using threads
pr_17_3	vector processing

18. More about software

pr_18_1	checkpoints

Using the software

Program compilation uses a command such as

```
gcc -O -o mdprog mdprog.c -lm
```

where *gcc* is the compile command for systems using the GNU C compiler, *-O* requests optimization, *-o mdprog* specifies the name of the executable program, *mdprog.c* is the program source file and *-lm* requests the C mathematical function library. Various levels of code optimization are available (such as *-O3*), although the details depend on the system[†].

The program source files and the corresponding input data files are to be found, respectively, in the directories *src* and *data* of the software distribution. The shell script *crun.sh*, shown below, combines the operations needed to compile and run any of the simulations. For example, typing

```
crun.sh 01 1
```

will compile program *src/pr_01_1.c*, and then run it using data contained in the file *data/pr_01_1.in*. The output is displayed and is also saved as the file *pr_01_1.out*. The listing of *crun.sh* is[‡]

```
#! /bin/tcsh
set f=src/pr_$1_$2
set b=`basename $f .c`
```

[†] The case study of §17.4 requires special treatment not covered here, and §17.5 may require an additional library for the thread functions. The reader should note that not all compilers optimize equally well and – at least in the past – some compilers have been reported to cause problems if too high a level of optimization is requested.

[‡] This requires just a minimal acquaintance with Unix shell programming.

```
gcc -O -I./src -o $b src/$b.c -lm
cp data/$b.in $b.in
$b | tee $b.out
```

Multiple runs with minor variations in the input data are most readily handled using shell scripts. For example, if the aim is to cover a series of temperature values, the following script will run a program *pr_01_1* for each of the required values, leaving all the output in the file *pr_01_1.out*.

```
#! /bin/tcsh
foreach t (0.4 0.6 0.8 1.0)
  cat > pr_01_1.in << EOD
    deltaT       0.005
    ...
    stepLimit    10000
    temperature  $t
  EOD
  pr_01_1 >> pr_01_1.out
end
```

References

[abr68]   Abramowitz, M. and Stegun, I., *Handbook of Mathematical Functions*, Dover, New York, 1968.

[abr86]   Abraham, F. F., Computational statistical mechanics: Methodology, applications and supercomputing, *Adv. Phys.* **35** (1986) 1.

[abr89]   Abraham, F. F., Rudge, W. E., and Plischke, M., Molecular dynamics of tethered membranes, *Phys. Rev. Lett.* **62** (1989) 1757.

[ada76]   Adams, D. J. and McDonald, I., Thermodynamic and dielectric properties of polar lattices, *Mol. Phys.* **32** (1976) 931.

[ada80]   Adams, D. J., Periodic, truncated-octahedral boundary conditions, in Ceperley, D., ed., *The Problem of Long-Range Forces in the Computer Simulation of Condensed Media*, Lawrence Berkeley Lab. Rept. LBL-10634, 1980, p. 13.

[ald57]   Alder, B. J. and Wainwright, T. E., Phase transition for a hard sphere system, *J. Chem. Phys.* **27** (1957) 1208.

[ald58]   Alder, B. J. and Wainwright, T. E., Molecular dynamics by electronic computers, in Prigogine, I., ed., *Transport Processes in Statistical Mechanics*, Interscience Publishers, New York, 1958, p. 97.

[ald59]   Alder, B. J. and Wainwright, T. E., Studies in molecular dynamics. I. General method, *J. Chem. Phys.* **31** (1959) 459.

[ald62]   Alder, B. J. and Wainwright, T. E., Phase transition in elastic disks, *Phys. Rev.* **127** (1962) 359.

[ald67]   Alder, B. J. and Wainwright, T. E., Velocity autocorrelation for hard spheres, *Phys. Rev. Lett.* **18** (1967) 988.

[ald70a] Alder, B. J., Gass, D. M., and Wainwright, T. E., Studies in molecular dynamics. VIII. The transport coefficients for a hard-sphere fluid, *J. Chem. Phys.* **53** (1970) 3813.

[ald70b] Alder, B. J. and Wainwright, T. E., Decay of the velocity autocorrelation function, *Phys. Rev. A* **1** (1970) 18.

[all83]   Alley, W. E., Alder, B. J., and Yip, S., The neutron scattering function for hard spheres, *Phys. Rev. A* **27** (1983) 3174.

[all87]   Allen, M. P. and Tildesley, D. J., *Computer Simulation of Liquids*, Oxford University Press, Oxford, 1987.

519

[all89]   Allen, M. P., Frenkel, D., and Talbot, J., Molecular dynamics simulation using hard particles, *Comp. Phys. Repts.* **9** (1989) 301.

[all93a]   Allen, M. P. and Masters, A. J., Some notes on Einstein relationships, *Mol. Phys.* **79** (1993) 435.

[all93b]   Allen, M. P. and Tildesley, D. J., eds., *Computer Simulation in Chemical Physics*, Kluwer Academic Publishers, Dordrecht, 1993.

[and80]   Anderson, H. C., Molecular dynamics simulations at constant pressure and/or temperature, *J. Chem. Phys.* **72** (1980) 2384.

[and83]   Anderson, H. C., Rattle: A 'velocity' version of the Shake algorithm for molecular dynamics calculations, *J. Comp. Phys.* **52** (1983) 24.

[ash75]   Ashurst, W. T. and Hoover, W. G., Dense-fluid shear viscosity via nonequilibrium molecular dynamics, *Phys. Rev. A* **11** (1975) 658.

[bar71]   Barker, J. A., Fisher, R. A., and Watts, R. O., Liquid argon: Monte Carlo and molecular dynamics calculations, *Mol. Phys.* **21** (1971) 657.

[bar88]   Barrat, J.-L., Hansen, J. P., and Pastore, G., On the equilibrium structure of dense fluids: Triplet correlations, integral equations and freezing, *Mol. Phys.* **63** (1988) 747.

[bar94]   Barker, G. C., Computer simulation of granular materials, in Mehta, A., ed., *Granular Matter: An Interdisciplinary Approach*, Springer, New York, 1994, p. 35.

[bee66]   Beeler Jr, J. R., The techniques of high-speed computer experiments, in Meeron, E., ed., *Physics of Many-Particle Systems: Methods and Problems*, Gordon and Breach, New York, 1966, p. 1.

[bee76]   Beeman, D., Some multistep methods for use in molecular dynamics calculations, *J. Comp. Phys.* **20** (1976) 130.

[ber77]   Berne, B. J., Molecular dynamics of the rough sphere fluid. II. Kinetic models of partially sticky spheres, structured spheres, and rough screwballs, *J. Chem. Phys.* **66** (1977) 2821.

[ber86a]   Berendsen, H. J. C., Biological molecules and membranes, in [cic86a], p. 496.

[ber86b]   Berendsen, H. J. C. and van Gunsteren, W. F., Practical algorithms for dynamic simulations, in [cic86a], p. 43.

[ber86c]   Berne, B. J. and Thirumalai, D., On the simulation of quantum systems: Path integral methods, *Ann. Rev. Phys. Chem.* **37** (1986) 401.

[ber98]   Bertsch, R. A., Vaidehi, N., Chan, S. L., and Goddard III, W. A., Kinetic steps for α-helix formation, *Proteins* **33** (1998) 343.

[bil94]   Billeter, S. R., King, P. M., and van Gunsteren, W. F., Can the density maximum of water be found by computer simulation?, *J. Chem. Phys.* **100** (1994) 6692.

[bin92]   Binder, K., ed., *Monte Carlo Methods in Condensed Matter Physics*, Springer, Berlin, 1992.

[bin95]   Binder, K., ed., *Monte Carlo and Molecular Dynamics Simulations in Polymer Science*, Oxford University Press, Oxford, 1995.

[bir94]   Bird, G. A., *Molecular Gas Dynamics and the Direct Simulation of Gas Flows*, Oxford University Press, Oxford, 1994.

[biz98]   Bizon, C., Shattuck, M. D., Swift, J. B., McCormick, W. D., and Swinnney, H. L., Patterns in 3D vertically oscillated granular layers: Simulation and experiment, *Phys. Rev. Lett.* **80** (1998) 57.

[boo91]   Boon, J.-P. and Yip, S., *Molecular Hydrodynamics*, Dover, New York, 1991.

[bro78]   Brostow, W., Dussault, J.-P., and Fox, B. L., Construction of Voronoi polyhedra, *J. Comp. Phys.* **29** (1978) 81.

[bro84]   Brown, D. and Clark, J. H. R., A comparison of constant energy, constant temperature and constant pressure ensembles in molecular dynamics simulations of atomic liquids, *Mol. Phys.* **51** (1984) 1243.

[bro86]   Brode, S. and Ahlrichs, R., An optimized molecular dynamics program for the vector computer Cyber 205, *Comp. Phys. Comm.* **42** (1986) 51.

[bro88]   Brooks, III, C. L., Karplus, M., and Pettit, B. M., *Proteins: A Theoretical Perspective of Dynamics, Structure, and Thermodynamics*, Wiley, New York, 1988.

[bro90a] Brooks, III, C. L., Molecular simulations of protein structure, dynamics and thermodynamics, in [cat90], p. 289.

[bro90b] Brown, D. and Clark, J. H. R., A direct method of studying reaction rates by equilibrium molecular dynamics: Application to the kinetics of isomerization in liquid n-butane, *J. Chem. Phys.* **92** (1990) 3062.

[cap81]   Cape, J. N., Finney, J. L., and Woodcock, L. V., An analysis of crystallization by homogeneous nucleation in a 4000-atom soft-sphere model, *J. Chem. Phys.* **75** (1981) 2366.

[car50]   Carlson, B. C. and Rushbrooke, G. S., On the expansion of a Coulomb potential in spherical harmonics, *Proc. Camb. Phil. Soc.* **46** (1950) 626.

[cat90]   Catlow, C. R. A., Parker, S. C., and Allen, M. P., eds., *Computer Modeling of Fluids, Polymers and Solids*, Kluwer Academic Publishers, Dordrecht, 1990.

[cha97]   Challacombe, M., White, C., and Head-Gordon, M., Periodic boundary conditions and the fast multipole method, *J. Chem. Phys.* **107** (1997) 10131.

[cic82]   Ciccotti, G., Ferrario, M., and Ryckaert, J.-P., Molecular dynamics of rigid systems in cartesian coordinates: A general formulation, *Mol. Phys.* **47** (1982) 1253.

[cic86a] Ciccotti, G. and Hoover, W. G., eds., *Molecular Dynamics Simulation of Statistical Mechanical Systems*, North-Holland, Amsterdam, 1986.

[cic86b] Ciccotti, G. and Ryckaert, J.-P., Molecular dynamics simulation of rigid molecules, *Comp. Phys. Repts.* **4** (1986) 345.

[cic87]   Ciccotti, G., Frenkel, D., and McDonald, I. R., eds., *Simulation of Liquids and Solids. Molecular Dynamics and Monte Carlo Methods in Statistical Mechanics*, North-Holland, Amsterdam, 1987.

[cla90]   Clark, J. H. R., Molecular dynamics of chain molecules, in [cat90], p. 203.

[cor60]   Corben, H. C. and Stehle, P., *Classical Mechanics*, Wiley, New York, 2nd edition, 1960.

[cun79]   Cundall, P. A. and Strack, O. D. L., A discrete numerical model for granular assemblies, *Géotechnique* **29** (1979) 47.

[daw84]   Daw, M. S. and Baskes, M. I., Embedded-atom method: Derivation and application to impurities, surfaces, and other defects in metals, *Phys. Rev. B* **29** (1984) 6443.

[del51]   de Laplace P. S., *A Philosophical Essay on Probabilities*, (transl.) Dover, New York, 1951.

[del80]   de Leeuw, S. W., Perram, J. W., and Smith, E. R., Simulation of electrostatic systems in periodic boundary conditions. I. Lattice sums and dielectric constants, *Proc. R. Soc. Lond. A* **373** (1980) 27.

[del86]   de Leeuw, S. W., Perram, J. W., and Smith, E. R., Computer simulation of the static dielectric constant of systems with permanent electric dipoles, *Ann. Rev. Phys. Chem.* **37** (1986) 245.

[des88]   de Schepper, I. M., Cohen, E. G. D., Bruin, C., van Rijs, J. C., Montfrooij, W., and de Graaf, L. A., Hydrodynamic time correlation functions for a Lennard-Jones fluid, *Phys. Rev. A* **38** (1988) 271.

[doo91]   Doolen, G. D., ed., *Lattice Gas Methods for PDEs*, North-Holland, Amsterdam, 1991.

[dua98]   Duan, Y. and Kollman, P. A., Pathways to a protein folding intermediate observed in a 1-microsecond simulation in aqueous solution, *Science* **282** (1998) 740.

[dul97]   Dullweber, A., Leimkuhler, B., and McLachlan, R., Symplectic splitting methods for rigid body molecular dynamics, *J. Chem. Phys.* **107** (1997) 5840.

[dun92]   Dunn, J. H., Lambrakos, S. G., Moore, P. G., and Nagumo, M., An algorithm for calculating intramolecular angle-dependent forces on vector computers, *J. Comp. Phys.* **100** (1992) 17.

[dun93]   Dünweg, B. and Kremer, K., Molecular dynamics simulation of a polymer chain in solution, *J. Chem. Phys.* **99** (1993) 6983.

[edb86]   Edberg, R., Evans, D. J., and Morriss, G. P., Constrained molecular dynamics: Simulations of liquid alkanes with a new algorithm, *J. Chem. Phys.* **84** (1986) 6933.

[edb87]   Edberg, R., Morriss, G. P., and Evans, D. J., Rheology of n-alkanes by nonequilibrium molecular dynamics, *J. Chem. Phys.* **86** (1987) 4555.

[ein68]   Einwohner, T. and Alder, B. J., Molecular dynamics. VI. Free-path distributions and collision rates for hard-sphere and square-well molecules, *J. Chem. Phys.* **49** (1968) 1458.

[erm80]   Ermak, D. L. and Buckholz, H., Numerical integration of the Langevin equation: Monte Carlo simulation, *J. Comp. Phys.* **35** (1980) 169.

[erp77]   Erpenbeck, J. J. and Wood, W. W., Molecular dynamics techniques for hardcore systems, in Berne, B. J., ed., *Modern Theoretical Chemistry*, Plenum, New York, 1977, vol. 6B, p. 1.

[erp84]   Erpenbeck, J. J. and Wood, W. W., Molecular dynamics calculations of the hard-sphere equation of state, *J. Stat. Phys.* **35** (1984) 321.

[erp85]   Erpenbeck, J. J. and Wood, W. W., Molecular dynamics calculations of the velocity autocorrelation function: Hard-sphere results, *Phys. Rev. A* **32** (1985) 412.

[erp88]   Erpenbeck, J. J., Shear viscosity of the Lennard-Jones fluid near the triple point: Green–Kubo results, *Phys. Rev. A* **38** (1988) 6255.

[ess94]   Esselink, K., Hilbers, P. A. J., van Os, N. M., Smit, B., and Karaborni, S., Molecular dynamics simulations of model oil/water/surfactant systems, *Colloids and Surfaces A* **91** (1994) 155.

[ess95]   Esselink, K., A comparison of algorithms for long-range interactions, *J. Comp. Phys.* **87** (1995) 375.

[eva77a] Evans, D. J., On the representation of orientation space, *Mol. Phys.* **34** (1977) 317.

[eva77b] Evans, D. J. and Murad, S., Singularity free algorithm for molecular dynamics simulation of rigid polyatomics, *Mol. Phys.* **34** (1977) 327.

[eva82] Evans, D. J., Homogeneous NEMD algorithm for thermal conductivity – application of noncanonical linear response theory, *Phys. Lett.* **91A** (1982) 45.

[eva83a] Evans, D. J., Computer 'experiment' for nonlinear thermodynamics of Couette flow, *J. Chem. Phys.* **78** (1983) 3297.

[eva83b] Evans, D. J., Hoover, W. G., Failor, B. H., Moran, B., and Ladd, A. J. C., Nonequilibrium molecular dynamics via Gauss's principle of least constraint, *Phys. Rev. A* **28** (1983) 1016.

[eva84] Evans, D. J. and Morriss, G. P., Non-Newtonian molecular dynamics, *Comp. Phys. Repts.* **1** (1984) 297.

[eva86] Evans, D. J. and Morriss, G. P., Shear thickening and turbulence in simple fluids, *Phys. Rev. Lett.* **56** (1986) 2172.

[eva90] Evans, D. J. and Morriss, G. P., *Statistical Mechanics of Nonequilibrium Liquids*, Academic Press, London, 1990.

[fer91] Ferrario, M., Ciccotti, G., Holian, B. L., and Ryckaert, J.-P., Shear-rate dependence of the viscosity of the Lennard-Jones liquid at the triple point, *Phys. Rev. A* **44** (1991) 6936.

[fey63] Feynman, R. P., Leighton, R. B., and Sands, M., *The Feynman Lectures on Physics*, vol. 1, Addison-Wesley, Reading, 1963.

[fin79] Finney, J. L., A procedure for the construction of Voronoi polyhedra, *J. Comp. Phys.* **32** (1979) 137.

[fin93] Fincham, D., Leapfrog rotational algorithms for linear molecules, *Mol. Simulation* **11** (1993) 79.

[fly89] Flyvberg, H. and Petersen, H. G., Error estimates on averages of correlated data, *J. Chem. Phys.* **91** (1989) 461.

[fri75] Friedman, H. L., Image approximation to the reaction field, *Mol. Phys.* **29** (1975) 1533.

[gal93] Gallas, J. A. C. and Sokolowski, S., Grain nonsphericity effects on the angle of repose of granular material, *Int. J. Mod. Phys. B* **7** (1993) 2037.

[gay81] Gay, J. G. and Berne, B. J., Modification of the overlap potential to mimic a linear site–site potential, *J. Chem. Phys.* **74** (1981) 3316.

[gea71] Gear, C. W., *Numerical Initial Value Problems in Ordinary Differential Equations*, Prentice-Hall, Englewood Cliffs, NJ, 1971.

[gei79] Geiger, A., Stillinger, F. H., and Rahman, A., Aspects of the percolation process for hydrogen-bond networks in water, *J. Chem. Phys.* **70** (1979) 4185.

[gei94] Geist, A., Beguelin, A., Dongarra, J., Jiang, W., Manchek, R., and Sunderam, V., *PVM3 User's Guide and Reference Manual*, Oak Ridge National Laboratory Technical Report ORNL/TM-12187, 1994.

[gel94] Gelbart, W. M., Ben-Shaul, A., and Roux, D., eds., *Micelles, Membranes, Microemulsions, and Monolayers*, Springer, New York, 1994.

[gib60] Gibson, J. B., Goland, A. N., Milgram, M., and Vineyard, G. H., Dynamics of radiation damage, *Phys. Rev.* **120** (1960) 1229.

[gil83]   Gillan, M. J. and Dixon, M., The calculation of thermal conductivities by perturbed molecular dynamics simulation, *J. Phys. C* **16** (1983) 869.

[gil90]   Gillan, M. J., The path-integral simulation of quantum systems, in [cat90], p. 155.

[gol80]   Goldstein, H., *Classical Mechanics*, Addison-Wesley, Reading, MA, 2nd edition, 1980.

[gra84]   Gray, C. G. and Gubbins, K. E., *Theory of Molecular Fluids*, vol. 1, Clarendon Press, Oxford, 1984.

[gre88]   Greengard, L., *The Rapid Evaluation of Potential Fields in Particle Systems*, MIT Press, Cambridge, MA, 1988.

[gre89a] Greengard, L. and Rokhlin, V., On the evaluation of electrostatic interactions in molecular modeling, *Chem. Scripta* **29A** (1989) 139.

[gre89b] Grest, G. S., Dünweg, B., and Kremer, K., Vectorized linked cell Fortran code for molecular dynamics simulations for a large number of particles, *Comp. Phys. Comm.* **55** (1989) 269.

[gre94]   Grest, G. S., Structure of many-arm star polymers in solvents of varying quality: A molecular dynamics study, *Macromolecules* **27** (1994) 3493.

[gro96]   Gropp, W., Lusk, E., Doss, N., and Skjellum, A., A high-performance, portable implementation of the MPI message passing interface standard, *Parallel Computing* **22** (1996) 789.

[han86a] Hansen, J.-P., Molecular dynamics simulation of Coulomb systems in two and three dimensions, in [cic86a], p. 89.

[han86b] Hansen, J.-P. and McDonald, I. R., *Theory of Simple Liquids*, Academic Press, London, 2nd edition, 1986.

[han94]   Hansen, D. P. and Evans, D. J., A generalized heat flow algorithm, *Mol. Phys.* **81** (1994) 767.

[hel60]   Helfand, E., Transport coefficients from dissipation in a canonical ensemble, *Phys. Rev.* **119** (1960) 1.

[hel79]   Helfand, E., Flexible vs rigid constraints in statistical mechanics, *J. Chem. Phys.* **71** (1979) 5000.

[her95]   Herrmann, H. J., Simulating granular media on the computer, in Garrido, P. L. and Marro, J., eds., *3rd Granada Lectures in Computational Physics*, Springer, Heidelberg, 1995, p. 67.

[hey89]   Heyes, D. M. and Melrose, J. R., Continuum percolation of 2D Lennard-Jones and square-well phases, *Mol. Phys.* **68** (1989) 359.

[hig76]   Higgins, R. J., Bulk viscosity of model fluids. A comparison of equilibrium and nonequilibrium molecular dynamics results, *Am. J. Phys.* **44** (1976) 772.

[hir54]   Hirschfelder, J. O., Curtis, C. F., and Bird, R. B., *Molecular Theory of Gases and Liquids*, Wiley, New York, 1954.

[hir98]   Hirshfeld, D. and Rapaport, D. C., Molecular dynamics simulation of Taylor–Couette vortex formation, *Phys. Rev. Lett.* **80** (1998) 5337.

[hir00]   Hirshfeld, D. and Rapaport, D. C., Growth of Taylor vortices: A molecular dynamics study, *Phys. Rev. E* **61** (2000) R21.

[hob31]   Hobson, E. W., *The Theory of Spherical and Ellipsoidal Harmonics*, Cambridge University Press, Cambridge, 1931.

[hoc74] Hockney, R. W., Goel, S. P., and Eastwood, J. W., Quiet high-resolution computer models of a plasma, *J. Comp. Phys.* **14** (1974) 148.

[hoc88] Hockney, R. W. and Eastwood, J. W., *Computer Simulation Using Particles*, Adam Hilger, Bristol, 1988.

[hol91] Holian, B. L., Voter, A. F., Wagner, N. J., Ravelo, R. J., Chen, S. P., Hoover, W. G., Hoover, C. G., Hammerberg, J. E., and Dontje, T. D., Effects of pairwise versus many-body forces on high-stress plastic deformation, *Phys. Rev. A* **43** (1991) 2655.

[hol95] Holian, B. L. and Ravelo, R. J., Fracture simulations using large-scale molecular dynamics, *Phys. Rev. B* **51** (1995) 11275.

[hon92] Hong, D. C. and McLennan, J. A., Molecular-dynamics simulations of hard-sphere granular particles, *Physica* **187** (1992) 159.

[hoo82] Hoover, W. G., Ladd, A. J. C., and Moran, B., High-strain-rate plastic flow studied via nonequilibrium molecular dynamics, *Phys. Rev. Lett.* **48** (1982) 1818.

[hoo85] Hoover, W. G., Canonical dynamics: Equilibrium phase-space distributions, *Phys. Rev. A* **31** (1985) 1695.

[hoo91] Hoover, W. G., *Computational Statistical Mechanics*, Elsevier, Amsterdam, 1991.

[hsu79] Hsu, C. S. and Rahman, A., Interaction potentials and their effect on crystal nucleation and symmetry, *J. Chem. Phys.* **71** (1979) 4974.

[hua63] Huang, K., *Statistical Mechanics*, Wiley, New York, 1963.

[jae96] Jaeger, H. M., Nagel, S. R., and Behringer, R. P., Granular solids, liquids, and gases, *Rev. Mod. Phys.* **68** (1996) 1259.

[jai91] Jain, A., Unified formulation of dynamics for serial rigid multibody systems, *J. Guid. Control Dyn.* **14** (1991) 531.

[jai93] Jain, A., Vaidehi, N., and Rodriguez, G., A fast recursive algorithm for molecular dynamics simulation, *J. Comp. Phys.* **106** (1993) 258.

[jai95] Jain, A. and Rodriguez, G., Base-invariant symmetric dynamics of free-flying manipulators, *IEEE Trans. Rob. Autom.* **11** (1995) 585.

[jam90] James, F., A review of pseudorandom number generators, *Comp. Phys. Comm.* **60** (1990) 329.

[jor83] Jorgensen, W. L., Chandrasekhar, J., Madura, J. D., Impey, R. W., and Klein, M. L., Comparison of simple potential functions for simulating liquid water, *J. Chem. Phys.* **79** (1983) 926.

[kar94] Karaborni, S., Esselink, K., Hilbers, P. A. J., Smit, B., Karthäuser, J., van Os, N. M., and Zana, R., Simulating the self-assembly of gemini (dimeric) surfactants, *Science* **266** (1994) 254.

[kar96] Karaborni, S. and Smit, B., Computer simulations of surfactant structures, *Current Opinion in Colloid and Interface Science* **1** (1996) 411.

[kle86] Klein, M. L., Structure and dynamics of molecular crystals, in [cic86a], p. 424.

[knu68] Knuth, D. E., *Fundamental Algorithms*, vol. 1 of *The Art of Computer Programming*, Addison-Wesley, Reading, MA, 1968.

[knu69] Knuth, D. E., *Seminumerical Algorithms*, vol. 2 of *The Art of Computer Programming*, Addison-Wesley, Reading, MA, 1969.

[knu73]   Knuth, D. E., *Sorting and Searching*, vol. 3 of *The Art of Computer Programming*, Addison-Wesley, Reading, MA, 1973.

[kol92]   Kolafa, J. and Perram, J. W., Cutoff errors in the Ewald summation formulae for point charge systems, *Mol. Simulation* **9** (1992) 351.

[kop89]   Koplik, J., Banavar, J. R., and Willemsen, J. F., Molecular dynamics of fluid flow at solid surfaces, *Phys. Fluids A* **1** (1989) 781.

[kre88]   Kremer, K. and Binder, K., Monte Carlo simulations of lattice models for macromolecules, *Comp. Phys. Repts.* **7** (1988) 259.

[kre92]   Kremer, K. and Grest, G. S., Simulations of structural and dynamic properties of dense polymer systems, *J. Chem. Soc. Faraday Trans.* **88** (1992) 1707.

[kus76]   Kushick, J. and Berne, B. J., Computer simulation of anisotropic molecular fluids, *J. Chem. Phys.* **64** (1976) 1362.

[kus90]   Kusalik, P. G., Computer simulation results for the dielectric properties of a highly polar fluid, *J. Chem. Phys.* **93** (1990) 3520.

[lan59]   Landau, L. D. and Lifshitz, E. M., *Fluid Mechanics*, Pergamon Press, Oxford, 1959.

[lan00]   Landau, D. P. and Binder, K., *A Guide to Monte Carlo Simulations in Statistical Physics*, Cambridge University Press, Cambridge, 2000.

[leb67]   Lebowitz, J. L., Percus, J. K., and Verlet, L., Ensemble dependence of fluctuations with application to machine computations, *Phys. Rev.* **153** (1967) 250.

[lee72]   Lees, A. W. and Edwards, S. F., The computer study of transport processes under extreme conditions, *J. Phys. C* **5** (1972) 1921.

[lev73]   Levesque, D., Verlet, L., and Kürkijarvi, J., Computer 'experiments' on classical fluids. IV. Transport properties and time-correlation functions of the Lennard-Jones fluid near its triple point, *Phys. Rev. A* **7** (1973) 1690.

[lev84]   Levesque, D., Weis, J. J., and Hansen, J.-P., Recent developments in the simulation of classical fluids, in Binder, K., ed., *Applications of the Monte Carlo Method in Statistical Physics*, Springer, Berlin, 1984, p. 37.

[lev87]   Levesque, D. and Verlet, L., Molecular dynamics calculations of transport coefficients, *Mol. Phys.* **61** (1987) 143.

[lev92]   Levesque, D. and Weis, J. J., Recent progress in the simulation of classical fluids, in [bin92], p. 121.

[lev93]   Levesque, D. and Verlet, L., Molecular dynamics and time reversibility, *J. Stat. Phys.* **72** (1993) 519.

[lie92]   Liem, S. Y., Brown, D., and Clarke, J. H. R., Investigation of the homogeneous-shear and nonequilibrium molecular dynamics methods, *Phys. Rev. A* **45** (1992) 3706.

[loo92]   Loose, W. and Ciccotti, G., Temperature and temperature control in nonequilibrium molecular dynamics simulations of the shear flow of dense liquids, *Phys. Rev. A* **45** (1992) 3859.

[mai81]   Maitland, G. C., Rigby, M., Smith, E. B., and Wakeham, W. A., *Intermolecular Forces*, Clarendon Press, Oxford, 1981.

[mak89]   Makino, J. and Hut, P., Gravitational N-body algorithms: A comparison between supercomputers and a highly parallel computer, *Comp. Phys. Repts.* **9** (1989) 199.

[mar72] Marsaglia, G., Choosing a point from the surface of a sphere, *Ann. Math. Stat.* **43** (1972) 645.

[mar92] Mareschal, M. and Holian, B. L., eds., *Microscopic Simulation of Complex Hydrodynamic Phenomena*, Plenum Press, New York, 1992.

[mck92] McKechnie, J. I., Brown, D., and Clarke, J. H. R., Methods for generating dense relaxed amorphous polymer samples for use in dynamic simulations, *Macromolecules* **25** (1992) 1562.

[mcq76] McQuarrie, D. A., *Statistical Mechanics*, Harper and Row, New York, 1976.

[med90] Medvedev, N. N., Geiger, A., and Brostow, W., Distinguishing liquids from amorphous solids: Percolation analysis of the Voronoi network, *J. Chem. Phys.* **93** (1990) 8337.

[mes64] Messiah, A., *Quantum Mechanics*, North-Holland, Amsterdam, 1964.

[mor85] Morriss, G. P. and Evans, D. J., Isothermal response theory, *Mol. Phys.* **54** (1985) 629.

[mor91] Morriss, G. P. and Evans, D. J., A constraint algorithm for the computer simulation of complex molecular liquids, *Comp. Phys. Comm.* **62** (1991) 267.

[mye88] Myers, D., *Surfactant Science and Technology*, VCH Publishers, New York, 1988.

[neu83] Neumann, M., Dipole moment fluctuation formulas in computer simulations of polar systems, *Mol. Phys.* **50** (1983) 841.

[nic79] Nicolas, J. J., Gubbins, K. E., Streett, W. B., and Tildesley, D. J., Equation of state for the Lennard-Jones fluid, *Mol. Phys.* **37** (1979) 1429.

[nos83] Nosé, S. and Klein, M. L., Constant pressure molecular dynamics for molecular systems, *Mol. Phys.* **50** (1983) 1055.

[nos84a] Nosé, S., A molecular dynamics method for simulations in the canonical ensemble, *Mol. Phys.* **52** (1984) 255.

[nos84b] Nosé, S., A unified formulation of the constant temperature molecular dynamics methods, *J. Chem. Phys.* **81** (1984) 511.

[orb67] Orban, J. and Bellemans, A., Velocity inversion and irreversibility in a dilute gas of hard disks, *Phys. Lett. A* **24** (1967) 620.

[par80] Parrinello, M. and Rahman, A., Crystal structure and pair potentials: A molecular dynamics study, *Phys. Rev. Lett.* **45** (1980) 1196.

[par81] Parrinello, M. and Rahman, A., Polymorphic transitions in single crystals: A new molecular dynamics method, *J. Appl. Phys.* **52** (1981) 7182.

[pay93] Payne, V. A., Forsyth, M., Kolafa, J., Ratner, M. A., and de Leeuw, S. W., Dipole time correlation functions of Stockmayer fluid in the microcanonical and canonical ensembles, *J. Phys. Chem.* **97** (1993) 10478.

[pea79] Pear, M. R. and Weiner, J. H., Brownian dynamics study of a polymer chain of linked rigid bodies, *J. Chem. Phys.* **71** (1979) 212.

[per88] Perram, J. W., Petersen, H. G., and de Leeuw, S. W., An algorithm for the simulation of condensed matter which grows as the 3/2 power of the number of particles, *Mol. Phys.* **65** (1988) 875.

[per96] Pérez-Jordá, J. M. and Yang, W., A concise redefinition of the solid spherical harmonics and its use in fast multipole methods, *J. Chem. Phys.* **104** (1996) 8003.

[pfa94]   Pfalzner, S. and Gibbon, P., A 3D hierarchical tree code for dense plasma simulation, *Comp. Phys. Comm.* **79** (1994) 24.

[pie92]   Pierleoni, C. and Ryckaert, J.-P., Molecular dynamics investigation of dynamic scaling for dilute polymer solutions in good solvent conditions, *J. Chem. Phys.* **96** (1992) 8539.

[pol80]   Pollock, E. L. and Alder, B. J., Static dielectric properties of Stockmayer fluids, *Physica* **102A** (1980) 1.

[pos93]   Pöschel, T. and Buchholz, V., Static friction phenomena in granular materials: Coulomb law versus particle geometry, *Phys. Rev. Lett.* **71** (1993) 3963.

[pos95]   Pöschel, T. and Buchholz, V., Molecular dynamics of arbitrarily shaped granular particles, *J. Phys. I France* **5** (1995) 1431.

[pow79]   Powles, J. G., Evans, W. A. B., McGrath, E., Gubbins, K. E., and Murad, S., A computer simulation for a simple model of liquid hydrogen chloride, *Mol. Phys.* **38** (1979) 893.

[pre92]   Press, W. H., Teukolsky, S. A., Vetterling, W. T., and Flannery, B. R., *Numerical Recipes in C: The Art of Scientific Computing*, Cambridge University Press, Cambridge, 2nd edition, 1992.

[pri84]   Price, S. L., Stone, A. J., and Alderton, M., Explicit formulae for the electrostatic energy, forces and torques between a pair of molecules of arbitrary symmetry, *Mol. Phys.* **52** (1984) 987.

[puh89]   Puhl, A., Mansour, M. M., and Mareschal, M., Quantitative comparison of molecular dynamics with hydrodynamics in Rayleigh–Bénard convection, *Phys. Rev. A* **40** (1989) 1999.

[que73]   Quentrec, B. and Brot, C., New method for searching for neighbors in molecular dynamics computations, *J. Comp. Phys.* **13** (1973) 430.

[rah64]   Rahman, A., Correlation of motion of atoms in liquid argon, *Phys. Rev.* **136A** (1964) 405.

[rah71]   Rahman, A. and Stillinger, F. H., Molecular dynamics study of liquid water, *J. Chem. Phys.* **55** (1971) 3336.

[rah73]   Rahman, A. and Stillinger, F. H., Hydrogen-bond patterns in liquid water, *J. Am. Chem. Soc.* **95** (1973) 7943.

[rai89]   Raine, A. R. C., Fincham, D., and Smith, W., Systolic loop methods for molecular dynamics simulation using multiple transputers, *Comp. Phys. Comm.* **55** (1989) 13.

[ral78]   Ralston, A. and Rabinowitz, P., *A First Course in Numerical Analysis*, McGraw-Hill, New York, 2nd edition, 1978.

[rap79]   Rapaport, D. C., Molecular dynamics study of a polymer chain in solution, *J. Chem. Phys.* **71** (1979) 3299.

[rap80]   Rapaport, D. C., The event scheduling problem in molecular dynamics simulation, *J. Comp. Phys.* **34** (1980) 184.

[rap83]   Rapaport, D. C., Density fluctuations and hydrogen bonding in supercooled water, *Mol. Phys.* **48** (1983) 23.

[rap85]   Rapaport, D. C., Molecular dynamics simulation using quaternions, *J. Comp. Phys.* **60** (1985) 306.

[rap87]   Rapaport, D. C., Microscale hydrodynamics: Discrete particle simulations of evolving flow patterns, *Phys. Rev. A* **36** (1987) 3288.

[rap91a]   Rapaport, D. C., Multi-million particle molecular dynamics I: Design considerations for vector processing, *Comp. Phys. Comm.* **62** (1991) 198.

[rap91b]   Rapaport, D. C., Multi-million particle molecular dynamics II: Design considerations for distributed processing, *Comp. Phys. Comm.* **62** (1991) 217.

[rap91c]   Rapaport, D. C., Time-dependent patterns in atomistically simulated convection, *Phys. Rev. A* **43** (1991) 7046.

[rap92]   Rapaport, D. C., Atomistic simulation of heat and mass transfer near the convection threshold, *Phys. Rev. A* **46** (1992) R6150.

[rap94]   Rapaport, D. C., Shear-induced order and rotation in pipe flow of short-chain molecules, *Europhys. Lett.* **26** (1994) 401.

[rap97]   Rapaport, D. C., An introduction to interactive molecular dynamics simulations, *Computers in Physics* **11** (1997) 337.

[rap98]   Rapaport, D. C., Subharmonic surface waves in vibrated granular media, *Physica A* **249** (1998) 232.

[rap02a]   Rapaport, D. C., Molecular dynamics simulation of polymer helix formation using rigid-link methods, *Phys. Rev. E* **66** (2002) 011906.

[rap02b]   Rapaport, D. C., Simulational studies of axial granular segregation in a rotating cylinder, *Phys. Rev. E* **65** (2002) 061306.

[ray72]   Ray, J. R., Nonholonomic constraints and Gauss's principle of least constraint, *Am. J. Phys.* **40** (1972) 179.

[ray91]   Ray, J. R. and Graben, H. W., Small systems have non-Maxwellian momentum distributions in the microcanonical ensemble, *Phys. Rev. A* **44** (1991) 6905.

[reb77]   Rebertus, D. W. and Sando, K. M., Molecular dynamics simulation of a fluid of hard spherocylinders, *J. Chem. Phys.* **67** (1977) 2585.

[rem90]   Remler, D. K. and Madden, P. A., Molecular dynamics without effective potentials via the Car–Parrinello approach, *Mol. Phys.* **70** (1990) 921.

[rod92]   Rodriguez, G. and Kreutz-Delgado, K., Spatial operator factorization and inversion of the manipulator mass matrix, *IEEE Trans. Rob. Autom.* **8** (1992) 65.

[ryc77]   Ryckaert, J.-P., Ciccotti, G., and Berendsen, H. J. C., Numerical integration of the cartesian equations of motion for a system with constraints: Molecular dynamics of n-alkanes, *J. Comp. Phys.* **23** (1977) 327.

[ryc78]   Ryckaert, J.-P. and Bellemans, A., Molecular dynamics of liquid alkanes, *Faraday Disc. Chem. Soc.* **66** (1978) 95.

[ryc90]   Ryckaert, J.-P., The method of constraints: Application to a simple n-alkane model, in [cat90], p. 189.

[sar93]   Sarman, S. and Evans, D. J., Self-diffusion and heat flow in isotropic and liquid crystal phases of the Gay–Berne fluid, *J. Chem. Phys.* **99** (1993) 620.

[sch73]   Schofield, P., Computer simulation studies of the liquid state, *Comp. Phys. Comm.* **5** (1973) 17.

[sch85]   Schoen, M. and Hoheisel, C., The shear viscosity of a Lennard-Jones fluid calculated by equilibrium molecular dynamics, *Mol. Phys.* **56** (1985) 653.

[sch86]   Schoen, M., Vogelsang, R., and Hoheisel, C., Computation and analysis of the dynamics structure factor for small wave vectors. A molecular dynamics study for a Lennard-Jones fluid, *Mol. Phys.* **57** (1986) 445.

[sch92]   Schmidt, K. E. and Ceperley, D., Monte Carlo techniques for quantum fluids, solids and droplets, in [bin92], p. 205.

[sch96]   Schäfer, J., Dippel, S., and Wolf, D. E., Force schemes in simulations of granular materials, *J. Phys. I France* **6** (1996) 5.

[smi91]   Smit, B., Hilbers, P. A. J., Esselink, K., Rupert, L. A. M., van Os, N. M., and Schlijper, A. G., Structure of a water/oil interface in the presence of micelles: A computer simulation, *J. Phys. Chem.* **95** (1991) 6361.

[smi92]   Smith, W. and Rapaport, D. C., Molecular dynamics simulation of linear polymers in a solvent, *Mol. Simulation* **9** (1992) 25.

[smi94]   Smith, P. E. and van Gunsteren, W. F., Consistent dielectric properties of the simple point charge and extended point charge water models at 277 and 300 K, *J. Chem. Phys.* **100** (1994) 3169.

[spr88]   Sprik, M. and Klein, M. L., A polarizable model for water using distributed charge sites, *J. Chem. Phys.* **89** (1988) 7556.

[spr91]   Sprik, M., Hydrogen bonding and the static dielectric constant in liquid water, *J. Chem. Phys.* **95** (1991) 6762.

[sta92]   Stauffer, D. and Aharony, A., *Introduction to Percolation Theory*, Taylor and Francis, London, 2nd edition, 1992.

[sti72]   Stillinger, F. H. and Rahman, A., Molecular dynamics study of temperature effects on water structure and kinetics, *J. Chem. Phys.* **57** (1972) 1281.

[sti74]   Stillinger, F. H. and Rahman, A., Improved simulation of liquid water by molecular dynamics, *J. Chem. Phys.* **60** (1974) 1545.

[sti85]   Stillinger, F. H. and Weber, T. A., Computer simulation of local order in condensed phases of silicon, *Phys. Rev. B* **31** (1985) 5262.

[str78]   Streett, W. B., Tildesley, D. J., and Saville, G., Multiple timestep methods in molecular dynamics, *Mol. Phys.* **35** (1978) 639.

[tan83]   Tanemura, M., Ogawa, T., and Ogita, N., A new algorithm for three-dimensional Voronoi tessellation, *J. Comp. Phys.* **51** (1983) 191.

[ten82]   Tenenbaum, A., Ciccotti, G., and Gallico, R., Stationary nonequilibrium states by molecular dynamics. Fourier's law, *Phys. Rev. A* **25** (1982) 2778.

[tho89]   Thompson, P. A. and Robbins, M. O., Simulations of contact-line motion: Slip and the dynamic contact angle, *Phys. Rev. Lett.* **63** (1989) 766.

[tho90]   Thompson, P. A. and Robbins, M. O., Shear flow near solids: Epitaxial order and flow boundary conditions, *Phys. Rev. A* **41** (1990) 6830.

[tho91]   Thompson, P. A. and Grest, G. S., Granular flow: Friction and the dilatancy transition, *Phys. Rev. Lett.* **67** (1991) 1751.

[tox88]   Toxvaerd, S., Molecular dynamics of liquid butane, *J. Chem. Phys.* **89** (1988) 3808.

[tri88]   Tritton, D. J., *Physical Fluid Dynamics*, Oxford University Press, Oxford, 2nd edition, 1988.

[tro84]   Trozzi, C. and Ciccotti, G., Stationary nonequilibrium states by molecular dynamics. II. Newton's law, *Phys. Rev. A* **29** (1984) 916.

[tuc94]   Tuckerman, M. E. and Parrinello, M., Integrating the Car–Parrinello equations. 1. Basic integration techniques, *J. Chem. Phys.* **101** (1994) 1302.

[van82]   van Gunsteren, W. F. and Karplus, M., Effect of constraints on the dynamics of macromolecules, *Macromolecules* **15** (1982) 1528.

[ver67]   Verlet, L., Computer 'experiments' on classical fluids. I. Thermodynamical properties of Lennard-Jones molecules, *Phys. Rev.* **159** (1967) 98.

[ver68]   Verlet, L., Computer 'experiments' on classical fluids. II. Equilibrium correlation functions, *Phys. Rev.* **165** (1968) 201.

[vog84]   Vogelsang, R. and Hoheisel, C., Structure and dynamics of supercritical fluid in comparison with a liquid. A computer simulation study, *Mol. Phys.* **53** (1984) 1355.

[vog85]   Vogelsang, R. and Hoheisel, C., Computation of low pressures by molecular dynamics based on Lennard-Jones potentials, *Mol. Phys.* **55** (1985) 1339.

[vog87]   Vogelsang, R., Hoheisel, C., and Ciccotti, G., Thermal conductivity of the Lennard-Jones liquid by molecular dynamics calculations, *J. Chem. Phys.* **86** (1987) 6371.

[vog88]   Vogelsang, R. and Hoheisel, C., Thermal transport coefficients including the Soret coefficient for various liquid Lennard-Jone mixtures, *Phys. Rev. A* **38** (1988) 6296.

[wal83]   Walton, O. R., Particle-dynamics calculations of shear flow, in Jenkins, J. T. and Satake, M., eds., *Mechanics of Granular Materials*, Elsevier, Amsterdam, 1983, p. 327.

[wal92]   Walton, O. R., Numerical simulation of inelastic, frictional particle-particle interactions, in Rocco, M. C., ed., *Particulate Two-Phase Flow*, Butterworth-Heineman, Boston, 1992, p. 884.

[whi94]   White, C. A. and Head-Gordon, M., Derivation and efficient implementation of the fast multipole method, *J. Chem. Phys.* **101** (1994) 6593.

[wig60]   Wigner, E. P., The unreasonable effectiveness of mathematics in the natural sciences, *Comm. Pure App. Math.* **13** (1960) 1.

[woo76]   Wood, W. W. and Erpenbeck, J. J., Molecular dynamics and Monte Carlo calculations in statistical mechanics, *Ann. Rev. Phys. Chem.* **27** (1976) 319.

[zim72]   Ziman, J. M., *Principles of the Theory of Solids*, Cambridge University Press, Cambridge, 2nd edition, 1972.

Function index

Index

Colophon

An abundance of freely available software was employed in preparing this book. Program development was carried out on computers running the Redhat distribution of the GNU/Linux operating system, C compiler, graphics libraries and assorted software tools, as well as the *NEdit* text editor. The manuscript was formatted using TEX, along with various LATEX macro packages, BIBTEX and *MakeIndex*, converted to PostScript form using *Dvips* and previewed with *Ghostscript*. Computer images and graphs were produced by specially written software.